Austin C. Apgar

Birds of the United States East of the Rocky Mountains

a manual for the identification of species in hand or in the bush

Austin C. Apgar

Birds of the United States East of the Rocky Mountains
a manual for the identification of species in hand or in the bush

ISBN/EAN: 9783337288211

Printed in Europe, USA, Canada, Australia, Japan

Cover: Foto ©berggeist007 / pixelio.de

More available books at **www.hansebooks.com**

BIRDS

OF THE UNITED STATES

EAST OF THE ROCKY MOUNTAINS

*A MANUAL FOR THE IDENTIFICATION OF SPECIES
IN HAND OR IN THE BUSH*

By AUSTIN C. APGAR

AUTHOR OF "TREES OF THE NORTHERN UNITED STATES," ETC.

————◦◦⚬❉⚬◦◦————

NEW YORK ∴ CINCINNATI ∴ CHICAGO

AMERICAN BOOK COMPANY

PREFACE

Much interest and enjoyment may be added to our lives by familiarity with those most beautiful, sprightly, and musical forms of life, — the birds. Yet few of us know or even see more than a very small part of the feathered songsters of our woods, fields, and waysides.

The object of this book is to encourage the study of birds by rendering it a pleasant and easy task. The introductory chapters explain briefly the meaning of technical terms used by ornithologists. These chapters are designed chiefly for reference, a much smaller vocabulary being employed in the body of the book.

The descriptions have been prepared with great care, and present several advantages over those in other books:

(1) They are short, being limited to points essential to the identification of the species.

(2) They consist generally of only two connected sentences, which can readily be recalled while looking at a bird. They are thus especially adapted for field use.

(3) Sufficient reference is made to the changes due to sex, age, and season, without describing in any particular species all the phases found in nature.

3

(4) They are adapted for the use of beginners in the study of birds, not for reference by ornithologists, who have access to more comprehensive works.

Keys, if properly arranged, furnish the easiest and most practicable method of enabling beginners to identify species. The Keys in this book were originally prepared as aids in discovering the names of birds by examination of their external features only. They were thus printed, and used by over a thousand students under the direct supervision of the author. Every difficulty encountered by the pupils suggested to the author changes to render the Keys more effectual; and now, after their final revision, they are so simply and carefully arranged that even a child can follow them with ease, and discover by their aid the names of birds both in the hand and in the bush.

Two series of Keys have been introduced:

(1) Keys to be used only with birds in the hand; that is, with prepared skins, mounted specimens, or recently killed birds. These place emphasis on the parts which change least with age, sex, or season, and give exact measurements of these parts.

(2) Keys to be used in the field for identifying the living birds that frequent our fields and groves. These emphasize such features as can be seen with the naked eye or through an opera glass, with the birds at some distance from the observer. In these Keys the birds are separated for convenience into groups, determined by their relation in size to our most familiar birds, the English sparrow and the robin.

The illustrations were drawn especially for this work by Miss Ada Collins Apgar and Mr. Richard B. Farley, and their scientific accuracy and careful execution add much to

the value and the interest of the book. The line under each
cut represents an inch, and can be used in measuring the
various parts. Its main purpose, however, is to show the
scale of the drawing. If the line is half an inch long, it
indicates that the illustration is one half as large as the
living bird; if the line is but one tenth inch, the scale is
but one tenth; etc.

The map on page 41 shows the territory covered by the
birds described in this book. Because of the migration of
birds, a book describing all the species of a given section
necessarily includes nearly all those of regions extending
hundreds of miles beyond. Hence the ground covered by
this book practically extends to Ontario, Quebec, etc.

In nomenclature and classification, the "Check List of
North American Birds," by the American Ornithologists'
Union, has been followed without any change, except a re-
versal of the order of the families, the higher classes of
birds being placed first. The numbers with the scientific
names in parenthesis are in accordance with those in the
Check List. These numbers will be found useful in com-
paring the descriptions with those in other books where the
same classification is followed; also in labeling specimens
of eggs, nests, or birds, without writing the full names.
The common name at the beginning of each description is
the one given in the Check List; the names at the end in
parenthesis are others in popular use.

Scientific names are marked to indicate the pronunciation.
The vowel of the accented syllable is marked with the grave
accent (`) if long, and with the acute (´) if short.

Through the kindness of the authorities of the Academy
of Natural Sciences, of Philadelphia, and of the American

Museum of Natural History, of New York, the large collections in both museums were placed at the disposal of the artists and the author. Thanks are due especially to Mr. Witmer Stone, Mr. Samuel N. Rhodes, Dr. J. A. Allen, and Mr. Frank M. Chapman for valuable advice and assistance.

AUSTIN C. APGAR.

State Normal School,
Trenton, New Jersey.

CONTENTS

———◆◇◆———

PART I

PART II

PART III

PART IV

PART I

*EXTERNAL PARTS AND THE TERMS NEEDED
FOR THEIR DESCRIPTION*

CHAPTER I

BIRDS AND THEIR FEATHERS

THERE is no group in Nature which can be defined so accurately and so easily as that of birds. *Birds are animals with feathers. All animals with feathers are birds.* Many other peculiarities might be mentioned; many statements might be made about the structure and the organs of birds, which would make us realize more comprehensively the differences between them and other animate forms. A complete definition is necessary for the ornithologist; but many years' work in botany and zoölogy in schoolrooms has convinced the author that such statements are beyond the comprehension of beginners, and that any attempt to force them on the pupils at the start results in loss of interest in the work. Full knowledge is a growth, hence the end, not the beginning, of the book is the place for a complete definition of birds.

The great external parts of birds are the head, the body, the tail, the wings, and the legs; these parts will be treated in subsequent chapters. The feathers form the covering, more or less complete, of all these parts. Feathers are the most wonderfully complex and perfect of skin growths. They not only protect the body from the effects of all atmospheric changes,

9

but form the best and lightest of all flying organs. Some study of the parts of feathers and a knowledge of the descriptive terms applied to them are important.

Parts and kinds of feathers. — Every feather[1] consists of the main *scape*, or stem, and the two *webs*. The scape has first the hollow portion, the *calamus* or *quill*, and then the four-sided solid portion, the *rhachis*, which extends to the tip of the feather. The rhachis bears on each of its sides lateral processes called *barbs*. These, with the rhachis, form the spreading portion of the feather, the *vane*. The calamus has an opening at the bottom through which the *pulp* penetrates, and another opening, the *superior aperture* or *umbilicus*, on the lower side where the calamus joins the rhachis.

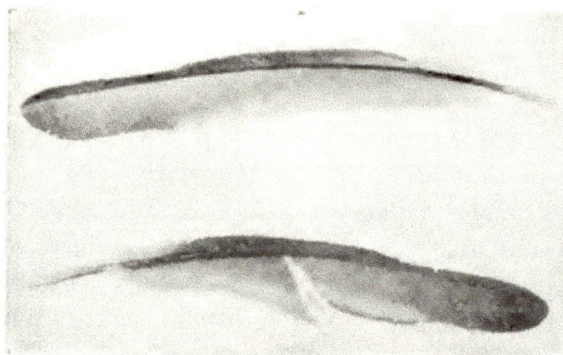

The barbs are narrow plates obliquely joining the rhachis, and tapering to points at their free ends, their edges being directed upward and downward when the vane is horizontal. On the sides of the barbs are minute processes, called *barbules*, branching from the barbs as the barbs branch from the rhachis. These barbules are often serrated and terminated by little hooks which interlock with hooks on the next barbule. (All these parts can be seen with the naked eye, or by the aid of a magnifying glass. With a microscope, the barbules will be found to divide again into *barbicels* or *cilia* and *hooklets*, forming a fringe to the barbules.) This gives firmness to the vane. If there is no interlocking of barbules, *downy*[2] feathers are formed. Sometimes the scape is very long, and the barbs are very short; such feathers are called *filament*[2] feathers, or *filoplumes*.

Many a feather [3] has, besides what is above described, another rhachis, on its lower side, called an *aftershaft*. This aftershaft joins the scape at the umbilicus, and has on its sides barbs and barbules about the same as those on the main rhachis. This part of the feather, even when present, is, in all of our birds, much smaller than the main vane. The figure shows a feather

from the back of the English sparrow, with an aftershaft, and, at the right, the aftershaft separated from the feather.

The description so far given is that of the *usual* feather, and, if the aftershaft is present, of a *complete* feather. There are, however, many modifications of these forms, concerning which some knowledge is important. First, as has already been said, the aftershaft is frequently wanting. Sometimes the barbs are found on only one side of the rhachis; this makes a one-sided vane. Frequently the barbs are lacking on both sides, thus changing the feather to a bristle,[4] as around the mouth, nostrils, and eyelids of most birds. Some-

times the barbs lack barbules on certain sections of their length, forming feathers with transparent portions. Sometimes the barbs are so far apart that there can be no locking of barbules, even when present; this causes the formation, in certain cases, of the most beautiful of plumes, as in the "aigrette" of the herons during the breeding season.

In review, it is well to recall the types of feathers spoken of in the foregoing pages, and to notice examples of each, as shown in the English sparrow.

1. The *typical* feather, or *pen* feather, where the interlocking of the barbs is complete, as in the great quills of the wing.

2. The *complete* feather, where there is an aftershaft as well as the main vane, as in the larger feathers of the back.

3. The *downy* feather, or *plume*[1] feather, where the stem is short and weak, the rhachis soft, and the barbs have long, slender, thread-like barbules without hooklets. These are abundant everywhere over the body of the sparrow, under and among the feathers which form the outer coating.

4. The *hairy*[1] feathers, where the stem is very long and slender and the vanes very small. These can readily be seen after plucking the feathers from the sparrow as, apparently, hairs scattered over the body. They are the parts singed off by the cook before preparing a bird for the oven.

5. The *bristly* feathers or *bristles*,[2] where the rhachis lacks vanes either throughout, or toward, the external end. These are abundant around the mouth of the sparrow.

Many feathers show in different portions two or even more of the above types. A complete feather may have a *downy* base, a *pennaceous* center, and a *bristly*[2] tip.

Location of different kinds of feathers. — The feathers which form the great bulk of the plumage of birds are called *contour* feathers. These usually consist of a perfect stem or quill at the base, an interlocked or pennaceous tip, and a downy portion between. They give outline, color, and most of the ornamental appendages of birds. Among the different birds there is a wonderful variety of contour feathers. They range from the almost fish-like scales of the penguins to the magnificent gorget of the hummingbirds. In their various modifications they form almost all the gorgeous crests, tufts, ruffs, and plumes which render the birds the most beautiful of animate forms. These contour feathers can all be moved by muscles situated under the skin. Many birds have thousands of these feather muscles, by the aid of which the feathers can be made to stand erect, as can readily be seen in the turkey when its tail is erected and its feathers ruffled up, giving the bird the appearance of great beauty and of twice its usual size.

Under these contour feathers and usually entirely hidden from view, but forming more or less of a complete covering to the body, there are the *downy*[1] feathers. These have the *plume-*

like structure throughout. They frequently consist of a stem without any rhachis, the barbs forming merely a tuft at the end of the quill.

Finally, there are among the contour feathers, coming from the same holes in the skin, long, slender, almost hair-like parts, *filament* feathers, or *hair*[1] feathers. These have little distinction of stem and rhachis, and almost no barbs at all, though sometimes there are a few small ones near the end of the rhachis.

Besides the foregoing, which can be found on nearly all birds, there are peculiar growths which are characteristic of certain groups, distinguishing them from others. Thus the herons and a few other birds have on their breast and hips downy feathers which continue to grow indefinitely; but as fast as they grow the ends crumble to powder, forming a whitish, greasy or dusty spot. These are called *powder-down tracts*, and are covered with *powder-down* feathers.

Very few birds have the feathers equally distributed over the skin. Most birds have the feathers closely placed on certain patches or bands of the body, while other spaces are either entirely bare (as the lower breast and belly of the English sparrow), or merely covered with down. The penguins and toucans have the skin almost entirely and evenly covered with feathers, but the great majority of birds have large open or naked spaces as far as the skin is concerned, though the plumage as a whole in most cases really covers the body completely. There are a few exceptions; thus the head and more or less of the neck are naked in such birds as the vultures, buzzards, etc.

The general marking or coloring of a bird depends upon the changes in the coloring of its individual feathers. *Mottled* plumage is given by *margined*[3] feathers; *streaked* plumage by *striped*[4] feathers; *spotted* plumage by *dotted*[5] feathers, and *barred* plumage by *cross-striped*[6] feathers.

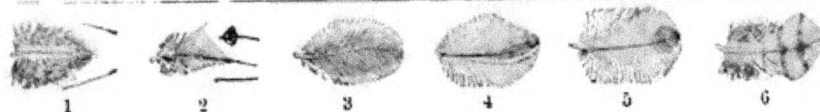

1   2   3   4   5   6

CHAPTER II

HEAD AND BODY

CERTAIN regions of the head and body have received special names, which are much used in descriptions. A few diagrams and definitions of these parts will be necessary.

The top of the head (see cut) is the *crown;* in front of this next the bill is the *forehead;* back of the crown is the *nape.* Above the eye there is a region often marked by a peculiar color; this is the *superciliary line,* in this book usually called the *line over eye.* A *line around the eye* has been called *orbital*

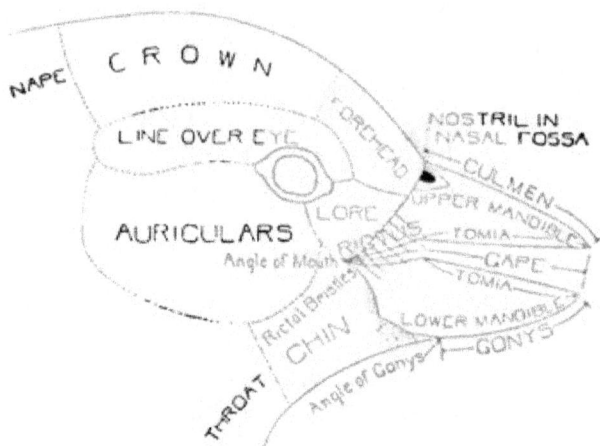

ring. A straight band extending from the eye to the bill is called the *lore;* this strip is bare of feathers on many swimming birds. Below and back of the eye, in the region of the ear, are the *auriculars.* This region, including a little below it, forms the *cheek.* The back corner of the mouth forms the *rictus.* This section is often bristly with hairs which are called *rictal bristles.*

The space just below the bill in front is the *chin* (see cut); below this, to about the bend of the closed wing (sometimes including the chin), is the *throat*. The greatest bulging portion of the body in front is the *breast*. From this backward, under the body, about to the legs in most birds, is the *belly*. Back of the position of the legs, in typical birds like the English sparrow, is the *anal region* (this is not marked on the diagram),

and still further back is the *crissum*, or *under tail coverts*. From the hind neck about half way to the tail is the *back;* next comes the *rump*, and then the *upper tail coverts*. The under and upper tail coverts are formed of those feathers which cover the stem portion of the *tail feathers*. By the side of the back there are often a number of enlarged feathers, and these form the *scapulars* or *shoulders*. Under the wings are the *sides* in front, and the *flanks* back of them. In the description of birds in Part II., the expression *back* or *upper parts* is often used, in a more general sense, to include all of the back, rump, etc. In the same way *below* is used to include nearly all the lower parts.

CHAPTER III

THE BILL

THE bills of birds, although equipped with neither lips nor teeth, have many offices. They are implements for cutting, handling, and carrying; they are organs of touch or feeling; they contain the nostrils for breathing and smelling. With the possible exception of the legs, no feature of birds is more varied in form, size, or appendages, or is more frequently used

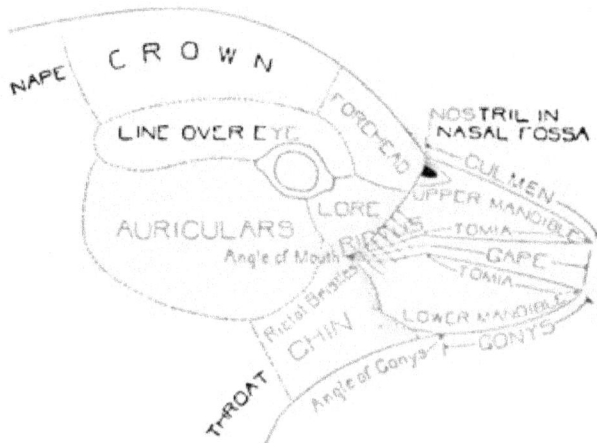

in systems of classification. Birds can often be classified into families by noting the peculiarities of the bill alone. It would therefore be well to study this chapter thoroughly before using the "Key to the Families of Birds."

Parts of the bill. — The two great parts are the *upper mandible* and the *lower mandible.* These consist of projecting skull bones covered by a horny material, usually comprising one piece for each jaw. Both jaws are movable. The lower jaw, in most birds, has a great range of movement, while that of the upper jaw is but slight. In parrots this is reversed, the upper jaw having the greater range. The ridge along the upper side of the upper mandible is the *culmen*, that along the lower side of the lower is the *gonys*. The gonys extends from the tip of the bill to where the central ridge forks; these two ridges at the base are called the *rhami* (singular *rhamus*). The *angle of the gonys* is between the gonys and the rhami.

The openings in the upper mandibles are the *nostrils*. These openings are frequently found in grooved portions of the bill; in such cases the groove is called the *nasal fossa* (plural *fossæ*). The *gape* is the whole opening of the mouth. Though *rictus* is sometimes used to mean the same thing, it is usually and more properly restricted to the back corner of the mouth as explained below. The term *commissure* is used to indicate the edges of the mouth when closed, and the *commissural point* or angle is the back angle of the mouth. The word *tomia* is used to indicate the cutting edges of the mandibles. The *rictus* proper extends from the basal end of the tomia to the corner of the mouth.

The covering of the bill. — The sheath-like covering of the mandibles is usually hard and horny as in the sparrows, but in many groups of birds it becomes, in part or as a whole, soft and skin-like, and is furnished with nerves of feeling. Most water birds, especially, have soft, leathery, or skin-like and very sensitive coverings to the bills, for feeling the food in the mud at the bottom of the water. A duck has a hard, so-called *nail* at the tip of the upper mandible.[1] A pigeon has a bill, soft at base and hard at tip, and a soft, swollen membrane at the top base of the upper mandible, roofing the nostrils.[2] Eagles, hawks, and parrots have a peculiar covering over the base of the upper mandible extending beyond the nostrils.[3] This covering is so peculiar that it is given a special name, the *cere*,[4] because it frequently has a waxy appearance. In the

parrots the cere is generally covered with feathers, and looks much like a part of the forehead, but as the nostrils open through it, it must be considered as part of the bill.

Positions of the nostrils. — The nostrils are frequently in a sort of hollow which has already been spoken of as a *nasal fossa*.[1] The groove, if long and narrow, is sometimes called the *nasal sulcus*.[2] Many birds show no evident fossae at all, but the nostrils open flush with the surface of the bill.[3]

There are a number of terms used to denote the position which the nostrils occupy in the upper mandible. They are almost universally *lateral*,[4] that is, on the sides of the mandible, away from the ridge; rarely they are *culminal*,[5] that is, together on the ridge of the culmen. The position on the sides, with reference to the width and length of the mandible, is defined as follows: *superior*[6] indicates a position above the central line between the culmen and the tomia, and *inferior*[7] below it; *basal*[8] indicates a position at the forehead; *sub-basal*[9] near it; *median*[10] half way between the base and the tip of the bill; *terminal*[11] nearer the tip than the base.

Kinds of nostrils. — The nostrils are usually *open* or *pervious*. Sometimes they are not distinctly open, in which case they are said to be *impervious*. Usually the two nostrils are separated by a partition; they are then said to be *imperforate*.[12] Rarely it is possible to see through the nostrils from side to side, as in the turkey buzzard, when they are said to be *perforate*.[13]

Forms of nostrils. — A *linear*[14] nostril is elongated and of

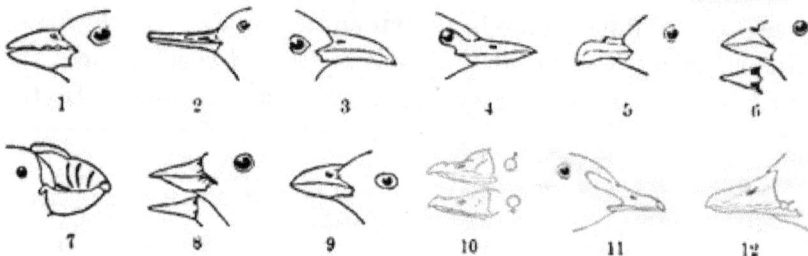

1   2   3   4   5   6

7   8   9   10   11   12

about equal width throughout; *clavate*, or *club-shaped*,[15] indicates an enlargement at the end nearer the tip of the bill, and
ovate[16] at the end nearer the base of the bill. An *oval*[17] nostril
is widest near the middle and wide for its length, while an
elliptical[18] one is narrower. If the nostril is about as wide as
long, it is called *circular*;[19] if there is a raised border to the
circular nostril, it is called *tubular*.[20]

Appendages to the nostrils. — The usual plan is to have the
opening through material like the covering of the rest of the bill,
of soft skin in the snipes, and of horn in the sparrows. Sometimes there is a special piece of about the same material as the
bill either above or below the nostril, called a *nasal scale*.[21] A
more frequent appendage consists of feathers proper or bristle-
tipped feathers. These are usually frontal feathers, more or
less changed into bristles, often entirely filling the nasal fossæ,
and frequently so covering the base of the bill as completely
to cover the nostrils.[22] This is well seen in the crow and in
the blue jay.

Sometimes the nostrils have special feathers of their own.
If these are separated and quite feather-like, they form *nasal
tufts*.[23] When not separated but extending from side to side,
they form a *ruff*.[24] Those feathers or bristles which are turned
forward are called *retrorse*.[22]

Other appendages to the bill. — The base of the bill is, in
hawks, etc., covered by a special membrane extending beyond
the nostrils; this is called a *cere*.[16] Something somewhat
cere-like, but consisting of an enlarged and *swollen mem-*

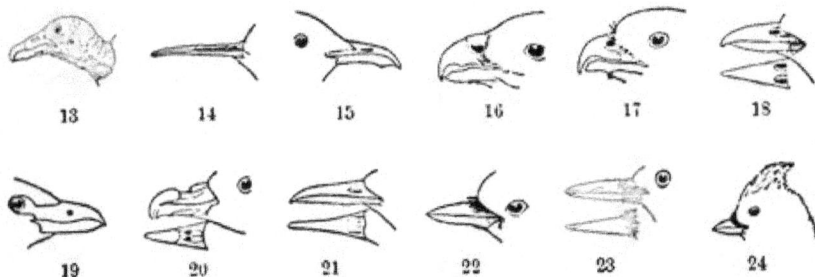

13   14   15   16   17   18

19   20   21   22   23   24

brane[1] extending over the nostrils, is found in the doves and
in the pigeons. The rictal portion of the mouth is frequently
fringed with longer or shorter hairs; these are called *rictal
bristles.*[2]

Shape of the gape. — The gape is *straight*[3] when the *commissural line*, formed by the tomia and the rictus, is straight.
This line may be *curved*,[2] *sinuate*,[4] or *angulate*.[5]

General size and form of the bill. — The length of the head is
used to measure the length of the bill. A *long*[6] bill is longer
than the head, a *short*[7] bill shorter, and one of *medium length*
is about as long as the head. A *compressed*[8] bill is one flattened sideways, so that its height is greater than its width. A
depressed[9] one is flattened up and down, or is wider than high.
A *straight*[10] bill is not only straight throughout its length, but
is also in line with the head. If not in line with the head, it
is said to be *bent*.[11] A *recurved*[12] bill is one that curves upward,
and a *decurved*[13] bill one that curves downward. A bill may
be decurved throughout, or it may have merely a *decurved tip*.[14]

The usual bill is sharp-pointed at the end and is called
acute.[15] If it has an abrupt and somewhat rounded end, it is
obtuse.[16] *Acuminate*[17] indicates not only an acute end, but a
slender bill as well. If very slender and sharp, it is called *attenuate*.[18] In some cases, as among the hummingbirds, still
more emphatic words are needed to denote slenderness and
sharpness. *Subulate*[19] is more emphatic than *attenuate*, and
acicular indicates the extreme limit in this direction.

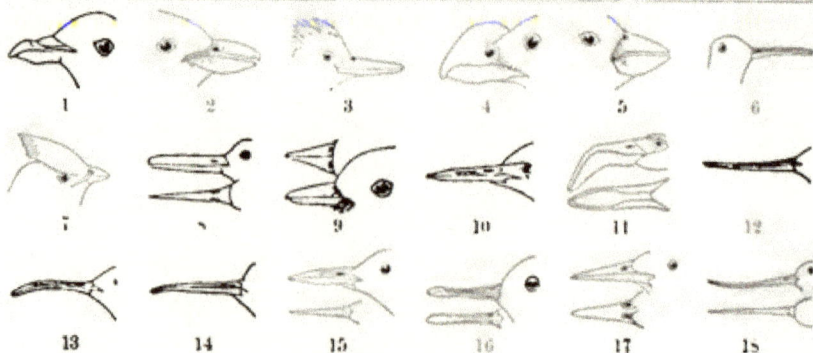

A bill is *hooked*[20] when the upper mandible is abruptly curved over the lower. In such cases the mandible often has teeth along its edge, and the word *dentate*[21] is used; if there are a number of teeth of about equal size, the word *serrate*[22] is used. *Spatulate*, or *spoon-shaped*,[23] indicates a bill much depressed as well as widened at the end. *Cultrate*, or *knife-shaped*,[24] indicates a much compressed bill with sharp edges. *Falcate*, or *scythe-shaped*, indicates a curved, cultrate one. In the crossbill, the upper and lower mandibles are *oppositely falcate*.[25] The ducks, geese, and a few other birds have a peculiar set of ridges just within the edges of the mandibles, in certain cases looking much like teeth; they are called *lamellæ*, and a bill that has them, *lamellate*.[26]

Besides the foregoing general terms, applying more or less to all bills, there are some special forms which have been given names that are frequently used in descriptions of birds. These need to be well fixed in mind. *Conirostral* indicates such a bill as the English sparrow has, — stout at base, conical in form, and with the gape so angulated as to bring the corners of the mouth down. Conirostral bills are *short*[5] in the sparrows and *long*[27] in the orioles. The swallows, etc., have *fissirostral*[28] bills. In this class of bills the culmen is very short, but the gape is both wide and deep, — about as wide as the head and so deep as to reach to the eyes. The creepers and the hummingbirds have *tenuirostral*[19] bills. The tenuirostral bill is slender, long, and has a rather short gape. The snipes have *longirostral*[6] bills; the bill is elongated, nearly equal in size throughout, and with the upper mandible grooved for the slit-like nostrils.

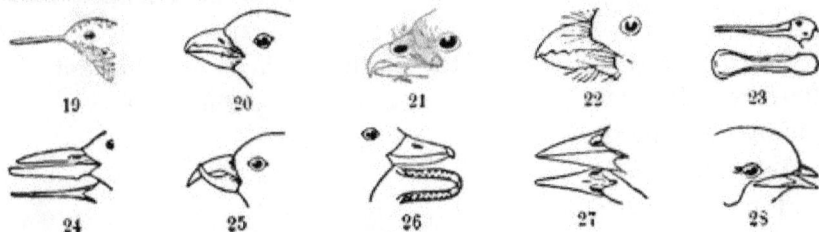

19   20   21   22   23

24   25   26   27   28

CHAPTER IV

WINGS

Use. — The general purpose of a wing is to be an organ of flight, and in most birds this is its principal use. In a few birds the body is too large and the wings are too small for this office. This is true in the ostrich and a few swimming birds. These use their wings to lighten their weight on the ground and possibly to aid them in running. In a few species, as in the penguins, the wings are not covered with feathers and quills, but with scale-like parts. In these, the wings act almost like the fins of fishes, and just like the paddles of whales and of porpoises, and enable the bird to move through the water almost if not quite as rapidly as any of the fishes.

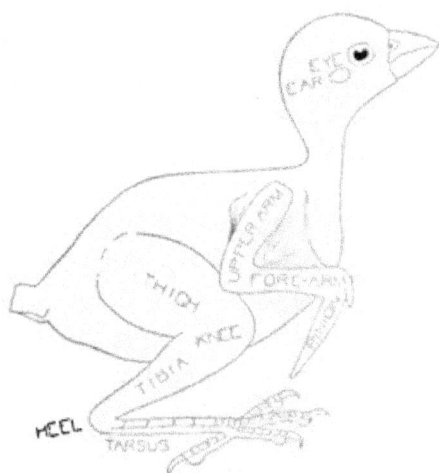

Some birds, as the divers, the dippers, etc., use their wings both for flight and for swimming. Many birds can use their wings as powerful weapons in fighting, and some have them fitted with strong and sharp spurs to render them the more useful for this office. Most birds make use of their wings to protect their young from enemies and from storms.

Parts. — The bones and the flesh of the wings consist of four readily seen parts, — the *upper arm*, the *forearm*, the *pinion*, and the *thumb*.

Feathers. — The *quills* or *remiges* are the stiffest, strongest, and most pennaceous (pen-like) of feathers, and form the spread

of the wing. These form the flight feathers proper. Their
number is smallest in the hummingbird (16) and very large in
the albatross (50 or more). Most of the other feathers are
small and very weak; they are used for covering up the bases
of the quills, both above and below, and for this reason are
called *wing coverts*. Besides the remiges and coverts of the
wing there is a third group of small quills, fastened to the thumb.
These quills form the *alula* or *little wing;* they are generally of
little use to the bird.

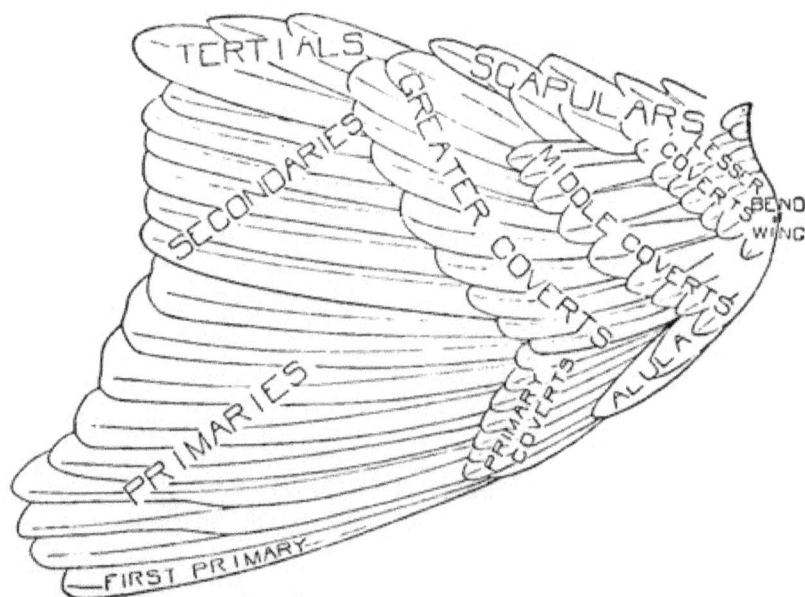

Quills. — The remiges or quills are readily divided into
three groups, according to the joint of wing to which they
are attached. Those fastened to the pinion are called *primaries;* those to the forearm, *secondaries;* and those to the
upper arm, *tertiaries* or *tertials.* This third term is generally
applied rather indifferently to the inner secondaries, those
attached to the elbow, which are frequently different in form,
size, and color from the other secondaries.

Primaries. — In number, the primaries are wonderfully uniform, being in almost all birds either *nine* or *ten.* Not only is

there great uniformity with birds in this number, but the position of a bird in a system of classification can often be determined most readily by the number of the primaries and the comparative length of the outer or *first primary.*

Secondaries. — The secondaries vary in number from only six in the English sparrow to upwards of forty in the albatross. These secondary quills are sometimes peculiarly colored; among some of the ducks they are very bright and iridescent. Such a colored spot on the secondaries is called a *speculum.*[1] Sometimes the secondaries are very much enlarged and brilliantly marked, as in the Argus pheasant, and sometimes of remarkable shapes, as in some tropical birds. The inner secondaries are much elongated in the larks and in the snipe, and in the grebes they are all so long as to cover the primaries completely when the wing is closed. In the chimney swift and in the hummingbirds they are peculiarly short.

Tertiaries or tertials. — The quills growing upon the upper arm — the true tertiaries — are not very evident upon most birds, but two or three of the inner secondaries are frequently conspicuous for either their length or their coloring; these are attached to the elbow and are the feathers which in the descriptions of the birds are generally called tertiaries. Sometimes conspicuously enlarged feathers on the shoulders, though not quills at all, are described as tertiaries. It is unfortunate that there is so little definiteness in the use of this term, but students will usually be right in considering any specially enlarged or peculiarly colored feathers about the shoulders of birds as being called tertiaries, as, for example, the enlarged inner secondaries of the larks, snipes, etc., and the peculiarly marked ones of the sparrows.

First primary and point of wing. — When quills are compared in length, the comparison refers to the position of their tips when the wing is closed. The first primary is the outer one, seen from below, and is often very short, as in the bluebird; frequently it is nearly as long as the longest; rarely it is the longest of all. Technically speaking, the expression *first* *primary* refers to the outer one of ten, as though we always considered the number to be ten; if there are only nine primaries, the first one is absent, and the series begins with the second. In other cases where there is a very short one beginning the series, the first primary is called *spurious*. In this book, which is written neither for anatomists nor ornithologists, but for beginners, no such technical use of the term will be attempted. The *first primary*[2] will always refer to the first apparent quill as seen from below at the outer edge of the wing. The *point of the wing* is frequently formed by about the third quill.[3] Sometimes, in what are called *rounded wings*,[4] the fifth or sixth forms it, while in the *pointed wings*[2] of the swallows it is formed by the first.

Coverts. — The feathers covering the bases of the primaries usually show imperfectly if at all on the closed wing, and are generally not mentioned in the descriptions of birds. The coverts fastened to the forearm on the upper side are the most important, and in many birds regularly form three series, as in the English sparrow.[5] The longest are called *greater coverts*. The next in size are called *middle coverts*. Each of these consists usually of a single row of feathers of nearly equal length. The last, called the *lesser coverts*, are generally small feathers in several rows. One or more rows of the coverts are apt to have their ends of a decidedly different color from the rest of the wing; these bands of color are called *wing bars*.[6] The English sparrow has one white wing bar formed by the tips of the middle coverts.[5] The under side of the wings has *under coverts*, but these are rarely mentioned.

The *first primary* and its length as compared with the others are important points to be determined in classifying most song birds. By raising the wing, if the bird is alive, or has been recently killed, the first primary will be readily seen.

SCAPULARS   QUILLS

1   2   3   4   5   6

If the bird is mounted, any raising of the wing should be prohibited as it would permanently injure the specimen. If the bird has been properly mounted for study, the wings will be spread enough to allow the first primary to show. If not, the feathers of the body can usually be pressed away from the wing by the tip of a pencil, enough to enable one to see it. The first primary is said to be *spurious*[1] when only about one third the length of the second, and *short*[2] when two thirds as long.

In many birds of prey and in many shore birds, more or less of the primaries are rather abruptly narrowed on their inner webs; such primaries are said to be *emarginate* or *notched*.[3] If not so abruptly narrowed, the word *attenuated*[4] is used. These words do not refer to the tip of the quill itself; it can be *rounded*, *acute*, or even *acuminate*.

Forms of wings. — The three great varieties of wings are the *long and pointed*[5] (swallows), *short and rounded*[2] (wrens), and the *ample*, or both long and broad (herons).

CHAPTER V

LEGS

Use. — The legs of birds serve many minor purposes in the different groups, besides the general one of locomotion. A large majority of birds perch on stems and hop (*leap* or *jump* would be more accurate words for the purpose) from twig to twig. The woodpeckers and many others climb up the surfaces of tree trunks: ducks swim: the grebes dive; and the parrots grasp and handle. In the use of the organ for locomotion there are wonderful differences in the various families. The

1   2   3   4   5

ostrich can run more rapidly than the horse, the barn fowls can walk and run, the bluebird can only leap or hop, while the auks can scarcely waddle.

The legs, like the bills, show a wonderful variety of modification in the different groups of birds. A careful study of either or both these parts will enable one to place any bird into its proper family. The use of the legs as a means of classification makes this chapter an important one, and it should be thoroughly studied before any attempt is made to determine the names of birds by the aid of the Key.

Parts. — The terms applied to the different parts of the legs of birds will be better understood by the student if he recalls what he learned in physiology about the bones of his own leg, and then compares the joints with those in the legs of a bird. In the sparrows and a large proportion of other birds, the space from the heel to the claws is all that shows of the leg (see cuts, pp. 15 and 22); these parts are called *tarsus* and *toes.* The *tibia* is entirely hidden by the feathers, and the *thigh* is so united with the skin of the body as to seem a part of it. In the grebes even the tibia is confined by the skin of the body.

The joint which bends forward in the hind limbs of all vertebrate animals is the *knee,* and the joint which bends backward is the *heel.* An examination of the horse's hind leg will show that it also has its heel as near the upper as the lower end of what appears to the eye as the leg, and that the knee is fastened to the body by the skin.

Covering of legs. — The thigh is feathered in all birds. The tibia is also feathered in most of the higher birds; but among wading birds there are on the tibia all stages of covering, from a completely feathered covering in the woodcock to one almost completely scaly in the stilts. The tarsus in most birds is scaly, but the grouse have it more or less completely feathered. Most of the owls have the tarsus fully feathered, and many of them the toes also. The barnyard fowls often have curious tufts of feathers on otherwise bare sections; some of the wild birds also have some odd tufts irregularly placed.

The parts of legs which are bare of feathers need close observation, as the kind and arrangement of the scaly covering of these parts have much to do with the classification of birds. The commonest arrangement is to have a distinct row of squarish scales down the front of the tarsus, as in the English sparrow. Sometimes such a row is also found down the back, as in most snipes; occasionally there is found a row down the outside of the tarsus, as in the flycatchers. These large, squarish scales are called *scutella*, and the tarsus is described as *scutellate* in front,[1] in front and behind,[2] or in front and along the outer side,[3] as the case may be. In the bluebird and in some others these front scales are so completely grown together as to look like a continuous covering; such a tarsus is said to be *booted*.[4] In many cases a portion of the tarsus, and in the geese the whole, is covered with small scales not very regularly arranged. These seem to form a fine network, and portions having such scales are said to be *reticulate*.[5]

The scutellate portions are different from the reticulate in another way. Scutella show as somewhat overlapping scales, and the whole forms a solid covering, but the small scales which form the reticulation are rather imbedded plates not touching at their edges, and the covering is apt to be more or less loose and pliable; rarely, these plates are elevated at their centers, and thus form tubercles, as in the fish hawk; such a leg is said to be *granulated*.[6] Sometimes a row of plates of any kind will be so roughened, in a regular way, as to be properly called *serrated*.[6]

The toes are almost invariably scutellate along the top. The tibia, when bare of feathers, has scales much like those of the tarsus, and of course the same words are used for their description. In some cases this part is covered with loose skin without any scales at all.

1   2   3   4

Length of leg. — The proportional length of leg and body of birds is extremely variable. The leg is very short in swallows and in all true swimming birds, medium in sparrows, longer in hawks, very long in the ostrich, and exceedingly long in cranes, stilts, and wading birds generally. The tarsus varies from about one thirtieth to one third the full length of the bird.

Number and arrangement of toes. — The usual number of toes is four, and among the birds of our region there are but few exceptions. The only other number represented in our fauna is three; but the ostrich has only two. When the toes are four in number they are arranged in three ways. The most common of all is shown in the sparrow, in which there are *three toes in front and one behind*.[1] In order to understand the modifications of this common plan, it is well to give names and numbers to the toes. The hind toe represents the great or inner toe of the human foot, and is called the *hallux* or first toe; the inner front toe is the second toe; the middle one the third toe; and the outer the fourth toe (see cut, p. 15). These, with few exceptions, have the following number of joints: the first toe two-jointed, the second three-jointed, the third four-jointed, and the fourth five-jointed. Some of our birds have the first toe absent, as will be shown hereafter; all the rest have joints as given, except the goatsuckers, which have but four joints to the fourth toe.

The second plan for the arrangement of four toes is shown in the woodpeckers, parrots, etc. In these there are *two in front and two behind*.[2] The first and fourth toes are behind, and the second and third in front.

The third plan is represented in the owls. In these, the first toe is permanently behind, the second and third permanently in front, and the fourth can be used either in front or behind,[3] and for this reason is called a *versatile* toe.

5

6

7

8

When there are but three toes, the usual arrangement is to have them all *three in front.*[1] This is the same as the arrangement in the sparrow, except that the first or hind toe is wanting, the three toes in front being the second, third, and fourth toes of the usual four-toed birds. Examples of this arrangement are found among the plovers. One of our woodpeckers lacks the first toe, and the fourth toe is thrown behind. This gives the last arrangement of three-toed birds; viz. *two in front and one behind.*[2]

It will thus be seen that the first toe is in many cases entirely wanting. From its absence to its reaching the length and strength of the front toes, there are all possible grades found in the feet of our birds. The kittiwake gull has the hind toe so small and wart-like (often without any claw), that it is readily overlooked by beginners in ornithology. Most of our plovers have just three toes, but the black-bellied plover shows a minute hind one. All of our barnyard fowl have a short hind toe, and in them, as in other birds with the first toe short, it is elevated[3] above the level of the front toes.

Appendages of toes. — The toes of birds have claw-like nails; these are called *claws* (or usually *nails* in this book), and vary much in strength, length, and curvature. They are so strong on birds of prey that they have the special name *talons.*[4] The hind claw is very long and almost straight[5] in the horned larks. In the grebes, the claws are much flattened[6] and resemble human nails. The herons and a few other birds have a curious saw-like ridge along the inner side of the middle claw; in these cases the claw is said to be *pectinate.*[7]

In many birds, the basal portions of some of the toes are more or less grown together. This growing together reaches the maximum in the kingfisher, where the outer and middle toes are united for half their length.[8]

  1   2   3   4   5   6   7

The principal union of toes is through their connection by a thin, movable membrane; this, whether small or large, is called *webbing*. In many families of birds, the three front toes have a distinct webbing at base only; if this webbing does not reach more than half way, the feet are *semipalmate*.[9] In the ducks, terns, etc., the front toes are webbed to the claws. This plan, which is so common, is called *palmate*.[10] A few of our birds have all four toes joined by a full webbing, and for this arrangement the word *totipalmate*[11] is used. Some birds with more or less webbing at the base of the toes have, in addition, a stiff, spreading membrane along the sides, sometimes lobed, sometimes plain; this plan is called *lobate*.[12] In the sea ducks, the front toes are palmate and the hind toes lobate;[13] in the grebes, the front toes are lobate. Many of the snipes have a narrow border along the edges of the toes, but not wide enough to be called lobate; these are said to be *margined*.[14]

CHAPTER VI

THE TAIL

Use. — The general office of the tail is to guide the bird in flight, but it is also used for other purposes. The woodpecker climbs trees, and the chimney swift climbs and rests on the sides of chimneys by its aid.

Kinds of feathers. — The feathers of the tail, like those of the wings, are of two sorts: *quill-like feathers* and *coverts*. The true tail feathers, or *rectrices*, are stiff, pennaceous, well-developed feathers having a strong quill and a broad, spreading vane, with rarely any plain aftershaft, or downy portion. The inner side of the vane is wider than the outer. The number

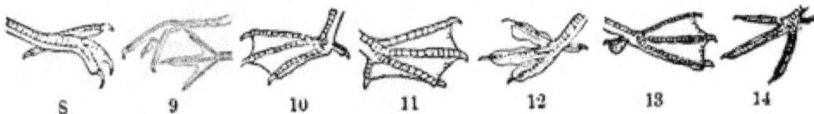

8   9   10   11   12   13   14

of tail feathers is almost always even, and varies from none to
upwards of thirty. This statement seems to indicate great
variation among birds with reference to the rectrices; in
reality there is but little variation, as a very large proportion
of birds have twelve, and the numbers eight, ten, twelve,
and fourteen, will include all except a few odd forms, most of
which are not found in the region covered by this book. The
rectrices have their bases covered, both above and below, by
short feathers called *upper tail coverts* and *lower tail coverts.*

Arrangement of rectrices. — The central pair of tail feathers
is above all the others, and each successive pair outward lies
under all the preceding ones.

Forms of rectrices. — A tail feather of the English sparrow
illustrates the usual form. It can be seen to widen gradually
toward the tip. The more important variations from this
type are the *lanceolate*,[1] where the vane is widest near the base,
and gradually narrows toward the tip; the *linear*,[2] where the
vane is narrow throughout; and the *filamentous*,[3] where it is
very narrow, as in the outer tail feathers of the barn swallow.

Varieties of tip and texture. — The usual tip is *rounded;*[4] if
very abruptly and squarely tipped, it is said to be *truncate;*
if obliquely and concavely cut off, *incised;* if regularly sharp-
pointed, *acute;* if abruptly sharpened, *acuminate.*[5] Most acuminate feathers are apt to be stiff and are used as an aid in
climbing; such feathers are said to be *rigid.* Some feathers,
while having the vane rounded, have the rhachis extending as
a hard point beyond it; in this case the feather is *spinous*[6] or
mucronate.

If the vane, instead of having its margin straight, has
its edge in rounded curves, it is said to be *crenulate.* Sometimes the rhachis curves upward at the center; in this case the
feather is said to be *vaulted* or *arched.* If the bending is side-

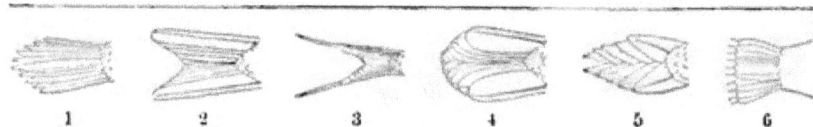

1   2   3   4   5   6

wise, it is described as *curved* outward or inward, according to
the side which shows the bulging outline.

Shape of the tail as a whole. — The usual shape of the tail is
like that of a fan, but there are many and very important modifications of this form. Some of these are characteristic of certain
groups and are much used in classification; thus most terns
can be separated from the gulls by this feature alone. If
the tail feathers are even in length, the tail is said to be even,
square, or *truncate.*[7] If the central pair is the longest, and
each successive outer pair is shorter, the tail is *graduated.*[5] If
each pair is shorter by a constant amount, the tail forms a
regular angle, and might, if at all common, be called an *angulated*[8] tail. A much commoner variety is said to have each
successive pair shorter by an increasing amount; this forms the
rounded tail;[4] sometimes each successive pair is shorter by a
decreasing amount, and this forms a *wedge-shaped* or *cuneate*
tail. If the central pair is excessively long, the tail is said to be
exserted;[9] when not so excessively elongated, it is *pointed.*

The opposite of graduation is very common among birds; that
is, each successive outer pair is longer than the preceding pair.
If this is true merely to an inappreciable extent, as in the English
sparrow, the tail is *emarginate;*[10] but when the difference is great
enough to make a very distinct angle, as in the chipping sparrow,
the tail is *forked;*[11] and the prefixing of the word *slightly* or *deeply*
tells how great the forking is. In deeply forked tails, like those
of the barn swallow, the outer rectrices are narrowed so as to be
filamentous. Such tails are said to be *forficate.*[3]

Sometimes there is a combination of the two plans above
given. If the middle pair is short, and about three pairs outward are successively longer, and the last two successively
shorter, the outer and middle pairs having about equal lengths,
a *doubly rounded*[12] tail is formed. If the middle pair is long

| 7 | 8 | 9 | 10 | 11 | 12 |

and the next two or three pairs successively shorter and the rest successively longer, a *doubly forked* tail is the result. This variety, though common among sandpipers, is so slight a forking that *doubly emarginate*[1] would be a better term.

In examining a tail to discover to which type it belongs, the student should be careful to spread the feathers but little. An emarginate tail might readily be made to appear square or even rounded by widely spreading it, and a truncate tail would always be changed to a rounded one.

The upper and lower tail coverts consist of numerous short feathers, and are never wanting, though the upper ones are often very short, as in the ruddy duck, and sometimes very long, as in the peacock, where the upper coverts, and not the rectrices, form the gorgeous tail of the male bird. In some of the storks the under coverts form the elegant plumes. The under tail coverts form the *crissum* of a bird.

CHAPTER VII

VOICE, MOVEMENT, AND MIGRATION

THE sounds made by birds are so peculiar, and so different from those that can be represented by letters, that any attempt to form such sounds into words is sure to prove more or less of a failure. The only successful way to learn a bird by its notes, is to see the bird while hearing it. Afterwards the sounds will reveal the bird. Beginners can hardly appreciate the variety of notes a single bird can make. Some have thought the only noises a catbird makes are those made when disturbed. The fine songs of birds are always made when undisturbed. Birds sing different songs at different seasons, but the finest of all are those made during the nesting time. A number of birds that seem to have no vocal powers during the greater part the year, sing sweetly in the spring. A still greater number, which merely chirp at other times, trill a long series of notes during mating time. It is practically only the male that sings; the female chirps. Nothing adds more to the enjoyment of nature than a knowledge of the notes, songs, and warblings of the birds. No teacher or book can give you more than a start toward the attainment of this knowledge. Two rules only can be given: (1) Learn to know birds. (2) Carefully observe them and listen to their songs.

As soon as you have learned to know birds, you will find among them many differences besides those of voice, form, and color. The places they frequent, — pond, marsh, meadow, upland, shrubbery, or forest, — in the water, on the ground, among the rocks, on the trunks of trees, or in the tree tops, — are as varied as their notes.

Their habits of sitting, their course in flight, their method of starting, their ways of coming to rest, are all peculiar to each bird.

Their solitary or social habits, their friendly or quarrelsome ways, are also well worthy of observation and study.

The way they flit their tails, the way they nod and twitch their heads, the way they use their feet, are other peculiarities that will aid you in recognizing them.

You will have to acquire this kind of knowledge out of doors. It cannot be taught in schoolrooms. It cannot be taught to any extent even by a teacher who accompanies his pupils on their trips. The teacher and books have done their work when they have given the names of the birds. The rest you must do for yourselves.

Among the most interesting of all the peculiarities of birds, are the migrations of a large proportion of them. Many live and nest in the far north, hundreds of miles beyond the limits of the United States, and go south to the Gulf States, in the winter, traveling more than a thousand miles to their new abode. These, for the northern United States, are but *birds of passage*. Others, while nesting in Canada and Labrador, spend their winters in the middle or the western states, and form for those sections *winter residents.* Still others nest with us and go south in winter to the Gulf States, or even to the West Indies and South America. These are *summer residents.* Some endure and even seemingly enjoy all the changes of climate any of our localities afford; these stay in the same place throughout the year. They form our *resident* birds. Doubtless many of those species which may be found at all seasons are somewhat *migratory;* that is, the individuals we have in the winter come from places somewhat further north, and those that are here in the summer find warmer places further south in the winter; but some birds, like the English sparrow, never migrate.

CHAPTER VIII

NESTS AND EGGS

There is no better or more useful work than to watch birds build their nests, hatch their eggs, and raise their young. After the student is able to recognize birds without difficulty, he is prepared to watch them and to learn all he can of their ways of living, their mating, their singing, their nesting, their eggs, their young, etc. It is not difficult to observe birds without disturbing them. An interest in living birds will soon lead the student to love them, and then he will be able to act when near them so as not to annoy or interrupt them in their work.

We have all read of men who could go among the most timid animals without disturbing them, and probably some of us have envied such people. But that power does not come spontaneously; it is gained only by careful attention to the peculiarities of the animals, the result of interest in their habits, which will lead to, if it does not begin with, an affection for them. Those who love birds find no great difficulty in studying their habits. A good opera glass will enable a person to see a bird as well as though it were at half the distance. When at a distance of fifty feet it can be seen as distinctly as with the naked eye at a distance of twenty-five feet. Most birds can be approached as near as fifty feet by a person who has no gun and who shows by his actions that he does not intend to harm them.

This book is written chiefly to help you to recognize birds, not to tell you all about them. But if you are interested in the study it will be a great pleasure to you to learn all you can about the birds that frequent your locality. Through book study alone no complete knowledge can be gained of birds or indeed of any animals. On the other hand, there are facts about the life history of migratory birds as well as the distribution of all birds, which can be learned only from the combined observations of many people, in many places, and so can be acquired only by reading. After you know a bird well enough to recognize it easily, it might be well for you to read a little about it, then watch it, listen to its song, examine its nest, observe all its habits. After that you will be ready to read with advantage and appreciation anything that has been written on the subject by creditable authors.

When examining birds' nests and eggs, do not handle them. It does no good, and may cause the bird to desert the nest. After the young birds have left the nest, you can without any harm secure it for your cabinet. In the chapter on preserving specimens you will find directions for cleaning eggs. Any extensive collections of eggs by students generally should not be encouraged; no eggs should ever be gathered without certain knowledge of the species of the bird. Such eggs are absolutely worthless for a collection. The variety of eggs which a single species lays is in many cases very great, and the number of species which lay similar eggs is also great, so there is no certain way of determining eggs except by observation of the birds. Your love for the birds, your feeling of horror at their useless destruction, and your desire for their protection and increase ought to make you slow to interfere with their nests and eggs. Single eggs of most birds can be carefully taken from nests, without special harm. More than this should never be appropriated except for the purpose of completing great collections, which can be studied by thousands of people. Such institutions as the American Museum of Natural History in New York City, the National Museum in Washington, and the Academy of Natural Sciences in Philadelphia, should of course be supplied with full sets of eggs with their nests. The educational value of such collections overbalances the injury done. But the usefulness of private collections is not great enough to justify the injury to the birds. A collection for the educational uses of a school, made by taking single eggs from nests, answers all the ordinary demands.

PART II

KEY, CLASSIFICATION, AND DESCRIPTION OF THE SPECIES

—oo:o:oo—

METHOD OF USING THE KEY

Caution. — In using the Key, never read any statements except those to which you are directed by the letters in parenthesis.

Rule. — First read *all* the statements following the stars (*) at the beginning of the Key; decide which one of these best agrees with the specimen you have. At the end of the chosen one you will find a letter in parenthesis ( ). Somewhere below, this letter is used two or more times. Read carefully *all* the statements following this letter; at the end of the one which most nearly states the facts about your specimen, you will again be directed by a letter to another part of the Key. Continue this process until instead of a letter there is a number and a name. The name is that of the Family to which your bird belongs. Turn to the descriptive part of the book where this family number, in regular order, is found. The headlines on the right-hand pages will show you which way to turn for the family sought. Under all Families of more than three species, another Key will enable you to determine the species.

The illustrations are as accurate as they could be made in black and white, but too much reliance must not be placed upon them. The student must remember that there are seasonal, sexual, local, and even individual differences as well as

39

the great variations for age. The whole description should be
read before deciding. The measurements of parts are very
important and should always be noted. Generally these measurements are only average ones, and some differences may
be allowed for. In order to judge of the amount to allow,
notice the extent of the variation in the length of the wing
as given in parenthesis. Other parts vary in about the same
proportion. Dimensions are always given in inches and such

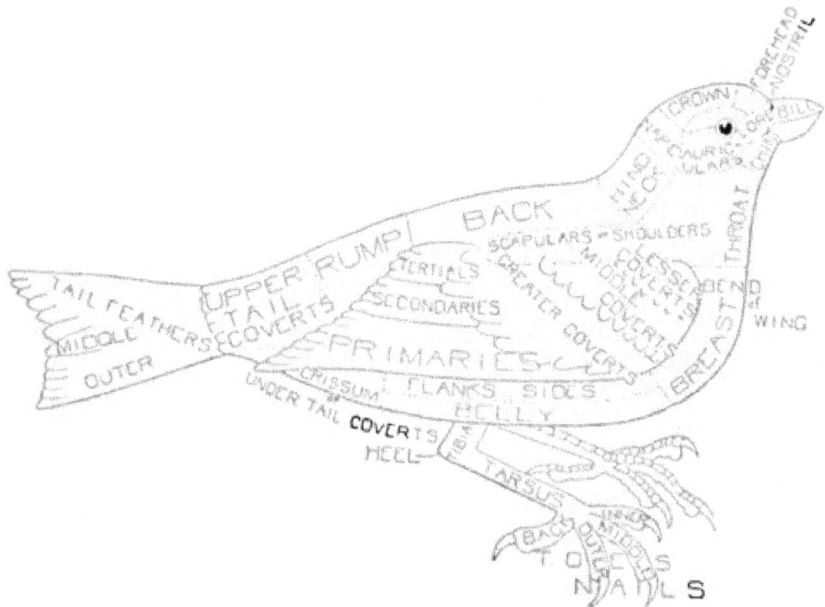

fractions thereof as are found on all common rulers. The
"*length*" of the bird is the distance from the tip of bill with
the neck extended to the end of the longest tail feather. In
mounted birds, allowance must be made for the curved neck.
The "*wing*" is the straight distance from the bend of the
wing to the tip of the longest primary. This can always be
accurately determined from any specimen; and so throughout
the book, in both keys and descriptions, great use is made
of this measurement. The "*tail*" is the length of the long-

est tail feather to the flesh in which it is fastened. This cannot be accurately measured without feeling (by placing the thumb and first finger above and below the tail coverts) for the fleshy mass to which all tail feathers are attached. The "*tarsus*" can be readily measured. It is the distance from the joint at the heel to the toes. The word "*culmen*" is almost always used in the book instead of "*bill*," because its measurement is more easy and certain. It is the *straight* dis-

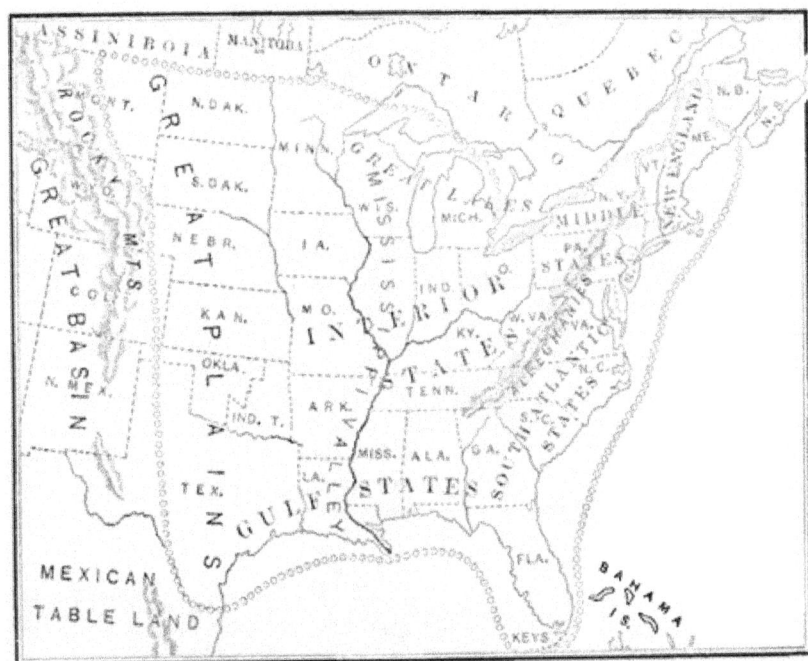

tance from the beginning of the bill at the forehead to the tip. The "*toe*" measures only the length to the base of the nail. If length to the tip of nail is intended, the expression "toe and nail" or "toe and claw" is used.

The name at the beginning of each description is the common name of the bird as decided upon by the American Ornithologists' Union; the names at the end in parenthesis are other names in more or less common use.

For convenience of reference the pictures of bird parts are repeated in the Key. The map on p. 41 shows the portion of the United States covered by the birds in this book.

KEY TO THE FAMILIES OF BIRDS

* Swimming birds: legs rather short; three front toes either with full webbing,[1] or with such membranes along their sides[2] as to take the place of webbing. (All birds with the front toes webbed to about the nails will be found in this group, excepting one very long-legged bird with the tarsus 12 inches or more long. All the illustrations in this book, from p. 279 to p. 348, are of swimmers.)

  Key to the Families of Swimming Birds, p. 48.

* Wading birds: legs in most cases much elongated; tibia always exserted, and in most cases more or less naked below (see p. 248); tail relatively short; few have the tail extending beyond the tip of the wings when closed; toes frequently with more or less webbing at base,[3] sometimes narrowly lobed along the sides. (Some species of these birds live habitually in dry localities, but their forms are very characteristic, and only a few ground birds, such as the meadow lark or quail, could possibly be placed in this group. All the illustrations from p. 227 to p. 275 are of waders.)

  Key to the Families of Wading Birds, p. 46.

* Birds fitted neither for swimming nor for wading. (A.)

  A   With only 2 toes in front, and in most species 2 behind;[4] eyes on the side of the head, as is usual with birds. (Q.)

  A.   Eyes in a facial disk, and thus directed forward instead of sidewise;[5] outer toe can be used either before or behind —

     — Nail of middle toe smooth on the inner side...............
  .... XXVII. **Horned Owls, etc.**

     — Nail of middle toe saw-like on the inner side[6] ....
  .... XXVIII. **Barn Owl.**

  A.   With **3 toes** permanently in **front and 1 toe** (rarely absent) behind;[7] eyes directed sideways. (The **vultures**, p. 212, are exceptions, as the outer toe can be used behind as with the owls: these are large birds, with the head and neck nearly bare of feathers.[9]) (B.)

1   2   3   4   5

6   7   8   9   10

B. Bill hooked and with a distinct membrane (cere) at the base, extending past the nostrils [8] —

— Head fully feathered. or nearly so.........XXIX. **Hawks, etc.**

— Head and neck naked or merely covered with hair [9]..........

....XXX. **Vultures.**

B. Bill without cere, and in most cases not strongly hooked. (**C.**)

  C. Hind toe short, small, inserted above the level of the others; [7] front toes with a plain webbing at base; [7] bill generally stout, short, and horny; [10] outer primaries of the wing curved and usually stiff; ground-living game birds —

    — Wing. 4–15 inches long...............XXXIV. **Grouse, etc.**

    — Wing over 16 inches long...............XXXIII. **Turkeys.**

  C. Bill straight, the horny tip separated from the base by a narrow portion; nostril opening beneath a soft, swollen membrane [11] (hard and somewhat wrinkled in mounted birds).......XXXI. **Pigeons.**

  C. Bill stout, straight, longer than the head; [12] feet with the outer and middle toes grown together for half their length; [13] tarsus very short.............................XXIII. **Kingfishers.**

  C. Bill very slender and long; [14] the smallest of birds; wings not over 2½ long in our species ....XIX. **Hummingbirds.**

  C. Bill with the top ridge or culmen very short, but the gape both wide and deep, reaching about to the eyes; [15] gape usually three times as long as the culmen. (**O.**)

  C. Not as above. (**D.**)

D. Inner secondary quills lengthened, nearly as long as the primaries in the closed wing; [16] nail of hind toe much lengthened and generally straightened; [17] the ground birds called "larks." (**N.**)

D. Inner secondaries not especially lengthened; the first primary short, [18] never more than ⅔ as long as the longest, usually less than ½ as long, sometimes barely noticeable on the under edge of the wing. [19] (**J.**)

D. With neither the inner secondaries very much lengthened nor the first primary much shortened; the first primary always more than ⅔ as long as the longest quill. (**E.**)

  E. Bill broad, depressed, wider than high at base, usually tapering to a point, which is often abruptly hooked. [20] (**H.**)

11

12

13

14

15

16

17

18

19

20

(E. Bill slender, about as high at base as broad, and regularly curved
   downward from the base to the very acute tip.[1]   The **Bahama
   Honey Creeper** (636. *Cœreba bahamensis*), of the family Cœrébidæ,
   has been found in southern Florida.)

  E. Bill higher than broad at base.[2]   (F.)

F. Bill stout at base, and with the gape so angulated as to bring the
   corners of the mouth downward;[2 3 4 5 6 1)] no lobes or nicks along
   the cutting edge of the upper mandible.   (G.)

F. Bill stout, with convex outline, and with lobes or nicks near the
   center of the upper mandible,[7] but not crossed at tip; wing, $3\frac{1}{4}$ long;
   tail even ....XII. **Tanagers.**

F. Bill stout, compressed, notched, and abruptly hooked near the tip;[8]
   plumage olivaceous; tail without either white or yellow blotches;
   wing, $3\frac{1}{4}$ or less long.................................VIII. **Vireos.**

F. Bill not as above, little, if at all, hooked; colors in most species
   bright and distinctly marked; tail feathers generally blotched with
   white.....................................VII. **Wood Warblers.**

  G. Upper ridge of bill extended backward so as to divide the feathers
     of the forehead;[2 5 6] no notch at tip of bill or bristles at the rictus
     (if any bristles can be seen they are less than $\frac{1}{3}$ of the length of
     the bill); bill not over $\frac{2}{3}$ as high at base as long, in most species
     less than $\frac{1}{2}$ as high........................XIV. **Blackbirds, etc.**

  G. Ridge of bill not especially extending upward on the forehead
     (except in a few very stout-billed birds with the bill as high as
     long); bill usually short, stout, and conical.[3 9 10 4] ....
     ....XIII. **Finches, etc.**

H. Rictal bristles absent; nostrils overhung with bristles; tail short,
   even, and tipped with a yellow band; head crested.[11]..X. **Waxwings.**

H. Rictal bristles numerous and long.[12]   (I.)

  I. Tarsus with a sharp ridge behind and a distinct row of square
     scales (scutella) merely down the front;[13] wing, 2-$2\frac{3}{4}$ long; no
     crest.....................VII. **Wood Warblers** (Flycatching).

  I. Tarsus rounded behind and with the scutella lapping round on the
     outside of the tarsus about to the back portion;[14] wing, $2\frac{1}{4}$-$5\frac{1}{2}$
     long; crest small or none .... XVIII. **Tyrant Flycatchers.**

J. Tarsus (booted) covered with a continuous plate along the front;[15]

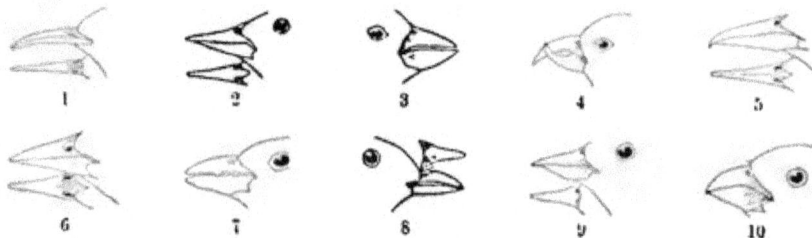

no distinct squarish scales, except near the toes (a very young bird of this group will show scutella, but they are gradually fused together as the bird grows older) —

— Wing, 2½ or less long ....II. **Kinglets.**
— Wing, 3 or more long ....I. **Thrushes, etc.**

(J. Tarsus (scutellate) covered with a row of rectangular scales in front and behind ;[16] wing, 7–9 long ; tail, 9–11 long..XXXII. **Curassows.**)

J. Tarsus (scutellate) covered with distinct rectangular scales only along the front.[13] [17] (**K.**)

  K. Bill stout, compressed, distinctly notched and hooked at tip ; nostrils and rictus with bristles ;[8] [18] no crest —

— Wing, 2–3½ long.............................VIII. **Vireos.**
— Wing, 3½ or more long ....IX. **Shrikes.**

  K. Bill, if hooked at tip, having the nostrils without bristles extending over them ; in most species the bill is not hooked. (**L.**)

  L. Tail feathers acute-pointed and somewhat stiff ;[19] bill slender and decurved ;[1] back mottled brown ; belly white........IV. **Creepers.**

  L. Tail feathers rounded at tip and the outer ones white ; bill slender and somewhat notched at tip ; back bluish-gray ; belly white ; wing, 2–2¼ long ....II. **Gnatcatchers.**

  L. Not as above, but with the tail feathers rounded at tip and soft. (**M.**)

  M. Bill long and stout; nostrils covered with bristly feathers[20] (excepting a western, dull, blue-colored jay); large birds, 10–25 long.. .... XVI. **Crows, etc.**

  M. Bill rather slender ; culmen more or less curved ; nasal feathers not directed forward over the nostrils ; tail rounded ; either small birds, 4–6½ long, with barred quills,[21] or large birds, 8–12 long, with quills not barred ......V. **Mocking Birds, Wrens, Thrashers, etc.**

  M. Bill neither notched at tip nor much decurved ; nostrils concealed by dense tufts of bristly feathers ; small birds, 4–7 long.......... ....III. **Nuthatches, etc.**

  M. Bill with the culmen about straight; wings about 5 long; tail about 2¼ long and square at tip...................XV. **Starlings.**

  N. Nostrils overhung with bristly feathers ; tarsus (scutellate) with a row of nearly square scales, behind as well as before ;[16] nail of hind toe longer than the toe and nearly straight ; bill not very slender ... .XVII. **Larks.**

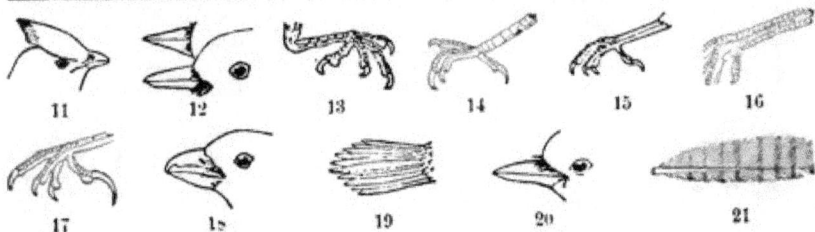

11   12   13   14   15   16

17   18   19   20   21

N. Nostrils exposed ; tarsus not scutellate behind ; nail of hind toe very long but curved ;[1] wing, 3–3½ long ....VI. **Pipits, etc.**

N. Nostrils exposed ; wing over 3¾ long ; breast with yellow.......... ....XIV. **Meadow Larks.**

N. Not as above ; some forms in..................XIII. **Finches, etc.**

   O. Plumage mottled browns and soft ; middle toe much longer than the side ones ; its nail (pectinated) with saw-like teeth on the inner side[2]................................XXI. **Goatsuckers, etc.**

   O. Plumage compact ; nail of middle toe not pectinated. (**P.**)

P. Tail of stiff feathers (in our common species the tail is rounded, with stiff shafts extending beyond the webs)[3]...............XX. **Swifts.**

P. Tail without stiff feathers and never rounded, often forked and without spinous tips ; head never crested........ ....XI. **Swallows.**

P. Tail nearly square ; its feathers tipped with yellow, head crested[4]... ....X. **Waxwings.**

   Q. Bill stout and decidedly hooked, higher at base than long ; bright-colored bird with yellow, orange, and green feathers ....XXVI. **Parrots.**

   Q. Bill stout and straight ;[5] tail feathers stiff and acute-pointed[6].... .... XXII. **Woodpeckers.**

   Q. Bill various, but always somewhat curved and without teeth along the cutting edge ; tail long, of round-tipped soft feathers......... ....XXV. **Cuckoos, etc.**

   Q. Bill short, broad, and decidedly toothed ;[7] tail long, of 12 broad feathers .... XXIV. **Trogons.**

Key to the Families of Wading Birds

* Toes in front webbed to the nails like the duck's ;[8] bill with teeth-like ridges, also like the duck's ; legs with the tarsus 12 inches or more long ; Florida............................... XLIX. **Flamingoes.**

* Nails of the toes excessively lengthened and nearly straight ; nail of the hind toe much longer than its toe ;[9] southern Texas .... XXXV. **Jacanas.**

* Birds with neither full-webbed toes nor nails lengthened and straightened. (**A.**)

1   2   3   4   5

6   7   8   9   10

A. Head with a horny shield on the forehead; [10] in other respects fully feathered.................................... XLII. **Rails, etc.**

A. Head with more or less of naked tracts (free from feathers but usually with some hairs) in front of the eyes or around the eyes; [11] [12] some species have the head entirely naked. (**H.**)

A. Head fully feathered and without horny shield. (**B.**)

B. Bill hard throughout and not sensitive (a peculiar smoothness of bill of dried specimens will show that the bill was hard in life). (**E.**)

B. Bill weak and soft, at least at base, often long and slender; if short, pigeon-like; hind toe always less than half the length of the inner one, sometimes absent; (dried specimens usually show the surface of the bills so roughened or dull in color as to indicate their soft condition when alive). (**C.**)

   C. Toes with lobed membranes along their edges as wide as the toes, sometimes wider; [13] tarsus much flattened sideways; body flattened below.......................................XLI. **Phalaropes**.

   C. Legs exceedingly long, the tarsus over 3½ long. XL. **Avocets, etc.**

   C. Tarsus less than 3½ long; toes with no wide membranes along their edges. (**D.**)

D. Bill usually shorter than the head, pigeon-like, the soft base separated by a narrow portion from the hard tip; [14] toes only three (one species has a hind toe ⅛ inch long); tarsus (reticulate) with rounded scales in front.............................. XXXVIII. **Plovers.**

D. Bill slender; nostrils narrow, exposed slits in elongated grooves extending from a half to nearly the full length of the bill; [15] tarsus (scutellate) with transverse and more or less square scales in front......
....XXXIX. **Snipes, etc.**

   E. Bill, 2⅜–8 long. (**G.**)

   E. Bill, ½–2⅜ long. (**F.**)

F. Tarsus, middle toe and nail, and bill each about 1 long; the bill nearly straight; wing about 6 long..........XXXVII. **Turnstones.**

F. Not as above; tarsus usually shorter than the middle toe and nail; if the tarsus is about 1 long, the wings are much less than 6 long.....
....XLII. **Rails, etc.**

   G. Bill nearly straight, much flattened sideways and very blunt at tip; [16] toes only three and webbed at base. XXXVI. **Oyster-catchers.**

   G. Bill somewhat curved downward; tarsus, 3½–6 long; wing, 10–14 long.......................................XLIII. **Courlans.**

   G. Bill about straight; tarsus, 6–12 long; wing, 16–25 long; young:
.... XLIV. **Cranes.**

     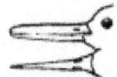

11   12   13   14   15   16

G. Bill very broad and flattened, twice as wide near tip as at the middle ;[1] young....................... ...XLVIII. **Spoonbills.**

H. Nail of middle toe (pectinated) with a fine, saw-like ridge on the inner edge ;[2] bill straight, acute, and with sharp cutting edges......
.... XLV. **Herons, etc.**

H. Nail of middle toe without saw-like teeth. (**I.**)

    I. Bill very broad and flattened, twice as wide near the tip as at the middle[1]................................XLVIII. **Spoonbills.**

    I. Bill narrow, about as wide as high, gradually and decidedly curved downward for its whole length[3]..................XLVII. **Ibises.**

    I. Bill either narrow, straight for half its length and then curved downward,[4] or else a very large bill (over 2 high at base) with the end curved upward[5].... .... XLVI. **Storks, etc.**

    I. Bill higher than broad, about straight, not very acute ;[6] very large birds over 40 long, with very long necks and legs. XLIV. **Cranes.**

Key to the Families of Swimming Birds

* Hind toe present and connected with the inner toe by a webbing; *i.e.* all four toes webbed.[7] (**E.**)
* The front toes bordered by broad membranes for their whole length —[8]
  — Diving birds with legs at the end of body ....LXIV. **Grebes.**
  — Legs near center of body. XLI. **Phalaropes,** or XLII. **Rails, etc.**
* The three front toes connected together by webbing.[9] (**A.**)

    A. Bill with teeth or ridges along the edges, easily seen from the lower side[10] ....L. **Ducks, etc.**

    A. Bill with the cutting edges even. (**B.**)

B. Legs inserted so far back along the body that the bird in standing has to take a vertical position (see p. 342) ; diving birds —
  — No hind toe ....LXII. **Auks.**
  — Hind toe present, short ....LXIII. **Loons.**

B. Legs so inserted that the body in standing takes nearly a horizontal position. (**C.**)

    C. Nostrils tubular, the tubes near together at the top of the bill ;[11] wings less than 13 long .... LVII. **Shearwaters, etc.**

    C. Nostrils tubular, the tubes on the sides of the bill near the base ; wings, 16–30 long .... LVIII. **Albatrosses.**

1   2   3   4   5

6   7   8   9   10

  C. Nostrils not tubular but slit-like. **(D.)**

D. Upper mandible decidedly hooked at tip and plainly made up of separate pieces, one forming a kind of roof to the nostrils;[12] tail dark-colored, with the middle feathers lengthened ....LXI. **Jaegers.**

D. Upper mandible not made up of separate pieces and at least as long as the lower mandible............................LX. **Gulls, etc.**

D. Bill flattened sideways and knife-like; the lower mandible longer than the upper one [13]...........................LIX. **Skimmers.**

  E. Bill straight or slightly curved. **(G.)**

  E. Upper mandible decidedly hooked at tip, hawk-like.[14] **(F.)**

F. Tail, 14–20 long and forked for half its length; space in front of eyes bare of feathers [15]........................LI. **Man-o'-War Birds.**

F. Tail, 5–10 long; bill less than 4; plumage dark. LIII. **Cormorants.**

F. Tail, 5–8 long; bill 8–15 long; plumage light........LII. **Pelicans.**

   G. Bill stout at base and slightly curved near tip;[16] wing, 14–22 long. ....LV. **Gannets.**

   G. Bill slender and nearly straight; wing, 12–14 long; neck very long...LIV. **Darters.**

   G. Bill stout, slightly curved; wing, 10–12 long...LVI. **Tropic Birds.**

ORDER I. PERCHING BIRDS (PÁSSERES)

This is the highest and much the largest order of birds; it contains nearly half of our birds (those east of the Rocky Mountains in the United States) and more than half of all known birds. In it are found the finest of the songsters.

The toes are four in number, three in front and one behind. The front toes are divided about to their bases and have no webbing or membrane along their sides. The hind toe is on a level with the rest and as long as the shortest front toe. The legs are slender, comparatively short, and so placed as to give the body, when at rest, a horizontal position. In size these birds range from very small to medium; from the size of a kinglet to that of a robin, or a little larger.

11   12   13   14   15   16

FAMILY I. THRUSHES, BLUEBIRDS, ETC. (TÚRDIDÆ)

This large family (300 species) is usually separated into several subfamilies.

The **Thrushes** are generally large, hopping birds, noted for their song, plain colors, and usually spotted breasts. The tail is nearly square tipped, of wide, soft feathers.  They are woodland birds of migratory habits; even when, as in the case of the robin, we have them throughout the year, it is probably true that those with us in the winter came from places farther north, and those which are found here in the summer wintered farther south.

Townsend's Solitaire

The first primary is a very short one;[1] bill rather long and slender; the upper mandible usually with a slight notch near the tip. Nostrils oval, the bristly front feathers nearly reaching but never concealing them; rictus with bristles;[2] tarsus booted.[3]

Key to the Species

* Tail about an inch shorter than the wings. **(A.)**
* Tail about as long as the wings and with its outer (under) feathers broadly tipped with white; bill peculiarly broadened at base and hooked at tip. **Townsend's Solitaire** (754. *Myadéstes townséndii*), which is pictured above, is sometimes found east of the Rocky Mountains, though its usual habitat is westward to the Pacific. It is a dull brownish-ash-colored bird with wings from 4 to 4¼ inches long.
* Tail slightly longer than the wings; no white on the tail, but the under tail coverts chestnut. The catbird might be looked for here, as its tarsus is somewhat booted. It will be found in Family V., p. 65.

A. Plumage more or less blue, rather brightly so on the tail ....
....9. **Bluebird.**

A. Head and tail quite dark colored, almost black ; outer (under) tail
feathers tipped with white ; breast brownish..6. **American Robin.**

A. Tail blackish, the outer feathers tipped with white ; a dark collar
across the breast ; western...................7. **Varied Thrush.**

A. Outer tail feathers white at base but broadly black tipped ; upper
tail coverts white.............................8. **Wheatear.**

A. Tail without white or blue ; breast spotted ; general color brownish.
(**B.**)

B. Upper parts reddish on head, shading to olive on rump and tail.....
....1. **Wood Thrush.**

B. Upper parts olive on head, shading to reddish on rump and tail......
....5. **Hermit Thrush.**

B. Upper parts from forehead to tip of tail of almost the same shade of
color. (**C.**)

   C. Upper parts reddish from head to tip of tail......... ....
....2. **Wilson's Thrush.**

   C. Upper parts olive throughout. (**D.**)

D. Throat, breast, and ring around eye a rich creamy-buff... ....
....4. **Olive-backed Thrush.**

D. No distinct buffy eye ring, and the throat
and breast nearly white, with only a slight
buffy tinge ; a grayish blotch in front of
the eye.........3. **Gray-cheeked Thrush.**

1. **Wood Thrush** (755. *Tŭrdus mus-
telīnus*). — A large, common, brownish-backed thrush, with white, heavily
spotted under parts, including the sides.
The crown is a bright cinnamon-brown,
and the back gradually changes in shade
to an olive-brown on the tail. It is not
at all confined to the woods, as its name
would indicate, but is often seen on
shaded lawns and in shrubbery. Its
power of song is very great, com-
paring well with that of any of the
thrushes.

Wood Thrush

Length, 8 ; wing, 4¼ (4–4½) ; tail, 3 ; tarsus,
1¼ ; culmen, ¾. Eastern United States ; breeding from Virginia and
Kansas northward, and wintering south to Central America.

2. **Wilson's Thrush** (756. *Túrdus fuscéscens*). — A large thrush, with a dull cinnamon-brown back, uniform in tint from head to tail. Its throat, belly, and sides are white; its breast buffy, delicately marked with triangular brownish spots. A retiring, though not particularly shy bird, inhabiting the dense woodlands, especially low, wet ones, and usually to be found nearer the ground than the wood thrush. Its notes are among the sweetest given by any bird, but it is impossible to write them in words or music. Its peculiarly weird song must be heard to be appreciated. (Veery; Tawny Thrush.)

Wilson's Thrush

Length, $7\frac{1}{2}$; wing, 4 ($3\frac{3}{4}$–$4\frac{1}{4}$); tail, 3; tarsus, $1\frac{1}{8}$; culmen, $\frac{5}{8}$. Eastern North America from Ontario southward; breeding from northern Ohio and New Jersey northward, and wintering mainly south of the United States. The **Willow Thrush**, a variety of the last (756a. *T. f. saliciola*), is a little larger, and has the upper parts less tawny, a russet-olive color, only a slight buff tint to the throat, and very few spots on the white breast. Length, $7\frac{3}{4}$; wing, 4; tail, $3\frac{1}{4}$; tarsus, $1\frac{1}{4}$; culmen, $\frac{5}{8}$. Rocky Mountain region, occasionally east to Illinois and possibly to South Carolina.

3. **Gray-cheeked Thrush** (757. *Túrdus alíciæ*). — A uniformly olive-backed thrush, with the middle of throat and belly white, the sides of throat and breast faintly buffy, spotted with triangular marks, and a whitish eye ring. In front of the eye there is a grayish blotch. A shy bird, of which but little is known, as it has been confused with the variety, Bicknell's Thrush, next given. (Alice's Thrush.)

Length, $7\frac{3}{4}$; wing, $4\frac{1}{8}$ ($3\frac{3}{4}$–$4\frac{3}{8}$); tail, $3\frac{1}{4}$; tarsus, $1\frac{1}{4}$; culmen, $\frac{1}{2}$. Eastern North America; breeding mainly north of the United States, and wintering south to Central America. **Bicknell's Thrush**, a variety of the last (757a. *T. a. bicknelli*), is somewhat brighter colored and smaller. Length, $6\frac{3}{4}$; wing, $3\frac{5}{8}$ ($3\frac{3}{8}$–$3\frac{7}{8}$); tail, $2\frac{5}{8}$; tarsus, $1\frac{1}{8}$; culmen, $\frac{1}{2}$. Breeding in the mountains of the northeastern states and Nova Scotia; migrating south in winter. Song very much like that of Wilson's Thrush.

4. **Olive-backed Thrush** (758ª. *Túrdus ustulátus swainsònii*). — A uniformly olive-backed thrush with the whole throat, breast, and eye ring a deep cream-buff, and the space in front of the eye the same color, instead of grayish as in the last species. A very shy bird, rarely seen, but often heard in notes similar to the hermit thrush's, though not so sweet. Its summer home is among the firs and spruces of the north.

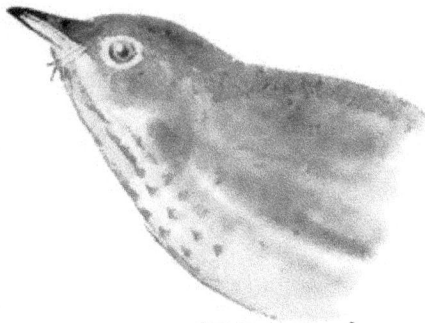

Olive-backed Thrush

Length, 7¼; wing, 3¾ (3½–4¼); tail, 2¾; tarsus, 1⅛; culmen, ½. Eastern North America, mainly in the mountains; breeding from northern New England northward, and migrating in winter to South America.

5. **Hermit Thrush** (759ᵇ. *Túrdus aonaláschkæ pallásii*). — A small thrush with olive-brown back changing abruptly to cinnamon-brown near the tail; the throat and breast somewhat buffy, with dark roundish spots abundant on the sides of the breast; middle of the belly white. A retiring, though not especially shy, bird, with about the sweetest and purest notes given by any of our feathered friends.

Hermit Thrush

Length, 7; wing, 3½ (3¼–3⅞); tail, 2¾; tarsus, 1⅛; culmen, ½. Eastern North America; breeding from the northern Alleghanies northward, and wintering from the northern states southward.

6. **American Robin** (761. *Mérula migratòria*). — A very common, large, red- or brown-breasted, slate-colored bird, with white on the throat, lower belly, and tips of the outer (under) tail feathers. The head and tail are much blacker than the back. The breast is unspotted except in very young birds. The robin is a noisy bird, but with less powers of song than any other of the thrushes.

American Robin

Length, 10; wing, 5¼ (4¾–5½) ; tail, 4¼ ; tarsus, 1¾ ; culmen, ⅞. Eastern North America ; breeding from Virginia northward, and wintering irregularly from Canada southward.

7. **Varied Thrush** (763. *Hesperocíchla nǽvia*). — A large slaty-backed bird with much orange-brown below and on the wings, and a dark collar across the breast. The tail is blackish, and the outer (under) feathers broadly tipped with white. This bird is about the size of the robin, belongs to the Pacific coast, but has been seen a few times in the Eastern States (New Jersey, New York, Massachusetts, etc.). (Oregon Robin.)

Varied Thrush

8. **Wheatear** (765. *Saxícola œnánthe*). — A small, northern, light-gray-backed, whitish-bellied bird, with black cheeks and wings. The fore-

head, upper tail coverts and basal half of the tail feathers are
white, the rest of the tail black. *Female* similar, but duller
and browner. *Young*
with much cinnamon-
brown and without the
cheek stripe. (Stone-
chat.)

Length, 6¼; wing, 4
(3¾–4¼); tail, 2¼; tarsus,
1; culmen, ½. An Old
World species breeding in
Labrador and straggling
southward to the United
States (Maine, Long Is-
land, New Orleans).

9. **Bluebird** (766. *Siália siális*). — A very
common, small, blue-backed, chestnut-breasted, white-bellied
bird. The *female* is more of a grayish-blue. Till the introduction of the English sparrow, this bird was to
be found everywhere around our homes. Its
sweet, joyous singing welcomed in the spring,
and its sadder notes of autumn told of
the dying year. From southern New
York and Illinois southward, it is to be
found throughout the year. In the
northern portion of its winter range
a few can generally be found living near cedar groves.

Wheatear

Length, 6¾; wing, 4 (3¾–4¼); tail,
2¾; culmen, ½. From the Rocky
Mountains eastward throughout the
United States, north to Ontario. The
Mountain Bluebird (768. *Siália
árctica*), a large bird without chestnut on the breast, and with a more
greenish-blue on the back, has been
occasionally seen east of the Rocky Mountains.

Bluebird

FAMILY II. KINGLETS, GNATCATCHERS, WARBLERS
(SYLVÍIDÆ)

This family includes a large subfamily (100 species) of Old World Warblers not represented in America, and two

small subfamilies represented in our fauna. The **Kinglets** are very small, musical, tree-loving, active, oliva-ceous birds, with, in the adult, some bright yellow or red on the crown, and a short, even or notched tail. The **Gnatcatchers** are very small, sprightly,

Golden-crowned Kinglet

ashy-colored, woodland birds, with long, graduated tails. Our species build very beautiful nests among the high branches of the trees.

Key to the Species

* Outer (under) tail feathers shortest [1] and white; tarsus scutellate; [2] colors gray.................................3. **Blue-gray Gnatcatcher.**
* Outer (under) tail feathers about the longest and without white; tarsus booted; [3] colors, olive-green with usually a yellow, orange, or ruby-colored spot on the crown. (**A.**)
  A. Crown patch bright-colored, bordered with black................
  ....1. **Golden-crowned Kinglet.**
  A. Crown patch, if present, ruby-colored, but without black........
  ....2. **Ruby-crowned Kinglet.**

1. **Golden-crowned Kinglet** (748. *Régulus satrápa*). — A very small, olive-green-backed, whitish-bellied bird, with a bright crown patch of gold or orange color, margined with black. The

1   2   3

male has orange and yellow; the *female*, only yellow. The kinglet is a fearless, nervous, quick-moving bird, found abundantly flitting among the most slender twigs of the trees at the proper season. The voice of the kinglet is marvelously rich and the singing unusually continuous for so small a bird.

Length, 4; wing, 2¼ (2–2¼); tail, 1¾; tarsus, ¾; culmen, ¼. North America in general; breeding from the northern states northward (in the mountains as far south as North Carolina), and wintering throughout most of the states, south to the Gulf or even into Central America.

2. **Ruby-crowned Kinglet** (749. *Régulus caléndula*). — This bird is like the last, excepting that there is no black on the head; the *female* even lacks the bright crown patch of color, and the *male* is apt to keep his bright red feathers hidden. The *female* and *young* appear just like warblers (the American warblers belong to Family VII.), but are de-

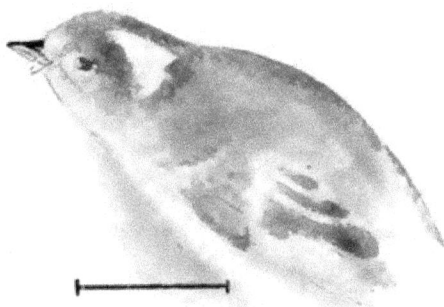

Ruby-crowned Kinglet

cidedly smaller than any of the olive-green-backed species, excepting those which have bright yellow below or conspicuous white blotches on the tail feathers, seen when the bird is in flight.

Length, 4¼; wing, 2¼ (2–2½); tail, 1¾; tarsus, ¾; culmen, ¼. North America in general; breeding mainly north of the United States (in the Rocky Mountains farther south), and wintering from the Carolinas south to Central America.

Blue-gray Gnatcatcher

3. **Blue-gray Gnatcatcher** (751. *Polióptila cœrúlea*). — A very small, bluish-gray bird,

with blackish wings and tail; the outer (under) tail feathers
are white, the forehead marked with a black border, and the
under parts lighter and duller than the back. The *female* is
without the black on the forehead. This is a bird usually
found among the upper branches of forest trees, and though
his song is sweet and varied, it can be heard but a little distance. His call note, a sharp *ting*, is readily heard.

Length, 4½; wing, 2½ (2–2¼); tail, 2½; tarsus, ½; culmen, ⅜. Middle
and southern sections of the eastern United States; breeding from Illinois
and New Jersey southward, and wintering from Florida to Central
America. It is rarely, though sometimes, found as far north as Maine
and Minnesota.

FAMILY III. NUTHATCHES AND CHICKADEES (PÁRIDÆ)

A family (100 species) of small birds, forming two widely
separated subfamilies. The **Nuthatches** are small, active, restless, creeping, short-tailed, long-winged birds, marked with
white, black, and brown colors. These noisy, but not musical,
sharp-billed birds are among the most nimble of creepers,
scrambling about in every direction, with the head downward
as often as in any other position. They derive their name
from the habit of wedging nuts into crevices of the bark, and
then hacking or hammering away with the bill till the shell
is broken. These nuts form only a small portion of their
food; generally they are insect eaters. The **Chickadees** are all
small, active, short-billed, long-tailed birds. Our species are
plain birds of white, black, and ashy colors. The *titmice*,
which are included in the subfamily, are conspicuously crested,
while the *chickadees* proper are without crest. The latter are
so called from the notes they utter, *chick-ā-dēē*.

Key to the Species

* Tail about as long as the wing and graduated;[1] bill less than a
  half inch long, and stout for its length. **(B.)**
* Tail about half the length of the wing and square; bill a
  half inch or more, long and slender. **(A.)**
  A. White below with rusty brown only on the under tail   ı
    coverts . . . . 1. **White-breasted Nuthatch.**

A. Under parts generally with much rusty brown ; crown black (*male*), or bluish-gray (*female*) ; a *white stripe* over eye . . . .
. . . . 2. **Red-breasted Nuthatch.**
A. Crown and sides of head brown, without stripes. . . . .
. . . . .3. **Brown-headed Nuthatch.**
B. Head conspicuously crested ;[2] throat and under parts nearly white, with rusty-brown sides. . . . . .4. **Tufted Titmouse.**
B. Head without crest ; throat black or dusky. (**C.**)
C. Top of head brown ; sides of body chestnut . . . .
. . . . .7. **Hudsonian Chickadee.**
C. Top of head black. (**D.**)
D. Greater wing coverts with whitish edges . . . . .5. **Chickadee.**
D. Greater wing coverts without whitish edges. .6. **Carolina Chickadee.**

1. **White-breasted Nuthatch** (727. *Sitta carolinénsis*). — A short-tailed, tree-creeping. bluish-backed. black-crowned, white-bellied bird. with brown blotches on the under tail coverts.
The sides of the head are white like the throat and breast, and the back neck black like the crown. The *female* has the black not so intense. The nuthatches are peculiar in their ability to run along tree trunks in all directions, with the head downward as often as upward. They are not singers, but have a call

White-breasted Nuthatch

note of *quank quank*, which they repeat with no reference to the position of their body.

Length, 6 ; wing, 3½ (3¼-3¾); tail, 2 ; tarsus, ¾ ; culmen, ¾. Eastern United States from Georgia to the Dominion of Canada; generally resident throughout. The **Florida White-breasted Nuthatch** (727b. *S. c. átkinsi*) is somewhat smaller and has the wing coverts and the quills very slightly, if at all, tipped with whitish. Wing less than 3½ ; tail, 1¼. It is found from South Carolina to Florida.

2. **Red-breasted Nuthatch** (728. *Sitta canadénsis*). — A short-tailed, tree-creeping, bluish-backed, brownish-red-breasted bird, with the black of the top and sides of the head separated by a

broad distinct white line over the eye. This is a more northern species than the last, and can easily be distinguished by the black line on the sides of the head and neck, and the generally brown under parts.

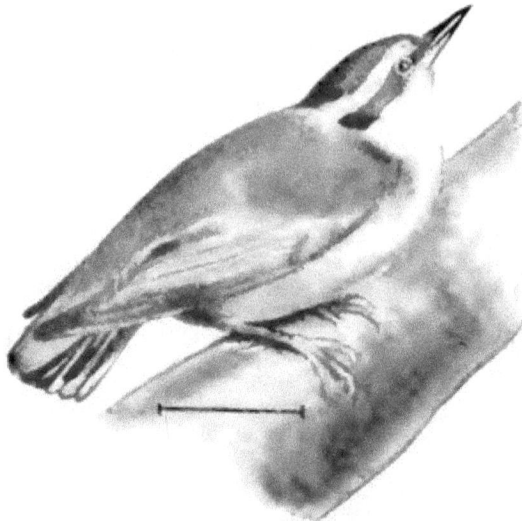

Length, 4⅝: wing, 2¾ (2½–2⅞); tail, 1½; culmen, ⅓. North America; breeding from northern New York, northern Michigan northward (farther south in the Alleghanies and Rocky Mountains), and wintering southwards to the Gulf.

Red-breasted Nuthatch

3. **Brown-headed Nuthatch** (729. *Sitta pusilla*). — A small, brown-crowned, bluish-backed, whitish-bellied nuthatch, with no white line over the eye, but with a whitish patch on the back neck. This is the nuthatch of the southern pine woods, where it is found associated with woodpeckers, but unlike them in their tree-top living habits, it scrambles up and down the trunks from the bottom

Brown-headed Nuthatch

to the top. All the nuthatches are much alike in habits, and are wonderfully nimble in their movements. Most creepers use the tail as an aid in supporting the body on perpendicular surfaces; but these birds make no such use of their short, square tails. The woodpecker's feet are strengthened by having

the outer toe turned backward; but the nuthatch's feet have only a slight enlargement of the nails.

Length, 4¾; wing, 2½; tail, 1¾; tarsus, ⅔; culmen, ⅓. South Atlantic and Gulf States, north to Maryland; accidentally to New York, Missouri, etc.

4. Tufted Titmouse (731. *Pàrus bícolor*). — A loud-voiced, conspicuously crested, gray bird of the woods, with some black on the forehead and brown on the sides.   Its loudest notes are a constant repetition of *peto peto*, sometimes changed to *de-de-de* in somewhat less ringing tones, producing a slight imitation of the notes of the chickadee. It is not at all shy, and so may be readily approached. (Crested Tit.)

Tufted Titmouse

Length, 6; wing, 3¼ (3-3½) ; tail, 3; tarsus, ⅞ ; culmen, ⅓.   Eastern United States north to northern New Jersey and southern Iowa ; casual in southern New England ; resident throughout. The **Black-crested Titmouse** (732. *Pàrus atricristàtus*) differs from the last species in having the whole crest, instead of only the forehead black. It is a somewhat smaller bird. Length, 5¼; wing, 2¾; tail, 2¾. Southeastern Texas and eastern Mexico.

5. Chickadee (735. *Pàrus atricapíllus*).—A small, black-capped, black-throated, ashy-backed bird, with the rest of the head and breast white; under parts buffy. This and the next species are much alike, but this has the greater wing coverts margined

Chickadee

with white.   Its common name expresses as closely as possible its whistled notes, *chick-à-dēē*. If its notes are well imitated, the bird will approach closely, or even alight on a person. (Black-capped Chickadee.)

Length, 5¼ ; wing, 2½ (2¾-2¼) ; tail, 2½ ; tarsus, ⅔ ; culmen, ¼. Eastern North America north of the Potomac and Ohio valleys to Labrador ; it migrates a little beyond its breeding range.

6. **Carolina Chickadee** (736. *Pàrus carolinénsis*). — A bird similar to the last, but smaller, and with the greater wing

Carolina Chickadee

coverts not margined with white. Though the notes of this species are somewhat different from those of the last, this difference is not so uniform as to render the printed form of much use to the beginner. It also calls itself a *chickadee*, though not so plainly. (Southern Chickadee.)

Length, 4½ ; wing, 2½ (2¼-2½) ; tail, 2¼ ; tarsus, ⅝ ; culmen, ₁⁵₆. Southeastern States north to New Jersey and Illinois, and west to Missouri and Texas ; practically breeding throughout.

7. **Hudsonian Chickadee** (740. *Pàrus hudsónicus*). — A small, brownish-ashy-backed bird, with grayish crown, wings, and tail, a distinct black throat patch, and brownish sides. The rest of the bird is white. This

Hudsonian Chickadee

northern chickadee has also peculiar notes, which need to be heard to be understood.

Length, $5\frac{1}{4}$; wing, $2\frac{1}{2}$ ($2\frac{3}{8}$–$2\frac{3}{4}$); tail, $2\frac{1}{2}$; tarsus, $\frac{5}{8}$; culmen, $\frac{3}{8}$. Northern North America from northern New England and northern Michigan northward; rarely south to Massachusetts.

FAMILY IV. CREEPERS (CERTHÍIDÆ)

A very small family (10 species) of Old World birds, represented in this country by the following:

1. **Brown Creeper** (726. *Cérthia familiáris americána*). — A small, tree-creeping bird, with mottled-brown back, white under parts, a slender decurved bill, and long, acute-pointed tail feathers. The tail is used as a partial support, as in the case of the woodpeckers; a common upward-creeping bird, with little fear of human observers. When the top is reached it suddenly drops to the bottom, and again begins its search for food.

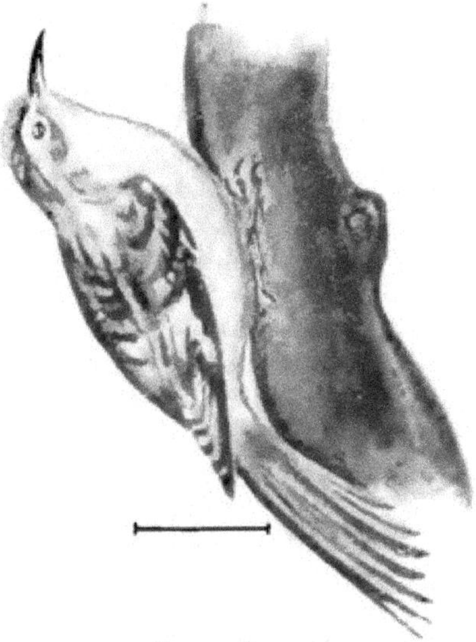

Brown Creeper

Length, $5\frac{1}{2}$; wing, $2\frac{1}{2}$ ($2\frac{3}{8}$–$2\frac{3}{4}$); tail, $2\frac{3}{4}$; tarsus, $\frac{5}{8}$; culmen, $\frac{3}{4}$. Eastern North America; breeding from Maine and Minnesota northward, and wintering as far south as the Gulf States.

FAMILY V. MOCKING BIRDS, WRENS, ETC. (TROGLODÝTIDÆ)

This family (150 species) of mainly American birds consists of two widely differing subfamilies. The **Mocking Birds** form a group of 40 species of American singing birds, of large size and plain colors, inhabiting mainly the bushy borders of the woods and other shrubbery. The tail in all cases is as long as the wings, and in one of our common species much longer. The bill is nearly as long as the head. The **Wrens** form a larger group (100 species) of small, mainly American, sprightly, fearless, excitable, plain-colored birds, with the plumage more or less extensively barred with narrow darker bands. The habit of holding the tail erect is very general. If these birds did not mingle so many of their characteristic scolding notes with their song, they would be considered very musical.

Key to the Species

* Birds under 6½ long, with wings under 3 long (Wrens). **(B.)**
* Birds over 8 long, with wings over 3 long (Mockers). **(A.)**
   A. Whole upper parts a rich reddish-brown; tail an inch longer than the wings; bill about an inch long............3. **Brown Thrasher.**
   A. Slate-colored bird, with chestnut on the under tail coverts.........
   ....2. **Catbird.**
   A. Wings blackish, with white bases to the primaries; outer tail feathers also white; bill, ¾ or less long.........1. **Mocking Bird.**
   A. Back ashy-gray; wings and tail darker and more brownish; bill, 1¼ or more long and decurved.[1] The Curve-billed **Thrasher** (707. *Harporynchus curviróstris*) of Mexico and New Mexico has been found in Texas.
B. Back with black, white, and brownish streaks, extending lengthwise. **(E.)**
B. Back without streaks extending lengthwise, or bars extending crosswise; a distinct white or whitish line over the eye. **(D.)**
B. Back without streaks, but with some cross bars; no very distinct whitish line over the eye. **(C.)**
   C. Under parts whitish; tail about as long as the wings...........
   ....6. **House Wren.**
   C. Under parts brownish, barred with black; tail a half inch shorter than the wings.............................7. **Winter Wren.**
D. Tail like the back in color, reddish brown; under parts buffy.......
   ....4. **Carolina Wren.**
D. Tail feathers, except the barred middle pair, blackish, tipped with grayish; under parts whitish...................5. **Bewick's Wren.**
   E. Crown as well as back streaked with white; bill under ½ long.....
   ....8. **Short-billed Marsh Wren.**
   E. Crown without white streaks, but a white line over eye; bill ½ or more long......................9. **Long-billed Marsh Wren.**

1. **Mocking Bird** (703. *Mimus polyglóttos*). — A large, ashy-colored, long-tailed bird, with much white on the center of wing and outer tail feathers. **This** is the most noted song-

bird of America, and as a mocker the most wonderful in the
world. He is to be found in woods, gardens, parks, and even
in the streets of
towns, always fear-
less and alert, and
with the power to
mimic almost any
sound in nature.
Mr. L. M. Lumis re-
ports having heard
one imitate thirty-
two different species
of birds in less than
a quarter of an hour.

Mocking Bird

Length, 10½ ; wing,
4½ (4–5) ; tail, 5 ; tarsus, 1¼ ; culmen, ¾. United States to Mexico ; rare
north of Maryland, though found in southern Ohio and Massachusetts ;
winters from Florida southward.

2. **Catbird** (704. *Galeoscoptes carolinénsis*). — A large, very
common, slate-colored bird, with a chestnut-colored patch
under the tail and almost black crown and tail. This gro-

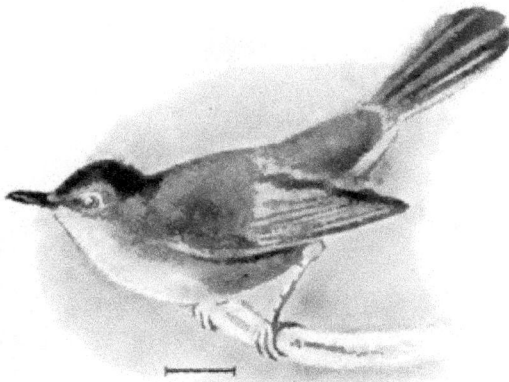

Catbird

tesquely active bird
can be found every-
where around our
orchards and shrub-
bery. It is a very
charming, but not
loud, singer, and a
good mimic; most
people know it only
by its cries when dis-
turbed.

Length, 8¾ ; wing, 3⅞
(3¾–3¾) ; tail, 4 ; tarsus,
1¼ ; culmen, ⅝. North
America, though common only east of the Rocky Mountains ; breeding
from the Gulf States northward, and wintering in the Southern States.

3. **Brown Thrasher** (705. *Harporhýnchus rúfus*). — A common, large, long-tailed, brown-backed bird, with the white under parts heavily spotted or streaked with dark-brown, except on the throat and middle of the belly. The wings, tail, and crown have the same rufous color as the back. It is an inhabitant of the ground or the lower growths along fences and the borders of the woods. It is a rich, sweet singer of its own notes, but not a mocker of the notes of other birds. When singing it usually perches on a twig in a prominent position as though it wished all to know how melodious a vocalist it is. (Brown Thrush.)

Brown Thrasher.

Length, 11¼; wing, 4¼ (4-4½); tail, 5¼; tarsus, 1⅜; culmen, 1. Eastern United States, west to the Rocky Mountains, north to Ontario; breeding throughout and wintering north as far as Virginia. Besides this species and the Curve-bill Thrasher given in the Key, there can be found in Texas **Sennett's Thrasher** (706. *Harporhýnchus longiróstris sénnetti*), a bird much like the brown thrasher, but with a darker-brown back, blacker spots on the lower parts, and a longer (1¼-1½) and somewhat decurved bill.

4. **Carolina Wren** (718. *Thryóthorus ludovicidnus*). — A nervous, scolding wren, distinctly barred,

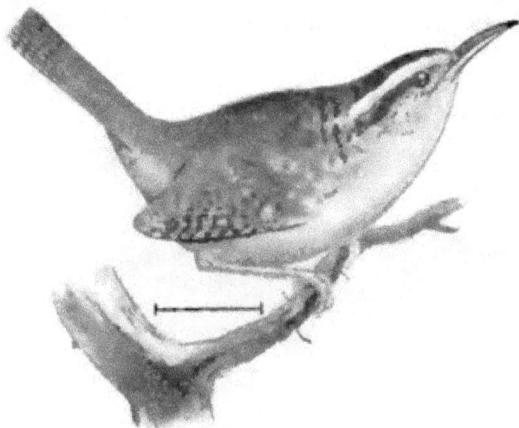
Carolina Wren

rich-brown, with long curved bill, a very distinct whitish line over the eye, and a tail the color of the back. It inhabits undergrowths in wet places, and has the ability to disappear

from sight and appear again with surprising quickness. Its
fear and its curiosity alternate in power over its actions. It
has been called *mocking wren* from the variety of its vocal
notes, some of which are so loud and ringing as to be com-
parable with those of the tufted titmouse. It is probably resident wherever found. (Mocking Wren.)

Length, 5½ ; wing, 2¾ (2½–2½) ; tail, 2¼ ; tarsus, ⅞ ; culmen, ½. Eastern United States, west to the Plains, and north to southern New York
and southern Michigan ;
resident or nearly so
throughout.

5. **Bewick's Wren**
(719. *Thryóthorus be-
wickii*). — This is a
slightly smaller, less
distinctly barred wren
than the last, with
a tail quite a little
darker than the back,
and without bars on the
primaries ; the outer
tail feathers are black,
tipped with grayish.

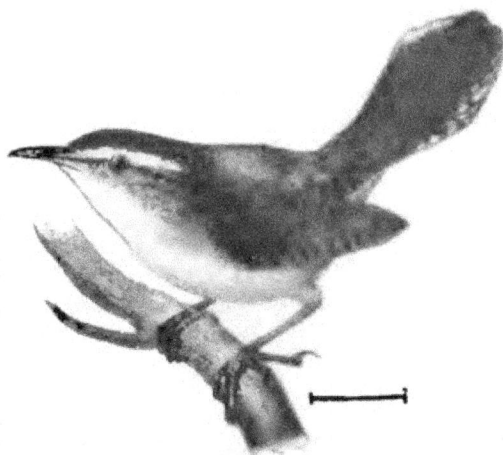

Bewick's Wren

This species is a sweet singer of clear, ringing notes, and
very fearless. It is found around outhouses, fences, etc., and
is in every way more deliberate in its movements than either
the house or Carolina wrens. Its long tail frequently leans
toward the head.

Length, 5¼ ; wing, 2¼ (2–2¼) ; tail, 2¼ ; tarsus, ¾ ; culmen, ½. Eastern
United States, west to Nebraska ; common in the Mississippi Valley ; rare
and local east of the Alleghanies and north of central New Jersey ; migratory along the northern border of its range. **Baird's Wren**, a form of
this species (719ᵇ. *T. b. leucogáster*), is found in Texas, Kansas, and westward to southern California. It is a more *ashy*-brown bird, with pure
white on the middle of the belly, and white specks on the sides of the
head.

6. **House Wren** (721. *Troglódytes ǎëdon*).— A dark-brown wren, with the tail decidedly more reddish than the back. The wings, tail, sides, and flanks are fully cross-barred with darker lines, and the under parts are whitish. As its name indicates, it likes to live near human habitations, returning to the same place year after year, and building its nest in the same hole in a log, bird box, or

House Wren

chink in an outhouse. It is active, irritable, noisy, and courageous. It is resident in the Southern States, and is there so numerous in winter as to overflow the settled regions, and so is found in the forests miles from any house.

Length, 5; wing, 2 (1⅞–2⅛); tail, 1⅝; tarsus, ⅔; culmen, ½. Eastern United States north to southern Ontario, and west to Indiana and Louisiana. It winters from South Carolina southward. The **Western House Wren** (721b. *T. a. áztecus*) is a variety of this species with less of red on the upper parts, and the back and rump are very distinctly barred with blackish. As a whole, it is a lighter colored bird. Interior United States from near the Pacific, eastward to Illinois.

Winter Wren

7. **Winter Wren** (722. *Troglódytes hiemális*).— A small, very short-tailed, cinnamon-brown wren, with more brownish under parts than any other species of ours. In its breeding range of the north, it is a very sweet singer; in other

localities, it merely gives its hearty *quip-quap* call notes. It lives in the woods, and can be found among the lower growths, and on and under old logs and stumps. Its quiet ways and dark colors render it difficult to be seen.

Length, 4; wing, $1\frac{7}{8}$ ($1\frac{3}{4}$–2); tail, $1\frac{1}{4}$; tarsus, $\frac{3}{4}$; culmen, $\frac{3}{8}$. Eastern North America; breeding from the northern United States northward (in the Alleghanies from North Carolina), and wintering from New York and Illinois southward.

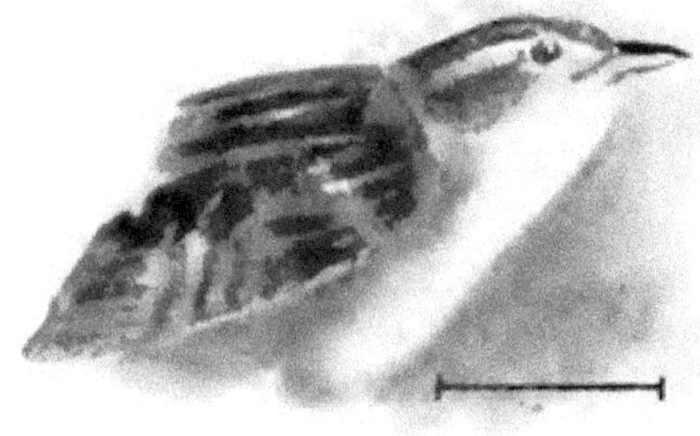

Short-billed Marsh Wren

8. **Short-billed Marsh Wren** (724. *Cistothòrus stellàris*). — A small, short-billed, marsh and meadow-living wren, with its whole back, including the crown, very distinctly streaked lengthwise with dark and light shades. The under parts are white, with buffy sides and breast. This, like many of the wrens, is so shy that it is much more frequently heard than seen.

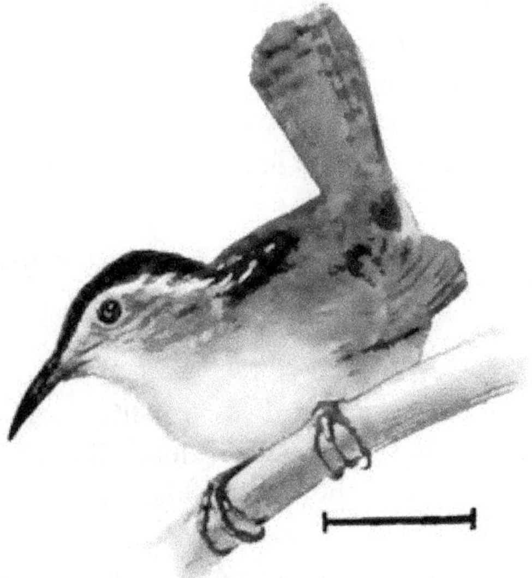

Long-billed Marsh Wren

Length, $4\frac{1}{4}$; wing, $1\frac{3}{4}$ ($1\frac{5}{8}$–$1\frac{7}{8}$); tail, $1\frac{5}{8}$; tarsus, $\frac{5}{8}$; culmen, $\frac{3}{8}$. The United States from the Plains eastward, and north to southern Michigan and southern Ontario. It winters in the South Atlantic and Gulf States.

9. **Long-billed Marsh Wren** (725. *Cistothòrus palústris*). — A long-billed, white-bellied wren, with a black back, striped lengthwise with white. The crown is unstreaked, and the wings, tail, and sides

are brown. This bird is common in reedy marshes, but because
of its shy, suspicious habits, needs careful, quiet searching.
Its grumbling notes can readily be heard, and if it were not
for its inquisitive nature, which leads it to expose itself for a
second or two to see its visitor, it would be impossible to
observe it.

Length, 5¼; wing. 2 (1¾–2¼); tail, 1¾; tarsus, ¾; culmen, ½. Eastern
United States north to Ontario; breeding throughout, and wintering (locally) from southern New England southward. **Worthington's** Marsh
Wren (725ᵇ. *C. p. griseus*), found along the coasts of South Carolina and
Georgia, is a lighter colored, more grayish and more faintly barred and
striped bird. Wing, 1¾; tail, 1½; bill, ½. **Marian's Marsh** Wren (725-1.
Cistothorus marianae) of western Florida is a darker bird than the long-
billed marsh wren, and has the sides and flanks of the same brown as the
rump. The under **tail** coverts and sometimes **the breast are** spotted with
black. Wing, 1⅞; tail, 1¾; bill, ½.

FAMILY VI. WAGTAILS. PIPITS (MOTACILLIDÆ)

A family (80 species) of mainly Old World, ground-living,
tail-wagging, walking or running birds, represented in our region by two species a little larger than the English sparrow.
They are usually to be found in open pastures, meadows, and
recently plowed fields. The tail is very long, usually as long
as the wings, and is moved in a peculiar see-saw way, as though
to enable the bird to balance itself. This habit is also common
among snipe, though they have short tails.

The bill is slender, acute, shorter than the head, and notched at tip.
The inner secondaries are lengthened, about as long as the primaries in
the closed wing; feet large, and the hind nail long and nearly straight.

1. **American Pipit** (697. *Anthus pensilvanicus*). — A common,
dark-olive-brown-backed, tail-twitching bird, with buffy under
parts marked on the breast with brownish dots like the
thrushes. This is a walking bird of the open fields, and in its
vacillating flight the white outer tail feathers can be seen.
The pipits are social birds seen in numbers (5–20), more or less
together while feeding, but usually flying in larger flocks when
startled. (Titlark; Wagtail.)

Length, $6\frac{1}{2}$; wing, $3\frac{3}{8}$ ($3\frac{1}{4}$–$3\frac{1}{2}$); tail, $2\frac{3}{4}$; tarsus, $\frac{7}{8}$; culmen, $\frac{1}{2}$. North America; breeding in the subarctic regions and higher mountains and wintering in the Gulf States to Central America.

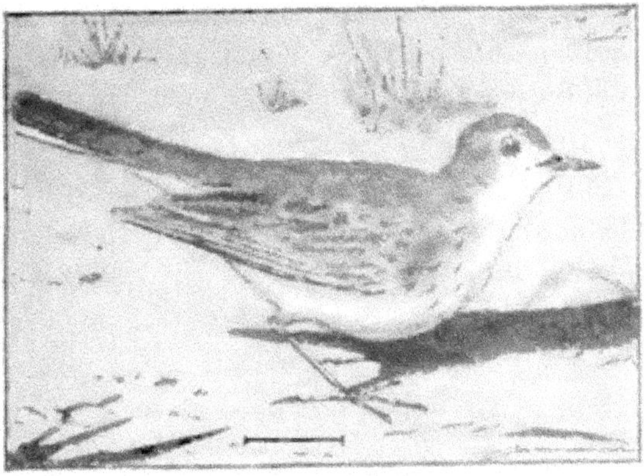

American Pipit

2. **Sprague's Pipit** (700. *Ánthus sprágueii*). — A bird in appearance very much like the last, but with the colors brighter and the markings more distinct. This species has the tarsus shorter than the hind toe and claw, while the preceding has it as long, sometimes longer. This has a tail always less than $2\frac{5}{8}$, the other greater, sometimes 3 long. While the two are so similar in form, size, and colors, they are wonderfully different in power of song. This is a sky-singing bird, like the skylark of Europe. Dr. Coues says: "No other bird music heard in our land compares with the wonderful strains of this songster; there is something not of earth in the melody, coming from above, yet from no visible source; . . . the whole air seems filled with the tender strains." (Missouri Skylark.)

Length, $6\frac{1}{2}$; wing, $3\frac{1}{4}$ (3–$3\frac{3}{8}$); tail, $2\frac{3}{8}$; tarsus, $\frac{7}{8}$; culmen, $\frac{1}{2}$. Interior plains of North America. Once recorded east of the Mississippi (in South Carolina).

FAMILY VII. WOOD WARBLERS (MNIOTÍLTIDÆ)

A family (100 species) of exclusively American brightly colored, small birds of woodlands and thickets. Their habits in gathering their insect prey vary greatly: some, like the vireos, search carefully for hidden insects, resting or crawling on leaf, on bark, or in flower; others flit from twig to twig, gathering the exposed insects, while still others are like the flycatchers, capturing most of their prey while on the wing.

Most warblers are tree living; some are only to be found in
the tops of tall trees, but many live in low bushes, while a
few are terrestrial. As a whole, the warblers do not deserve
their name, as their vocal powers are inferior, though a few
species are remarkable singers. But few species are as large
as the English sparrow, and only one, the chat, is larger.
Numbers 1 and 11 are *creepers;* Nos. 2–10 are *worm eaters;*
these are usually creepers along the smaller twigs; Nos. 12–28
are the usual or *typical warblers;* Nos. 29–35 are *ground warblers;* Nos. 37–40 are *flycatching warblers;* while No. 36 is a
large, heavy-billed, aberrant form placed in this family only
because it belongs nowhere else.

Key to the Species

* Large, over 6½ long; bill rather stout and compressed;[1] under parts
  bright yellow, abruptly changing to white at about the middle of the
  length from chin to tail......36. **Yellow-breasted Chat.**

* Bill depressed, broader than high at base, notched and
  slightly hooked at tip; rictal bristles nearly or quite half
  the length of the bill.[2] (R.)

* Bill slender and not depressed; rictal bristles small[5] or
  none.[3] (A.)

  A. No bright yellow or orange anywhere, at most a slightly
  yellowish tinge. (L.)

  A. Yellow nearly everywhere; inner web of under tail
  feathers yellow, outer web dusky; no white blotches
  on under tail feathers... ......13. **Yellow Warbler.**

  A. Whole head and neck bright yellow; wings ashy, with
  neither white nor yellow wing bars; inner web of under
  tail feathers mostly white..2. Prothonotaria Warbler.

  A. *Whole* head and neck *not* bright yellow; under tail feathers blotched
  with white. (**E.**)

  A. Under tail feathers with no white blotches, but of about the same
  color on both webs; no distinct wing bars. (B.)

B. Tail and wings of about equal length, each about 2 inches (1⅞–2¼);
  back, wings, and tail olive-green.......35. **Maryland Yellow-throat.**

B. Tail about ½ inch (⅜–⅝) shorter than the wing. (D.)

B. Tail over ½ inch (⅝–1) shorter than the wing. (C.)

  C. Head, neck, and breast bluish-gray (or in the *female* and *young,*
  grayish-brown); other upper parts olive-green; belly yellow; a
  well-marked white line around the eye..33. **Connecticut Warbler.**

C. A clear yellow line extending from the bill over the eye and curving round back of the eye ; under parts bright yellow. ....32. **Kentucky Warbler.**

C. On account of the very short tail and the yellowish tint to the olive of the head, see........................10. **Tennessee Warbler.**

D. Head, neck, and throat bluish-gray (*male*) ; head and neck grayish (*female*) ; *no white ring around eye ;* belly yellow ....34. **Mourning Warbler.**

D. Top and sides of head bluish-gray, changing to olive-green on the back (or in the *female* only sides of head brownish-gray) ; breast yellow changing to nearly white on the lower belly.................... ....8. **Nashville Warbler.**

D. A large black breast patch surrounded by yellow (*male*); bend of wing yellow (*female*) ; crown black (*male*) ; grayish (*female*) ; Gulf States, accidental in Virginia..........5. **Bachman's Warbler.**

E. Bluish-gray above with a golden-green patch in the *middle* of the back ; two white wing bars [4] ....11. **Parula Warbler.**

E. Rictus without evident bristles (less than $\frac{1}{16}$ long if any) ; bill very acute.[3] (**K.**)

E. Rictus with evident bristles ; bill usually not very acute and usually with a slight notch near tip.[5] (**F.**)

F. Wing bars or wing patch white. (**H.**)

F. Wing bars if present not white (sometimes in the young yellowish-white). (**G.**)

  G. Wing bars yellowish and belly yellow (young have the wing bars very indistinct) ; back usually spotted with chestnut ; wing. 2¼ or less..28. **Prairie Warbler.**

  G. Wing bars yellow (yellowish-white in young); belly pure white ; sides usually with more or less chestnut........................ ....19. **Chestnut-sided Warbler.**

  G. Wing bars brownish and inconspicuous ; white blotches square and on the tips of the under tail feathers ; crown more or less distinctly marked with chestnut................27. **Palm Warbler.**

  G. Wing bars inconspicuous ; whole under parts pale yellow ; back ashy without any tint of green or olive ; wing. 2½ or more ....25. **Kirtland's Warbler.**

H. Rump and crown patch yellow (crown patch somewhat obscure in winter) ; sides of breast also generally yellow ; throat *white*........ ....15. **Myrtle Warbler ;** throat *yellow*....16. **Audubon's Warbler.**

H. Rump and belly yellow ; white blotches on the *middle* of nearly all the tail feathers ; crown not yellow, usually clear ash.............. ....17. **Magnolia Warbler.**

H. Rump and sides of neck usually yellow ; bill very acute and distinctly decurved near the tip..............12. **Cape May Warbler.**

H. Rump not yellow ; bill not very acute. (**I.**)

I. Throat yellow or orange ; crown with a small or large yellow or orange spot ; under tail feathers with outer edge white edged as well as white blotches on the inner web........................
....22. **Blackburnian Warbler.**

I. Sides of head bright yellow ; inner web of under tail feathers entirely white ; outer web white at base.........................
.........................24. Black-throated **Green Warbler.**

l. Upper parts, chin, throat, breast, bill. and feet black ; sides of head yellow. In southern Texas the Golden-cheeked **Warbler** (666. *Dendroica chrysoparia*) can be found. Its habitat extends to Central America. (See p. 87.)

I. Not as above ; throat more or less yellow. (**J.**)

J. White tail blotches large and *oblique* near the end of two or three under tail feathers ; no sharp markings anywhere..26. **Pine Warbler.**

J. Throat definitely yellow ; belly white ; back not greenish..........
....23. **Yellow-throated Warbler.**

K. Wings with *white wing bars ;*[1] back bright olive-green ; eye with a black line extending across it
....6. Blue-winged **Warbler.**

K. Wing coverts yellow, forming a *yellow wing patch*
................7. Golden-winged Warbler.

K. A black throat patch surrounded by yellow (*male*) ; bend of wing yellow (*female*) ; a *yellow wing bar* formed only of the lesser coverts.................................5. **Bachman's Warbler.**

K. The supposed hybrids, Brewster's and Lawrence's Warblers might be looked for here. They seem in markings and habits intermediate between 6 and 7. Brewster's has the throat and breast white, and Lawrence's has a large black patch on the breast. About a dozen specimens of **Lawrence's** have been found, chiefly in New Jersey and southern Connecticut, and many of Brewster's from southern New England to Michigan. (See p. 79.) No. 10 (Tennessee Warbler) might also be looked for here, as its tail feathers are sometimes marked with white.

L. Under tail feathers **without** white blotches ; wings without **wing bars.** (**N.**)

L. Under tail feathers blotched with **white.** (**M.**)

  M. Body nearly everywhere streaked with black **and white,** including the crown, which has a middle streak of white.............. ...
...............................1. Black and White Warbler.

  M. Crown black ; all other parts much streaked with black and white ; back with some ashy........ ....21. Black-poll Warbler (*male*).

  M. Upper parts olive-green more or less streaked with black ; under parts more or less yellowish and somewhat streaked on breast and sides.... ....21. Black-poll Warbler (*female*).

  M. Under parts, especially the **crissum, buffy** ; crown and throat usually chestnut............................ .20. **Bay-breasted Warbler.**

M. Two white wing bars;[1] entire upper parts sky-blue (*male*) or dull greenish, brightest on the head (*female*), under parts white (tinged with pale yellow in the *female*).... .. ....18. **Cerulean Warbler.**

M. Slightly yellowish-white wing bars; sides with some chestnut markings................................19. **Chestnut-sided Warbler.**

M. No wing bars but a white blotch on the primaries near the base (very small in *female* and young); upper parts grayish-blue (*male*) or olive-green (*female*).........14. **Black-throated Blue Warbler.**

N. Crown with two black stripes separated by a broader one of buff; two other black stripes back of the eyes; under parts buffy, unspotted................................4. **Worm-eating Warbler.**

N. Head brown, a whitish line over eye; under parts white, grayer on sides and not definitely spotted............3. **Swainson's Warbler.**

N. Not as above, and the tail ³⁄₄ inch or more shorter than the wing. (**P.**)

N. Tail not over ½ inch shorter than the wing. (**O.**)

  O. Upper parts somewhat ashy with more or less of an orange-brown patch on the crown (except in the young); under parts dull, sometimes with dusky streaks on the breast; a yellowish or white ring around the eye.....................9. **Orange-crowned Warbler.**

  O. A small white patch on the base of the primaries, near and partly hidden by the coverts; upper parts olive-green, with a brownish tinge on the tail ....14. **Black-throated Blue Warbler.**

P. Upper parts yellowish-olive; under parts dull white more or less tinged with yellowish but without definite spots; a whitish line over the eye and *white* under tail coverts........10. **Tennessee Warbler.**

P. Conspicuously spotted or streaked below, thrush-like; back brown, brownish-olive, or dusky: head striped, at least a distinct line over the eye. (**Q.**)

  Q. Crown with an orange-brown stripe bordered with black lines..... ....29. **Oven-bird.**

  Q. No central stripe on crown, but a whitish to buffy line over the eye; under parts, including the throat, tinged with yellow and very fully streaked with black.....................30. **Water Thrush.**

  Q. Line over the eye conspicuously white; under parts slightly buffy tinted, and the black streaks do not extend over the throat or middle of the belly.... ....31. **Louisiana Water Thrush.**

R. Without bright yellow, but with more or less of flame color or dull yellow on wings and tail..................40. **American Redstart.**

R. Breast bright yellow. (**S.**)

  S. Under tail feathers with white blotches......37. **Hooded Warbler.**

  S. Under tail feathers without white blotches on the inner webs; no wing bars. (**T.**)

T. Above bright olive-green; crown black without streaks (black cap sometimes lacking in *female* and *young*).....38. **Wilson's Warbler.**

T. Above bluish-ash; a necklace of black (*male*) or dusky (*female*) spots across breast ....39. **Canadian Warbler.**

1. **Black and White Warbler** (636. *Mniotilta varia*). — A black and white streaked warbler, with a broad white stripe on the top of the head and no yellow anywhere. *Female* with some brownish on the sides and fewer black stripes on the lower parts. This is a silent bird, common in woodlands, *creeping* over twigs and branches, often hanging from the lower surfaces, hunting industriously for insect food. (Black and White Creeper.)

Black and White Warbler

Length, 5¼; wing, 2¾ (2¾-2⅞); tail, 2¼; tarsus, ⅞; culmen, ½. Eastern North America. Breeds from Virginia north to Hudson Bay, and winters from the Gulf States south to northern South America.

2. **Prothonotary Warbler** (637. *Protonotaria citrea*). — A very pretty warbler, with the whole head, neck, upper back, and under parts a rich orange. The rest of the upper parts gradually change through greenish to bluish to ashy, and the lower parts to almost white on the crissum, and large white blotches on the under tail feathers. The *female* has the yellow paler. It is found most frequently in the low growths near and over the water, where it is more like a creeper in its habits than like a flycatcher. Its usual notes are clear, penetrating *peet, tweet, tweet, tweet*, given without change of pitch. (Golden-headed Warbler.)

Prothonotary Warbler

Length, $5\frac{1}{2}$; wing, $2\frac{7}{8}$ ($2\frac{3}{4}$–3); tail, $2\frac{1}{4}$; tarsus, $\frac{3}{4}$; culmen, $\frac{1}{2}$. Eastern United States, north to Virginia and southern Michigan; south in winter to northern South America; breeding throughout its United States range.

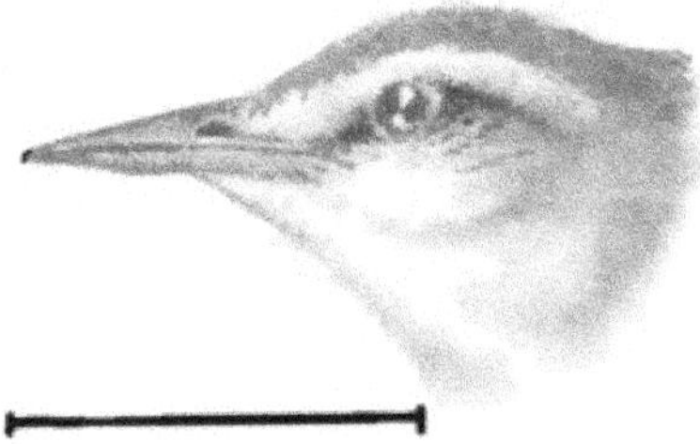
Swainson's Warbler

3. **Swainson's Warbler** (638. *Helinaia swainsonii*). — A brownish warbler, with whitish under parts, inclined to yellow on the middle, and grayish on the sides. This ground warbler of the Southern States is a beautiful singer of loud, rich, yet tender notes of most penetrating quality.

Length, 6; wing, $2\frac{3}{4}$; tail, $1\frac{7}{8}$; tarsus, $\frac{3}{4}$; culmen, [illegible]. Southeastern United States, north to southern Virginia and southeastern Missouri, and south in winter to central Mexico.

4. **Worm-eating Warbler** (639. *Helmitherus vermivorus*). — An olive-green-backed and creamy-bellied warbler, with a buffy head, distinctly marked with four black lines, two on the crown and two through the eyes; no white on back, wings, or tail, all being of about the same shade of olive-green. A rare, shy bird, found usually in the dense undergrowth of wooded hills and ravines.

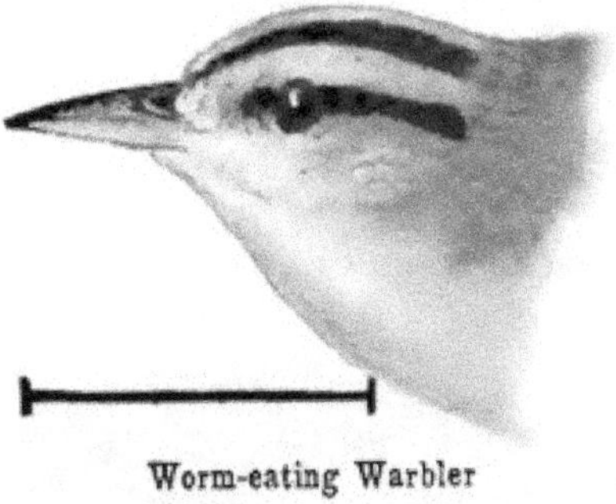
Worm-eating Warbler

Length, $5\frac{1}{2}$; wing, $2\frac{3}{4}$ ($2\frac{1}{2}$–$2\frac{7}{8}$); tail, $2\frac{1}{8}$; tarsus, $\frac{5}{8}$; culmen, $\frac{5}{8}$. Eastern United States north to southern New England; west to Nebraska and Texas; in winter south to northern South America.

5. **Bachman's Warbler** (640. *Helminthophila bachmanii*). — A rare, southern warbler, having forehead, throat, and belly yellow, with a large conspicuous patch of black on the breast; the rest of the bird bright olive-green, shading to grayish on the wings, and with a white-blotched tail. *Female* with the yellow

of the forehead and the black of the breast not so distinct.
but with the bend of the wing yellow; breeding range and
habits unknown.

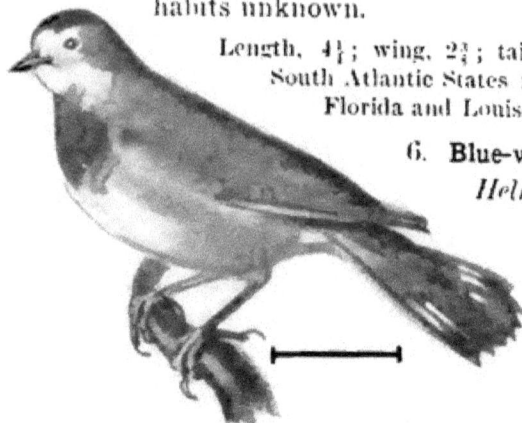

Length, 4¼; wing, 2¾; tail, 1¾; culmen, ¼ nearly.
South Atlantic States from southern Virginia to
Florida and Louisiana; Cuba in winter.

6. **Blue-winged Warbler** (641.
Helminthóphila pínus). —
A yellow warbler,
with slaty-blue wings
and tail: the yellow
of the upper parts
changes to olive-
green on the rump.
There is a black line
through the eye, and

Bachman's Warbler

the wing bars are yellowish. *Female* with less yellow. This
is a common. creeper-like warbler, found mainly in the undergrowth of woods. In its creeping movements it is slow and deliberate, and may occa-
sionally be seen hang-
ing head downwards.
This is more of a
singer than most of
the warblers.

Length, 4¾; wing, 2¾;
tail, 2; tarsus, ¾; cul-
men, ⁷⁄₆. Eastern United
States from southern
New England and south-
ern Minnesota south-
ward; west to Nebraska
and Texas; in winter
south to Central America.

Blue-winged Warbler

7. **Golden-winged
Warbler** (642. *Helminthóphila chrysóptera*). — A warbler with a
yellow crown, yellow wing coverts, a black patch around and

below the eye, and another on the breast, with two conspicuous white stripes, a narrow one above the eye patch and a wider one between the eye and breast patches. Upper parts grayish, below white. The under tail feathers blotched with white. *Female* with the head less distinctly marked, and with grayish instead of black. An insect-

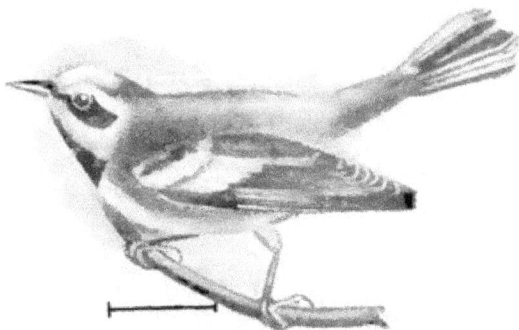

Golden-winged Warbler

eating creeper in the lower growths at the borders of woods.

Length, 5; wing, 2¼; tail, 2; culmen, ⅜. Eastern United States north to southern New England and southern Minnesota; breeding from its northern limit south to New Jersey and Indiana, and in the mountains to South Carolina. In winter south to northern South America. The last two species are supposed to interbreed and form two named hybrids. **Brewster's Warbler** (*H. leucobronchialis*) and **Lawrence's Warbler** (*H. lawrencei*). The former has broad yellow wing bars and white breast, the latter white bars and a black throat patch. Different specimens show great variety of plumage, but all are intermediate between Nos. 6 and 7. The Key, page 74, gives other facts about these puzzling birds.

8. **Nashville Warbler** (645. *Helminthóphila rubricapilla*). — An olive-green-backed warbler, with all the lower parts bright yellow, lighter on the belly. The top and sides of the head are

gray, with a more or less concealed chestnut patch on the crown. No white bars on wing or white blotches on tail. The *young* are duller and have brownish washings on head, back, and sides. An inhabitant of open woods and fields.

Nashville Warbler

Length, 4¾; wing, 2⅜ (2¼–2½); tail, 1⅞; culmen, ⅜. From the Plains eastward and northward to the fur countries; breeding from the northern United States northward, and wintering as far south as Central America.

9. **Orange-crowned Warbler** (646. *Helminthóphila celàta*). —
A rare warbler, with the whole upper parts nearly uniform
olive-green except the more or less concealed orange-brown
crown patch. The lower parts greenish-yellow, with slight dusky
streaks on the breast. *Young* lack the crown patch, and are in
all respects duller birds, but with a white ring around the eye.

Length, 5; wing, 2½ (2¾-2¼); tail, 2; culmen, ½. Eastern North
America from Mackenzie River south through the Rocky Mountains;
wintering in the South
Atlantic and Gulf States
and Mexico. Rare north
of Virginia in the Eastern
States.

10. **Tennessee War-
bler** (647. *Helminthó-
phila peregrìna*). — A
rare warbler, with
the lower parts white,
more or less tinged

Tennessee Warbler

with yellow, and the upper parts bright-olive-green, changing abruptly to bluish-gray on the head. No white wing
bars, but the inner web of the under tail feathers generally
white at tip. *Female* has the crown tinged with greenish and the
under parts more dis-
tinctly yellowish. The
breast of this species
is pale greenish yellow
with no streaks, and the
under tail coverts white.

Length, 4¾; wing, 2½;
tail, 1¾; culmen, ½. Eastern North America; breed-
ing from northern New
York to Hudson Bay, and
wintering from Mexico to
northern South America.

Parula Warbler

11. **Parula Warbler** (648. *Compsóthlypis americàna*). — A
greenish-yellow-backed, yellow-breasted, grayish-blue warbler,

with white wing bars and belly. This bird generally has a
darkish, more or less reddish band across the breast. The
greenish-yellow of the back forms a central patch. The *female*
sometimes lacks the dark-reddish breast band. (Blue Yellow-
backed Warbler.)

Length, 4¾; wing, 2¼ (2¼-2¾); tail, 1¾; culmen, ⅜. United States
from the Plains eastward, north to Canada; breeding throughout, and
wintering from Gulf States south to eastern Mexico and West Indies.
Sennett's Warbler (649. *Compsóthlypis nigrilòra*), of western Texas,
southward, is a similar bird, but the parula has a white spot on each
eyelid, which is lacking in Sennett's warbler; and in front of the eyes
(lores) of the parula there is a dusky spot. In Sennett's warbler this
spot is intensely black, and this black crosses the front of the head just
above the bill.

12. **Cape May Warbler** (650. *Dendroìca (drò-ca) tigrìna).* — A
rare but beautiful warbler, with black-streaked, olive-green
back, chestnut cheeks, black crown, and yellow rump; the
under parts are yellow, heavily streaked with black, but
changing to white
on the crissum. The
wing coverts form a
large white patch,
and the under tail
feathers have white
patches near the tips
on the inner webs.
The *female* lacks the
white wing patch,
but has a narrow,
white wing bar; the
back is somewhat

Cape May Warbler

grayish, the rump less yellow, but there is a yellow line over
the eye. The *young female* has almost no yellow below. This
is a warbler of the tree tops.

Length, 5¼; wing, 2¾ (2½-2⅞); tail, 2; culmen, ⅜. North America
from the Plains eastward, north to Hudson Bay Territory; breeding
from northern New England northward, and wintering in the West Indies.

13. Yellow Warbler (652. *Dendroica æstiva*). — This is the yellow warbler in fact as well as name, having some shade of

yellow throughout, and forming our only canary-colored wild bird. The under parts are somewhat streaked with reddish, and the under tail feathers are yellow on the inner webs and dusky on the outer. The *female* is less brightly yellow, and the

Yellow Warbler

under parts are less streaked. This is a common inhabitant of our gardens and orchards, and is often thought to be an escaped canary: its slender bill shows that it is a different species. (Summer Yellow-bird; Golden Warbler.)

Length, 5; wing, 2½ (2⅜–2⅝); tail, 2; culmen, ⅜. North America throughout, except the southwest; breeding in nearly its whole range, and wintering south to northern South America.

14. Black - throated Blue Warbler (654. *Dendroica cærulés-cens*). — A common, grayish-blue-backed, white-bellied warbler, with black sides of head and throat, and irregular patches of black along the sides of the body. The

Black-throated Blue Warbler

bases of the primaries form a white patch on the wings. The *female* has the upper parts olive-green and the lower parts yellowish, and in the main lacks the black throat, while the white wing patch is much reduced.

Length, $5\frac{1}{4}$; wing, $2\frac{1}{2}$; tail, $2\frac{1}{8}$; culmen, $\frac{5}{16}$. North America from the Plains eastward; breeding from northern New York northward (in the Alleghanies south to Georgia), and wintering in the tropics.

15. **Myrtle Warbler** (655. *Dendroica coronàta*). — A common, large, streaked, bluish- and black-backed warbler, with distinct patches of yellow on crown, rump, and sides of breast, and a white throat and lower belly. There are two white wing bars, white blotches on the under tail feathers, and heavy black marks on the breast. The *female* has browner upper parts, and fewer black marks on the breast. The yellow on the crown and sides of the breast are much reduced in *young* and *winter* birds. (Yellow-rumped Warbler.)

Myrtle Warbler

Length, $5\frac{3}{4}$; wing, $2\frac{7}{8}$ ($2\frac{3}{4}$–3); tail, $2\frac{1}{4}$; culmen, $\frac{3}{8}$. North America, but rare west of the Rocky Mountains; breeding from northern United States northward, and wintering from southern New England and the Ohio Valley, southward to Central America.

16. **Audubon's Warbler** (656. *Dendroica àuduboni*). — A western warbler similar to the last, but with yellow on the throat instead of the white of that species. The wing bars blend together into a wing patch.

Length, $5\frac{1}{2}$; wing, 3 ($2\frac{3}{4}$–$3\frac{1}{4}$); tail, $2\frac{1}{4}$; culmen, $\frac{3}{8}$. Western United States eastward to the western borders of the Plains; accidental in Pennsylvania and Massachusetts.

17. **Magnolia Warbler** (657. *Dendroica maculòsa*). — A gray-crowned, black-backed, yellow-rumped warbler, with the breast and throat yellow; heavily streaked on the breast and sides with black. The wing coverts form a large white patch: the middle of the under tail feathers is white, and the end third

black. *Female* similar, but duller. Both the *female* and *young* have the white tail blotches on the middle of the feathers. (Black and Yellow Warbler.)

Magnolia Warbler

Length, 5; wing, 2¾ (2¼-2½); tail, 2; culmen, ⅜. North America from the Rocky Mountains eastward; breeding from northern New York northward (southward in the mountains to Pennsylvania), and wintering south of the United States to Central America.

18. **Cerulean Warbler** (658. *Dendroica cærulea*). — A warbler with bright blue upper parts, white lower parts, and many black streaks on the sides; wing bars and much of the under tail feathers white. *Female* with greenish tint to the back and yellowish tint to the belly. It lives in the tops of the forest trees. Its song is very much like that of the parula warbler.

Length, 4½; wing, 2¾ (2¾-2½); tail, 1¾; culmen, ⅜. Eastern United States and southern Ontario west to the Plains; rare east of the Alleghanies; in winter, south to northern South America.

19. **Chestnut-sided Warbler** (659. *Dendroica pensylvánica*). — A chestnut-sided, yellow-crowned warbler, with mottled black and olive back and white under parts; wing bars yellowish and cheeks white, outlined with

Cerulean Warbler

black. *Female* similar, but duller; the *young* has the back somewhat streaked with black on a ground that is yellowish-olive, and the under parts silky-white; the sides are sometimes blotched with chestnut; an inhabitant of bushy borders.

Chestnut-sided Warbler

Length, 5; wing, 2½ (2⅜–2⅝); tail, 2; culmen, ⅜. Eastern United States from the Plains, including southern Ontario; breeding from northern New Jersey and central Illinois northward (southward to Georgia in the mountains), and wintering in the tropics.

20. **Bay-breasted Warbler** (660. *Dendroica castánea*). — A brownish-ashy-backed warbler, with chestnut crown and brownish breast and sides; forehead and cheeks black; wing bars and belly white; the under tail feathers have the white patches at their tips. *Female* with the crown somewhat olive, the under parts not so white, and less rufous on the breast and sides. This is a beautiful warbler, living in its summer home, among the tree tops. (Autumn Warbler.)

Length, 5½; wing, 2⅞ (2¾–3); tail, 2⅛; culmen, ⅜. Eastern North America from Hudson Bay southward; breeding from northern New England and northern Michigan northward, and wintering in Mexico and Central America.

Bay-breasted Warbler

21. **Black-poll Warbler** (661. *Dendroica striáta*).— A common, very much streaked, mainly black and white warbler, with distinct black cap and white cheeks. The *male* has grayish and the *female* olive-green tints on the back, including the crown, thus obliterating the black cap. The *female* is less distinctly

streaked. The *young* is even less streaked than the female, has greenish-yellow tinting on the under parts, and almost no markings. It is found in orchards, gardens, and open, especially evergreen woods.

Length, 5½; wing, 2⅞; tail, 2¼; culmen, ⅜. North America from the Rocky Mountains eastward; breeding from northern New England northward, and wintering south to northern South America.

Black-poll Warbler

22. Blackburnian Warbler (662. *Dendroica blackburniæ*). — A warbler, with orange-colored throat, breast, and center of crown, black upper parts mottled with lighter, and white belly; wing coverts and under tail feathers with much white. The back of the *female* is brownish-olive, streaked with black; the orange of the *male* is replaced by yellow, and the white of the wing coverts forms two wing bars. The *young* has the crown patch nearly absent. It lives mainly in the upper branches of evergreens. (Orange-throated Warbler; Hemlock Warbler.)

Blackburnian Warbler

Length, 5½; wing, 2⅔ (2½–2¾); tail, 2; culmen, ⅜. North America from eastern Kansas eastward; breeding from Massachusetts and Michigan northward (farther south in the mountains), and wintering south to Peru.

23. Yellow-throated Warbler (663. *Dendroica dominica*). — A yellow-throated, gray-backed, white-bellied warbler, with black cheeks; white wing

Yellow-throated Warbler

bars distinct, and white blotches near the tips of the under tail feathers. A southern warbler, with some of the habits of a "creeper" among the tree tops.

Length, 5¼; wing, 2⅜ (2⅜-2¾); tail, 2¼; culmen, ½. Southern United States; breeding from Virginia southward, and wintering from Florida southward; accidental in New York and Massachusetts. The **Sycamore Warbler** (663ᵃ. *D. d. albilora*) is a variety very much like the yellow-throated, but that species has a yellow line in front of the eye and a white line over it, while the sycamore has the line in front of the eye white.

24. Black-throated Green Warbler (667. *Dendroica virens*). — A common, olive-green-backed, black-breasted warbler, with whitish belly and yellow sides of head. Two white wing bars and the under tail feathers with much white, including the base of the outer web; black streaks on the sides. *Female* with much less of black on throat and breast, and some yellowish. The *young* may entirely lack black on the throat. During the breeding season, its home is in the tops of coniferous trees; when migrating it can be found in the growths anywhere.

Length, 5; wing, 2¼ (2¼-2⅜); tail, 2; culmen, ⅜. Eastern North America from the Plains; breeding from Connecticut and Illinois north to Hudson Bay (in the mountains south to South Carolina), and wintering south to Central America. The **Golden-cheeked Warbler** (666. *Dendroica chrysoparia*), a species found in southern Texas, has black upper parts, yellow sides of the head and neck, yellowish-white belly, black chin, throat, and breast, and black streaking on the side of the body. There are two white wing bars and a black stripe through the eye and extending some distance back of it. The tail has large white patches. This bird is much like No. 24, and probably the *female* varies as in that species.

25. Kirtland's Warbler (670. *Dendroica*

Kirtland's Warbler

kirtlandi). — An extensively black-spotted warbler with brownish-ashy back, bluish-gray head, and light yellow under parts.

The sides, breast, and back are the most fully spotted; the space in front of the eye and the sides of the throat are almost wholly black. There are no white wing bars, but the under tail feathers have white blotches near their tips. This is one of the rarest of the warblers.

Length, $5\frac{3}{4}$; wing, $2\frac{3}{4}$ ($2\frac{5}{8}$-$2\frac{7}{8}$); tail, $2\frac{1}{2}$; culmen, $\frac{3}{8}$. Breeding home unknown; migrates through the Mississippi Valley and the southeastern United States.

26. **Pine Warbler** (671. *Dendroica vigórsii*). — A bright olive-green-backed warbler, with the under parts bright yellow except near the tail, where the yellow is gradually changed to white. Sometimes there is a touch of ashy color both on the back and on the belly. The wing bars are whitish, the under tail feathers have white blotches near their tips, and the sides are sometimes streaked with black. *Female* similar but less bright, the upper parts somewhat brownish, and the lower parts yellow only on the breast. As its name indicates, it is nearly always to be found among the pines; in summer up in the trees; in winter mainly on the ground.

Pine Warbler

Length, $5\frac{1}{2}$ (5-$5\frac{3}{4}$); wing, $2\frac{7}{8}$ ($2\frac{3}{4}$-3); tail, $2\frac{1}{4}$; culmen, $\frac{3}{8}$. United States from the Plains eastward, north to New Brunswick; wintering in the South Atlantic and Gulf States.

27. **Palm Warbler** (672. *Dendroica palmárum*). — A warbler, with the upper parts grayish-brown, the lower parts yellow,

The sides are streaked with chestnut, and the crown has a
chestnut patch, very distinct in the breeding season. In winter
the crown patch is rendered more or less indistinct by brownish tips to the feathers.
In summer there is a yellow line over
the eye; in winter this is white,
and a ring around the eye is
also white. The under parts
in winter are rather yel-
lowish than yellow.
This is the Palm War-
bler of the Mississippi
Valley; in winter in
the South Atlantic and

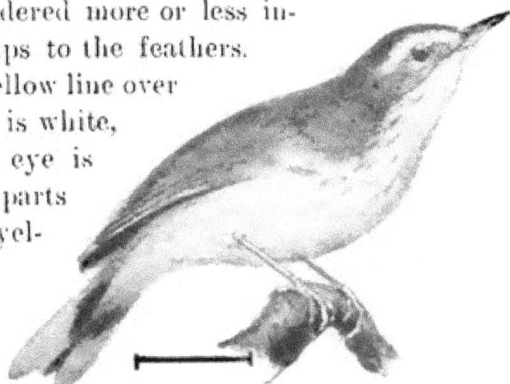

Palm Warbler

Gulf States; occasionally seen in other localities (Red-poll
Warbler). **Yellow Palm Warbler** (672ª. *D. p. hypochrysea*). — A
warbler, with the upper parts dark olive-green, the lower parts
entirely bright yellow, the crown chestnut, and the sides
streaked with chestnut; over and around the eye there is a
yellow line. There are no white wing bars, but the under tail
feathers have white blotches near their tips. In winter the
chestnut crown is partly concealed by brownish. The yellow
of the under parts of this variety is much the brighter and
more uniform. (The Palm Warbler always shows whitish on
the belly.) This is an active warbler of the open field rather
than of the woods, and in winter in the south it is a common
town and village bird. It has a tail-wagging habit which is
very characteristic. (Yellow Red-poll.)

Length, 5¼; wing, 2¾ (2¼-2¾); tail, 2¼; tarsus, ¾; culmen, ¾. Atlantic States; breeding from eastern Maine to Hudson Bay, and wintering in
the South Atlantic and Gulf States. (The Palm Warbler is found north
to Great Slave Lake, and winters south to Mexico.)

28. **Prairie Warbler** (673. *Dendroica discolor*). — A small
olive-green-backed warbler with the under parts bright yellow,
streaked with black on the sides. The center of the back is
marked in the adult with a brownish patch, and the under tail

feathers have large white patches at their tips, even the outer
webs having white at their bases.   There is a yellow line over
the eye, and a black crescent-shaped mark under
the eye. The *female* sometimes lacks the
brown patch of the back, and the *young*
usually has the whole upper parts
ashy in shade. A shy inhabitant
of bushy fields and pastures.

Length, 4¾; wing, 2¼; tail, 2;
culmen, ¼. United States
from the Plains eastward;
breeding from Florida
north to Michigan and
southern New England,
and wintering from Flor-
ida to the West Indies.

Prairie Warbler

29. Oven-bird (674.
Seiùrus aurocapíllus). — A rather small, thrush-like, olive-green-
backed bird, with white under parts, spotted with black on the
breast and sides; the crown is marked with brownish-orange,
bordered with black stripes; wings and tail unmarked. This is
a walking, tail-wagging bird, found mainly in wooded ground,
except in early spring, when
it may often be seen in the
shrubbery and gardens loudly
singing its notes, which seem
to say *teacher*, TEACHER,
TEACHER. (Golden-crowned
"Thrush.")

Length, 6¼; wing, 3 (2¾–3¼);
tail, 2¼; tarsus, ⅞; culmen, ½.
Eastern North America; breeding
from Kansas and Virginia northward, and wintering from Florida
south to Central America.

Oven-bird

30. Water-Thrush (675. *Seiùrus noveboracénsis*). — A small,
olive-backed, thrush-like bird, with the under parts yellowish,

streaked everywhere with black. Over the eye there is a distinct buffy line. This is a walking, tail-wagging bird, like the last, but as its name indicates, it prefers localities near the water, though it is sometimes found in dry places. It is not a shy bird. (Water Wagtail.)

Length, 5¾; wing, 3 (2¾-3½); tail, 2⅛; tarsus, ⅞; culmen, ½. Illinois and eastward; breeding from northern Illinois and northern New England northward, and wintering from the Gulf States to northern South America. **Grinnell's Water-Thrush** (675ᵃ. *S. n. notábilis*) is very much like the last, but larger, on the average, and with the upper parts darker and the under parts whiter. This is the western variety, and is found from Illinois to California north into British America, and winter-

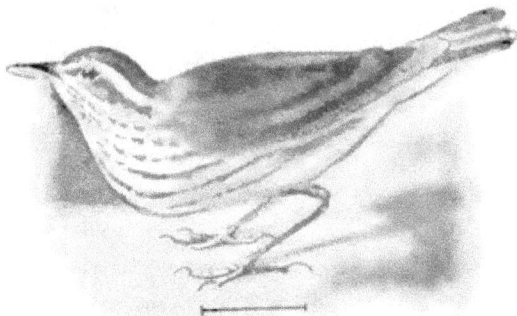

Water-Thrush

ing from the Gulf States to South America. During migrations it has been found in Virginia and even in New Jersey.

31. Louisiana Water-Thrush (676. *Seiúrus motacilla*). —This bird is much like No. 30, but the streakings on the lower parts do not include the throat and middle of belly, the line over the eye is white and conspicuous, and the under parts are tinged with buff color rather than yellow. This is a much shyer bird, more fond of the water, and a noted songster, sometimes singing while on the wing. (Larger-billed Water-Thrush.)

Length, 6¼; wing, 3½ (3-3¼); tail, 2¼; tarsus, nearly 1; culmen, ½+. United States from the Plains eastward, north to central New England; wintering south of the United States to Central America.

32. Kentucky Warbler (677. *Geóthlypis formòsa*). — An olive-green-backed warbler, with all the lower parts and a line over the eye bright yellow. The crown, and a blotch under the eye, extending along the side of the throat, are black. There are no wing bars or tail patches. The *female* has the dark sections more grayish. It is an inhabitant of dense, especially wet, woods, and may be found on the lower growths or *walking* on the ground. It is a loud, clear singer.

Length, 5½; wing, 2⅜ (2⅜–3); tail, 2; tarsus, ⅞; culmen, ½. United States, from the Plains eastward; breeding from the Gulf States to southern Michigan, and wintering south of the United States to Central America.

Kentucky Warbler

33. **Connecticut Warbler** (678. *Geóthlypis ágilis*). — An olive-green-backed, yellow-bellied warbler, with much bluish-gray on the head, neck, and breast. The wings and tail are almost the exact tint of the back, and have no bars or blotches, but there is a white ring round the eye. The *female* has the crown the same as the back, and the throat and breast grayish-brown. This active, sprightly bird is a migrant over most of the eastern United States, and is usually found near the ground in low shrubbery.

Length, 5½; wing, 2⅞ (2⅝–3); tail, 2; tarsus, ¾; culmen, ⅜. Eastern North America; breeding north of the United States, and wintering south of it to northern South America.

34. **Mourning Warbler** (679. *Geóthlypis philadélphia*). — A warbler, with bluish-gray head, olive-green upper parts, and yellow belly. The bluish-gray changes to black on the breast, and the wings and tail are unmarked. The *female* has a head only slightly grayer than the back, and the breast is also only grayish. This shy

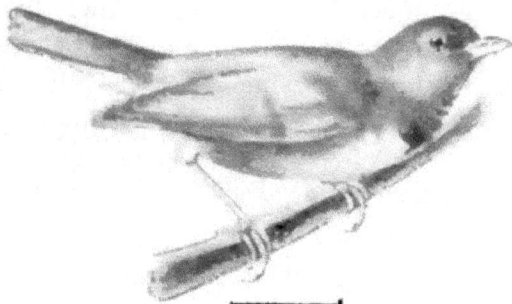

Mourning Warbler

bird lives mainly in the low bushes, and receives its common name from the appearance of crape on the head. It frequently perches on low limbs and sings its clear, whistling notes.

Length, 5½ ; wing, 2¼ (2¼–2½); tail, 2 ; tarsus, ⁊ ; culmen, ⅜. North America, from the Plains eastward ; breeding from the mountains of Pennsylvania and northern Michigan northward, and wintering south of the United States to northern South America.

35. **Maryland Yellow-throat** (681. *Geóthlypis tríchas*). — A bright, yellow-breasted, olive-green-backed warbler, with a peculiar, distinctly out-lined, black mask across the forehead and over the cheeks; wings and tail short and unmarked. *Female* with less distinct mask, and sometimes none. This is a common, bush-living. sprightly bird. which chirps and sings throughout the summer.

Maryland Yellow-throat

Length, 5 ; wing, 2¼ (1⅞– 2¼); tail, 2 ; tarsus, ¾ ; culmen, ⅜. Eastern United States; breeding from Georgia to southern Labrador, and wintering in the South Atlantic States to Central America (even as far north as Massachusetts). The **Florida Yellow-throat** (681ᵇ. *G. t. ignòta*) has the under parts a deeper yellow, the upper parts browner, and the black mask larger. Florida and Georgia. The **Western Yellow-throat** (681ᵃ. *G. t. occidentàlis*) is a larger and brighter colored bird, the bright yellow of the breast extending almost to the anal regions, and the black mask bordered behind by a grayish white band. The wings and tail are each about 2¼ long. From the Mississippi Valley to the Pacific coast.

36. **Yellow-breasted Chat** (683. *Ictèria vírens*). — A large, bright-yellow-breasted, white-bellied, olive-green-backed bird, with a white line over the eye, no wing bars, and a stout bill. This is a bright-colored, noisy dweller of bushy thickets, much more readily heard than seen. Its notes are indescrib-

able in their taunting, mocking, and ventriloquistic qualities. When disturbed in its medley, it merely repeats the complaining call notes of *chŭt chŭt*.

Length, 7¼; wing, 3 (2¾–3¼); tail, 3⅛; culmen, ¼. United States from the Plains eastward; breeding north to Ontario, and wintering south to Central America.

37. Hooded Warbler (684. *Sylcània mitràta*). — A beautiful, black-hooded, olive-green-backed, yellow-bellied, flycatching warbler with yellow forehead and cheeks, and almost completely white under tail feathers. The hood is made up of a crown

Yellow-breasted Chat

piece connected on the sides of the neck with a large throat patch. The *female* has a less distinctly outlined hood. This is a restless bird, generally found among the lower trees or higher shrubs of dense wet woods. It is a sweet singer.

Length, 5½; wing, 2⅝ (2¼–2¾); tail, 2¼; culmen, ⅜. United States from the Plains eastward; breeding from the Gulf of Mexico northward to

Hooded Warbler

southern New England and southern Michigan, and wintering south of the United States to Central America.

38. **Wilson's Warbler** (685. *Sylcània pusílla*). — A yellow-faced, bright olive-green-backed, yellow-bellied, flycatching war-

bler with a distinct black cap, but no wing bars or tail blotches. *Female* similar, but usually lacks the black cap. It is generally to be found among low bushes near the water, and acts much like the true flycatchers in its habit of darting in and out by short flights, in search of its insect prey. The flycatchers proper almost invariably return to the same twigs from which they darted; the warbling flycatchers do not. (Green Black-capped Warbler.)

Length, 4¾; wing, 2¼ (2-2⅜); tail, 2; culmen, ¼+. North America from the Rocky Mountains eastward; breeding mainly north of the United States, and wintering south to Central America.

Wilson's Warbler

39. Canadian Warbler (686. *Sylvania canadénsis*). — A gray-backed, flycatching warbler with all the lower parts yellow, except a necklace of black spots across the breast, and white under tail coverts. It is without either wing bars or tail blotches, but has spots of black on the crown, black sides of neck. and a yellow spot in front of the eye. *Female* lacks the black of the head, and the necklace is made up of dusky spots. It is generally to be found in the same localities as No. 38 and has about the same habits. It is a loud but sweet singer.

Canadian Warbler

Length, 5½; wing, 2⅔; tail, 2¼; culmen, ⅜. North America. from the Plains eastward; breeding from northern New York northward (farther south in the mountains), and wintering south of the United States to northern South America.

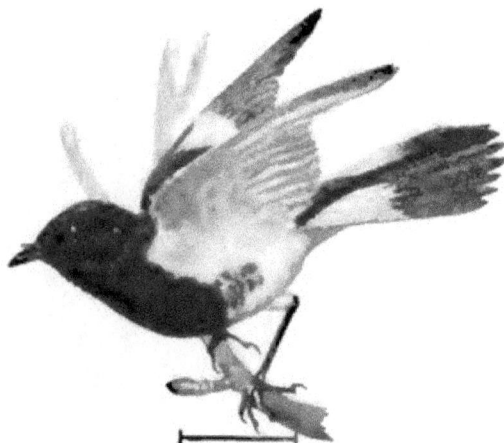

40. American Redstart (687. *Setóphaga ruticilla*). — A small, very lively, dark-colored, brilliantly-marked, flycatching warbler, with bright orange or flame color at base of tail, middle of wings, and under the wings. The belly is nearly white, the bill is very broad, and the rictal bristles fully half as long as the bill. The *female* is a brownish-gray bird with dull-yellow markings replacing the orange of the male. This is one of the most beautiful and active of the warblers, and is to be found abundantly in most woodlands and shrubberies.

American Redstart

Length, 5½ (4¾-5¾); wing, 2¼ (2¼-2¾); tail, 2¾ ; tarsus, ⅝ ; culmen, ⅜. North America, west casually to California ; breeding from North Carolina and Kansas north to Labrador, and wintering south of the United States to South America.

FAMILY VIII. VIREOS (VIREÓNIDÆ)

A family (50 species) of exclusively American, small, olive-backed birds of woods and thickets, with narrow, stout, notched and hooked bills.[1] Our largest species is about the size of the English sparrow. The vireos are insect-eating birds, but unlike many warblers and all the true flycatchers, they gather their prey while perching. With rather slow movements they patiently search over and under leaves, on twigs and bark, for spiders, beetles, caterpillars, etc. All our species are good singers, and some are noted for their vocal powers. Dr. Coues thus speaks of them : "Next after the warblers, the greenlets [vireos] are the most delightful of our forest birds, though their charms address the ear and not the eye. . . . In the quaint and curious ditty of the white-eye, in the earnest, voluble strains of the red-eye, in the tender secret that the warbling vireo confides in whispers to the passing breeze — he is insensible who does not hear the echo of thoughts he never clothes in words." They build beautiful basket-like nests, which are suspended from forked twigs, sometimes near the ground and sometimes from the highest parts of forest trees. The vireos are usually to be found on trees or bushes, very rarely on the ground. Any of the common names given can end in *Greenlet* as well as *Vireo*.

Key to the Species

* Rather stout species with distinct white or whitish wing bars.[2] (C.)
* Rather slender species with no distinct wing bars. (A.)
   A. The first primary less than one inch long ;[3] under parts white with yellowish on the sides (if there is almost no yellowish on sides, look for 2. Red-eyed Vireo, as it has been found with a short first primary).....................................5. **Warbling Vireo.**
   A. The first primary much over one inch long. (B.)
   B. Under parts yellowish, brightest on the breast ; crown ashy, without a bordering black line over the whitish eye line. 4. **Philadelphia Vireo.**
   B. Under parts mainly white with almost no yellowish ; crown ashy-gray margined with blackish just above the white line over the eye. ....2. **Red-eyed Vireo.**
   (or 1. **Black-whiskered Vireo,** if in Florida, and if there is a dusky streak on the side of the throat.)
   B. Sides bright olive-yellow, and the under tail coverts clear sulphur-yellow (Texas).......................... 3. **Yellow-green Vireo.**
   C. First primary nearly as long as the second ; breast bright yellow ....6. **Yellow-throated Vireo.**
   C. First primary less than one inch long.[3] (D.)
   D. Top and sides of head grayish-blue, with a distinct white line around the eye.............................7. **Blue-headed Vireo.**
   D. Crown olive, much like the back ; a distinct yellow line over the eye. ....9. **White-eyed Vireo.**
   D. Crown ashy ; a distinct white line around the eye (Western). ....10. **Bell's Vireo.**
   D. Crown and sides of head black (Kansas to Texas)................. ....8. **Black-capped Vireo.**

   1   2   3

1. **Black-whiskered Vireo** (623. *Vireo calidris barbátulus*). — A species found in southern Florida, West Indies, and south in winter to Central America. It is very much like the next species, but has a narrow dusky streak extending from the bill, below the eyes, along the side of the throat.

Length, 5¾ ; wing, 3⅛ (3–3¼) ; tail, 2¼ ; culmen, ⅝.

2. **Red-eyed Vireo** (624. *Vireo olivàceus*). — A very common, small, olive-green-backed, white-bellied vireo, with a black-margined slaty-gray crown, white line over the eye, and no wing bars. The dark border to the crown gives emphasis to the white line over its red eye. This common inhabitant of trees has been called the preacher, because of its tireless singing. In the words of Wilson Flagg, "We might suppose him to be repeating moderately, with a pause between each sentence, 'You see it — you know it — do you hear me? — do you believe it?' All these strains are delivered with the rising inflection at the close, and with a pause, as if waiting for an answer."

Red-eyed Vireo

Length, 6¼ ; wing, 3¼ (3–3½) ; tail, 2¼ ; tarsus, ¾ ; culmen, ⅝. North America, from Utah eastward ; breeding from the Gulf States to Hudson Bay, and wintering from Florida to South America.

3. **Yellow-green Vireo** (625. *Vireo flavoviridis*). — A vireo of western Texas and southward to Peru ; very much like the last species, but with the under parts, especially the sides, flanks, and under tail coverts, much more brightly yellow.

Length, 6½ ; wing, 3 (2¾–3¼) ; tail, 2¼ ; culmen, ⅝.

4. **Philadelphia Vireo** (626. *Vireo philadélphicus*). — A small, light, olive-green-backed, grayish-crowned vireo with all under

parts light greenish-yellow and a whitish line over the eye. There are no wing bars. This is a smaller and rarer bird than No. 2, but with similar habits and song.

Length, 5; wing, $2\frac{5}{8}$ ($2\frac{1}{2}$–$2\frac{3}{4}$); tail, 2; tarsus, [illegible]; culmen, [illegible]. Eastern North America; breeding from New Hampshire north to Hudson Bay, and wintering south of the United States to Central America.

5. **Warbling Vireo** (627. *Vireo gilvus*).—An olive-green-backed, white- or whitish-bellied vireo, without wing bars and with some yellowish on the sides. This, as its name indicates, is a good, continuous "warbler" of rich notes. It lives mainly among the tops of tall trees, so that it can be heard more easily than seen.

Length, $5\frac{3}{4}$; wing, $2\frac{7}{8}$ ($2\frac{5}{8}$–3); tail, $2\frac{1}{4}$; tarsus, [illegible]; culmen, $\frac{1}{2}$. North America in general; breeding nearly throughout, and wintering in eastern Mexico.

6. **Yellow-throated Vireo** (628. *Vireo flavifrons*).—An olive-green-backed, bright-yellow-breasted, white-bellied vireo, with two distinct white wing bars. It is an inhabitant of the tree tops. Its notes are deep, rich, and varied, and occasionally it shows a power of song which is surprising in its fine and intricate quality. The bird looks in color much like the yellow-breasted chat, though decidedly smaller.

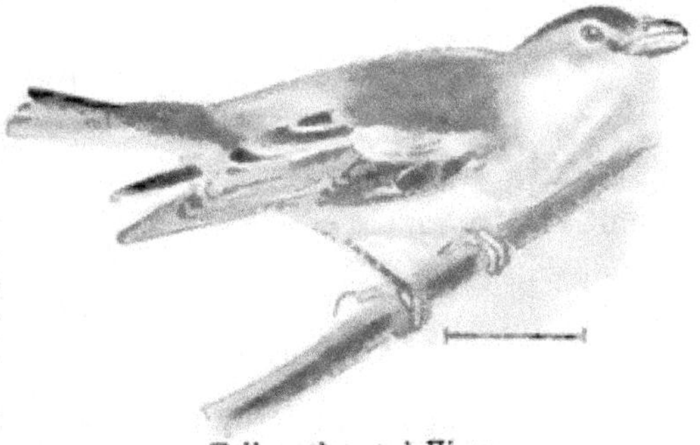

Yellow-throated Vireo

Length, $5\frac{3}{4}$; wing, $3\frac{1}{8}$ (3–$3\frac{1}{4}$); tail, $2\frac{1}{4}$; tarsus, $\frac{5}{8}$; culmen, [illegible]. Eastern United States; breeding from Florida north to Ontario, and wintering from Mexico to Central America.

7. **Blue-headed Vireo** (629. *Vireo solitarius*).—An olive-green-backed, bluish-headed vireo, with the lower parts, ring around eye, and two wing bars white. It is, like the vireos in general, an inhabitant of the woods and a fine singer. (Solitary Vireo.)

Length, $5\frac{5}{8}$; wing, $2\frac{7}{8}$ ($2\frac{3}{4}$–3); tail, $2\frac{1}{8}$; tarsus, $\frac{3}{4}$; culmen, $\frac{3}{8}$. North America, from the Plains eastward; breeding from New England to Hudson Bay, and wintering in Mexico to Central America. The **Mountain Solitary Vireo** (629^c. *V. s. alticola*), of the higher southern Alleghanies, is a similar bird but larger, and with the entire upper parts a nearly uniform dark lead-color with almost no tinge of green. Wing, $3\frac{1}{8}$ (3–$3\frac{1}{4}$); tail, $2\frac{1}{4}$.

Blue-headed Vireo

8. **Black-capped Vireo** (630. *Vireo atricapillus*). — A small, rare, Texan, black-headed, olive-green-backed, white-bellied vireo, with olive-shaded sides.

Length, $4\frac{1}{4}$; wing, $2\frac{1}{8}$; tail, 2; tarsus, $\frac{3}{4}$; culmen, $\frac{1}{2}$. Central and western Texas, north to southwestern Kansas.

9. **White-eyed Vireo** (631. *Vireo noveboracensis*). — A small, olive-green-backed, white-bellied vireo, with much yellowish on wings, tail, and sides of head, including two distinct wing bars. A ring around the eye, and line from eye to bill yellow. This is a saucy inhabitant of underbrush, with mocking bird powers, which it uses in intricate "medleys" of bird notes.

White-eyed Vireo

Length, 5; wing, $2\frac{3}{8}$; tail, 2; tarsus, $\frac{3}{4}$; culmen, $\frac{1}{2}$. United States from the Rocky Mountains eastward; breeding north to southern New England and Minnesota, and wintering from Florida as far south as Central America.

10. **Bell's Vireo** (633. *Vireo béllii*). — A stout, western, thicket-living, olive-green-backed, yellow-sided vireo, with the ring around the eye, and line from eye to bill white or nearly so. The belly is white, and the crown ashy-gray, gradually changing to a bright olive-green on the rump. This is a smaller bird than the last.

Length, 4¾; wing, 2¼ (2–2¼); tail, 1⅞; tarsus, ¾; culmen, ⅜. Great Plains, to the upper Mississippi Valley, eastward to western Indiana, and southwestward to Mexico.

FAMILY IX. SHRIKES (LANIIDÆ)

A family (200 species) of mainly Old World birds, repre-sented in America by two species of large-headed, strong-bodied gray birds, with black wings and tail, and white belly. Their size is not much less than that of our robin. The hawk-like bill[1] enables them to capture their prey, which consists of mice, small birds, insects, etc. Their greatest peculiarity is the habit of [1] impaling their food upon thorns, the barbs of wire fences, etc. In watching for their prey, the shrikes almost always occupy the outside twigs of bushes or trees or other exposed situations. When an insect, a small bird, a mammal, or a reptile is seen, they will dart through a distance of many feet and secure it.

Key to the Species

* Breast generally with distinct wavy cross lines; black on the sides of the head *not* connected by a black line on the forehead..........
....1. **Northern Shrike.**
* Breast usually with no distinct wavy cross lines; black on the sides of the head connected by a black line across the forehead. ....
....2. **Loggerhead Shrike.**

1. **Northern Shrike** (621. *Lànius boreàlis*). — A gray-backed, white-bellied bird, with black wings and tail. There are black blotches on the side of the head, which are not joined together by a black line across the forehead, and narrow wavy bars across the breast. It is a low-flying bird, with a peculiarly straight course till it is ready to alight, when it makes a short

upward turn and perches on the outside twigs of the tree or
bush. It may be distinguished by the terror it causes among
the small birds in its vicinity. (Butcher-bird.)

Length, 10; wing, 4½ (4¾–4¾); tail, 4¾; tarsus, 1;
culmen, ¾. Northern North America; breeding
north of the United States, and wintering in the
Middle States.

2. **Loggerhead Shrike** (622. *Lanius lu-
doviciánus*). — A bird similar to the
last, but with more black on the
sides of the head, connected
across the forehead by
a narrow black stripe.
There are fewer wavy
lines, or almost none,
across the breast. Both
these species are noted

Northern Shrike

for the habit of impaling their prey — grasshopper, lizard,
snake, or bird — on thorns. Both of these birds sing in the
springtime. The notes of northern shrike are very musical,
and resemble some-
what those of the cat-
bird, but those of the
loggerhead are too
harsh to be pleasant.

Length, 9; wing, 3⅞
(3¾–4); tail, 4; tarsus,
1; culmen, ½. United
States, from the Plains
eastward, and north to
northern New England;
breeding from the Gulf
States north to southern
New Jersey and the

Loggerhead Shrike

Great Lakes. The **White-rumped Shrike** (622ª. *L. l. excubitorides*), a
variety found from the Plains to the Pacific, has, as its name indicates,
the upper tail coverts more or less distinctly whitish.

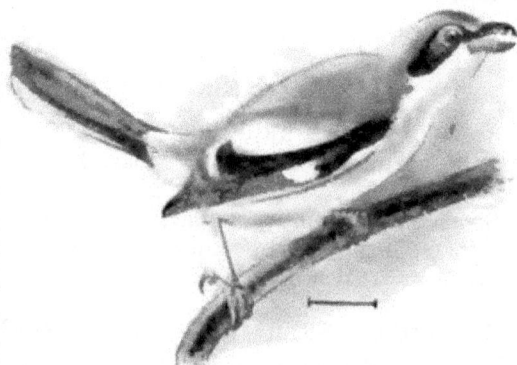

FAMILY X. WAXWINGS (AMPÉLIDÆ)

This very small family of birds includes two of our crested, smooth-plumaged, rich grayish-brown species, with short, square, yellow-tipped tails and long wings. The waxwings practically have no song, and their notes are so quietly uttered as to be by many unnoticed. The name "waxwing" is derived from the fact that the secondary wing quills, and sometimes the tail feathers, are tipped with horny appendages resembling red sealing wax.

Key to the Species

* Under tail coverts chestnut; wing bar white..1. **Bohemian Waxwing.**
* Under tail coverts white; no wing bar...........2. **Cedar Waxwing.**

1. **Bohemian Waxwing** (618. *Ampelis gárrulus*). — A rare, distinctly crested, rich brown-backed, grayish-bellied bird, with the under tail coverts chestnut and the tail feathers tipped with yellow; having a white wing bar, white tips to the secondary quills, and a brownish breast. The forehead, chin, and line through the eye are black. (Northern Waxwing.)

Bohemian Waxwing

Length, 8; wing, 4½ (4⅜-4⅝); tail, 2¼; tarsus, ¾; culmen, ⁷⁄₁₆. Northern parts of the northern hemisphere; breeding north of the United States, and wintering rarely south to Pennsylvania, Illinois, and Kansas.

2. **Cedar Waxwing** (619. *Ampelis cedròrum*). — A common, distinctly crested, rich brown-backed, yellowish-bellied waxwing, with the under tail coverts white, and all the tail feathers tipped with yellow. There is no wing bar. The breast is like the back and the forehead; the chin and the line over the eye are black. These smooth-plumaged birds move, excepting in the

breeding season (May to August), in small flocks, and when on the wing fly close together in a straight line on about a level with the tree tops. They are chatterers rather than singers. (Cedar-bird; Cherry-bird.)

Length, 7¼; wing, 3¾ (3½-3⅞); tail, 2¼; tarsus, ¾; culmen, ⅜. North America; breeding from Virginia and Kansas northward (farther south in the mountains), and wintering throughout the United States south to Central America.

Cedar Waxwing

FAMILY XI. SWALLOWS (HIRUNDÍNIDÆ)

This family comprises eighty species of long-winged, small birds. They spend most of the time in the air in pursuit of their food, which consists almost entirely of insects. Many have forked tails; few have colors other than black and white; many have glossy, and some, iridescent plumage. On account of their weak, small feet, they usually perch on very slender twigs, or by preference on telegraph wires. The top of the bill is very short, but the mouth is both wide and deep, reaching about to the eyes.[1]

Key to the Species

* Wing over 5 long ....1. **Purple Martin**, or 2. **Cuban Martin.**
* Wing, 5 or less long. **(A.)**
  A. Tail, 3 or more long, the notch more than an inch deep; the under tail feathers with white blotches; throat chestnut; back lustrous steel-blue ....4. **Barn Swallow.**
  A. Tail, 2¼ or less long; back with metallic luster. **(C.)**
  A. Tail, 2¼ or less long; back brownish, without luster. **(B).**
B. Breast brownish; belly and throat white.........6. **Bank Swallow.**

B. Throat and breast brownish ; belly white. 7. **Rough-winged Swallow.**
B. All under parts white........ ....5. **Tree Swallow.**
  C. Throat chestnut or black ; upper tail coverts *reddish*............
  ....⌐.........3. **Cliff Swallow.**
  C. All under parts white......................5. **Tree Swallow.**

1. **Purple Martin** (611. *Prógne sùbis*). — A large, shining, blue-black swallow, with a notched tail. The *female* is not so glossy on the back ; and her throat, breast, and sides are brownish-gray, and her belly white. It nests in boxes, gourds, etc., near human habitations, and is very common throughout the Southern States.

Length, 8 ; wing, 5¾ (5¼-6¼) ; tail, 3¼ ; forked, ¾ ; culmen, ½. North America from Mexico to Ontario, wintering from Mexico to South America.

2. **Cuban Martin** (611. 1. *Prógne cryptoleùca*). — A Florida and Cuban species, very much like the last in habits and appearance, but if the belly feathers are opened, there will be found a broad, white spot on each. The *female* has the neck, chest, and sides a sooty-brown, changing abruptly to the white of the belly and under tail coverts.

Purple Martin

Length, 7½ ; wing, 5¼ (5¾-5½); tail, 3¼. Southern Florida and Cuba.

3. **Cliff Swallow** (612. *Petrochélidon lùnifrons*). — A steel-blue-backed swallow, with a white forehead, much chestnut on the neck, pale brownish above the

Cliff Swallow

tail, and a white belly. Tail very slightly notched. This bird builds gourd-shaped mud nests under the eaves of buildings and on rocks. (Eave Swallow.)

Length, 5½; wing, 4¼ (4–4½); tail, 2; culmen, ¼. North America; breeding from the Potomac and Texas northward into the Arctic regions, and wintering in Central and South America.

4. Barn Swallow (613. *Chelidon erythrogástra*). — A common, chestnut-bellied, steel-blue-backed swallow, with a deeply forked tail. The breast is dark chestnut, but the other under parts are lighter; the under tail feathers are white blotched. It nests in barns, using mud and grass for building.

Barn Swallow

In flying, it keeps nearer the ground than most swallows.

Length, 7; wing, 4¾ (4½–5); tail, 3–5; culmen, ¼. North America; breeding from Mexico to the Arctic regions, and wintering in Central and South America.

5. Tree Swallow (614. *Tachycineta bícolor*). — A steel-blue-backed swallow, with all the under parts pure white. The back sometimes has a tinge of green, and the wings and tail are blackish. The *young* has brownish-gray upper parts. The tail is very slightly forked. The nests are found mainly in

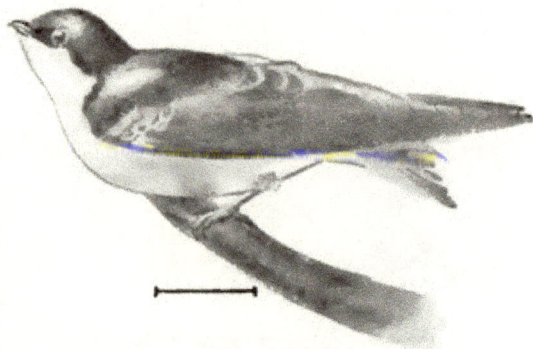

Tree Swallow

hollows in trees, but some are built in boxes, like the martins. (White-bellied Swallow.)

Length, 5¾; wing. 4¾ (4½–5); tail, 2¾ ; culmen. ⅓. North America ; breeding from the Ohio Valley northward, and wintering from the Gulf States to Central America.

6. Bank Swallow (616. *Clivicola ripària*). — A small, common, dull, brownish-backed swallow, with white

Bank Swallow

throat and belly. and a broad band of grayish-brown on the breast. The tail is slightly notched. There is a curious tuft of feathers above the hind toe. This bird breeds in great colonies in appropriate sandy banks, and if the locality is suitable. is very abundant. (Sand Martin.)

Length, 5; wing. 4 (3½–4½) ; tail, 2 ; culmen, a little over ⅓. Northern hemisphere ; breeding from the Gulf States northward, and wintering from Central to South America.

7. Rough-winged Swallow (617. *Stelgidópteryx serripénnis*). — A dull, brownish-gray swallow, with white only on the

Rough-winged Swallow

lower belly; tail slightly notched. The *adult* has recurved hooklets on the outer edge of the first primary. The *young* lack these, and have the breast somewhat tinged with chestnut.

In general appearance much like the last, but slower in its flight. It nests in sand banks or among the timbers of bridges.

Length, 5½; wing, 4⅜ (4–4⅜); tail, 2½; culmen, ¾. Southern Ontario and Connecticut southward ; breeding throughout.

FAMILY XII. TANAGERS (TANÁGRIDÆ)

This is a large family (300 species) of tropical, tree-living birds, with brilliant colors and generally weak voices. Our three species are stout-billed,[1] migratory birds. The males are mainly bright red, and without the crest which is so conspicuous on the cardinal grosbeak.

Key to the Species

* *Male* red, with black wings and tail, wings without wing bars ; *female*, olive, with most under parts greenish-yellow ....2. **Scarlet Tanager**.
* *Male* red throughout ; *female*, yellowish-olive, with the under parts buffy-yellow... 3. **Summer Tanager**.
* With yellow or yellowish wing bars ....1. **Louisiana Tanager**.

1. **Louisiana Tanager** (607. *Piránga ludoviciána*). — An extreme western species which has been recorded from a few of the Eastern States. It is a crimson-headed, yellow-bodied tanager, with the back, wings, and tail black ; wing with two yellow bars. *Female*, much like the female of No. 3, but with two light-colored wing bars. Size like that of the other tanagers. Western United States, from the Plains to the Pacific.

2. **Scarlet Tanager** (608. *Piránga erythromélas*). — A common, summer, red-bodied bird, with

Scarlet Tanager

black wings and tail, and no wing bars. *Female*, olive-green above, greenish-yellow below ; blackish wings and tail. This brilliantly colored bird is found in dense woods, singing its robin-like carol in the tree tops.

Length, 7; wing, 3¼ (3½–3¾); tail, 3; culmen, ⅜. The United States, from the Plains eastward; breeding from Virginia to New Brunswick, and wintering from Mexico to South America.

3. **Summer Tanager** (610. *Piránga rúbra*). — A common, summer, red bird of the south, without either crest on head or black on wings or tail. *Female*, brownish-olive above and buffy-yellow below. This is a sweet singer in open woods, with notes which resemble those of the last species. Its call notes are very peculiar, and have been written *chicky-tucky-tuck*. (Summer Red Bird.)

Length, 7½; wing, 3¾ (3½–4); tail, 3; tarsus, ¾; culmen, ⅝. Eastern United States; breeding from Florida to New Jersey, wandering to Nova Scotia, and wintering in Mexico to South America.

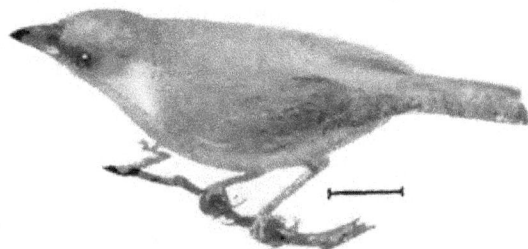

Summer Tanager

FAMILY XIII. FINCHES, SPARROWS, AND GROSBEAKS (FRINGÍLLIDÆ)

This is the largest of the families of birds (550 species), and comprises medium to small forms to be found everywhere (except in Australia) at all seasons of the year. The family has never been successfully divided into groups, and the student, in working with these forms, will have greater difficulty in determining species than anywhere else among birds. All have somewhat short, conical bills, with the corners of the mouth abruptly bent downward.[1][2][3][4][5] Most of our small species have plain colors arranged more or less in a streaky manner; these

1   2   3   4   5

are popularly called sparrows. About a dozen of the large species have very heavy, stout bills, and are called grosbeaks.[13] Some are bright colored, others have bright markings of red or yellow; these often have names to indicate their colors. Some have the nail of the hind toe peculiarly elongated and straightened;[4] these constitute the longspurs. Others, the crossbills, have the bill remarkably curved and crossed at tip.[2] Others, as the juncos and towhees, have the plumage unstreaked, but with masses of different colors on different portions of the body. None of our species equal the robin in size, though a few come near it. The English sparrow is about the average, there being about twenty species smaller, twenty larger, and about twenty like it in size. The painted bunting, the smallest species (except Sharp's seed-eater of Texas), is about the size of the kinglets. The singing power varies wonderfully; some hardly sing at all, while others are noted songsters. Some of our favorite cage birds — the canary for example — belong to this family. Nearly all are seed-eaters, and for this reason are not so migratory as the insect-eaters of other families; the migration of birds being more due to lack of food than to inability to stand the cold. The streaked species are mainly inhabitants of the ground, while the brighter colored ones are more generally to be found among the trees.

Key to the Species

* Mandibles long and much curved, their points crossed at tip.[2]
  — Without wing bars......................4. American **Crossbill**.
  — With white wing bars..............5. **White-winged Crossbill**.
* Bill very stout, as high at base as the culmen is long; top and bottom of bill usually much curved.[13] (**X.**)
* Bill neither very stout (at least not so high at base as long) nor the points crossed at tip. (**A.**)
  A. Rather evenly colored birds; there may be large patches of different colors, but they are not sharply spotted or streaked either above or below; some are somewhat mottled, but not in any very definite manner. (**T.**)
  A. Decidedly spotted or streaked either above or below. (**B.**)
B. Upper (middle) tail feathers especially narrow and sharp-pointed, much more so than the under ones.[5] (**Q.**)

B. All tail feathers rather narrow and acutely pointed, and in many cases stiff.⁶ (**L.**)

B. Tail feathers neither especially narrow nor especially sharp-pointed, and in no cases stiff. (Nos. 36–37 have narrow but not acute tail feathers.) (**C.**)

 C. Wing, 4 or more long; under parts white, sometimes with brownish markings ....12. **Snowflake.**

 C. Wing, 3–4 long; no yellow anywhere. (**K.**)

 C. Wing, 3–4 long; some distinct yellow on bend of wing and head.

  — Some yellow on breast also...................52. **Dickcissel.**

  — No yellow on breast...........30. **White-throated Sparrow.**

  And under that species..........**Golden-crowned Sparrow.**

 C. Wing, 3 or less long. (**D.**)

D. With a spot of bright red on the crown..........6 and 7. **Redpolls.**

D. With some distinct yellow somewhere. (**I.**)

D. With neither distinct red nor yellow anywhere. (**E.**)

 E. Tail rounded; breast without distinct streaks; crown dark chestnut or streaked; no whitish wing-bars......40. **Swamp Sparrow.**

 E. Tail rounded; breast sharply streaked. (**H.**)

 E. Plumage not streaked below; tail somewhat notched. (**F.**)

F. Crown slate-color, ashy-brown, or liver-brown; a distinct white or buffy wing bar..11. **European House Sparrow** and **E. Tree Sparrow.**

F. Crown grayish with a light central stripe; a white line over the eye.

  — Rump brownish ....33. **Clay-colored Sparrow.**

  — Rump slate-gray....................32. **Chipping Sparrow.**

F. Crown chestnut. (**G.**)

 G. Crown bright chestnut; a narrow black line back of the eye and some black on the forehead..............32. **Chipping Sparrow.**

 G. Crown bright chestnut; a reddish-brown line back of the eye and a black or blackish spot on the breast.........31. **Tree Sparrow.**

 G. Crown dull chestnut; no black on the forehead; a whitish eye ring ....34. **Field Sparrow.**

H. Back, sides, breast, and tail coverts much streaked; crown with a faint, pale, medium line.......................38. **Song Sparrow.**

H. Everywhere sharply streaked; crown not chestnut; a buffy band across breast ....39. **Lincoln's Sparrow.**

 I. Wing quills and under tail feathers yellow at base; tail notched; under parts white, heavily streaked with black....9. **Pine Siskin.**

 I. Breast at least with some yellow; tail notched and the under tail feathers white blotched.................8. **American Goldfinch.**

 1   2   3   4   5   6

I. Breast with yellow; tail slightly double-rounded and the under feathers not white blotched.....................52. **Dickcissel**.

I. Tail somewhat longer than the wings; bend of wing yellow. (**J.**)

J. Head striped and two of the stripes white with yellow in front; a white throat patch....................30. **White-throated Sparrow**.

J. No yellow on head; upper tail feathers not barred...............
....36. **Bachman's Sparrow**.

J. No yellow on head; upper tail feathers barred [1]....................
....................................37. **Cassin's Sparrow**.

K. Rump, tail, and wings with much rusty-red; large arrow-shaped spots on the white breast....................41. **Fox Sparrow**.

K. *Male* with much red; *female* olive-brown; tail an inch shorter than the wings and notched at tip [2]....................3. **Purple Finch**.

K. Under parts pure white, except black spots on the breast; tail rounded,[3] and the under feathers black, with white tips...........
..............................27. **Lark Sparrow**.

K. Wing with a conspicuous, light-colored (white or whitish) patch; tail about square [4]....................53. **Lark Bunting**.

K. Tail rounded; wing, 3¼ or less long; crown pure white or pale brownish, margined with darker....29. **White-crowned Sparrow**.

K. Tail rounded; wing, 3½–3½; center of crown more or less black...
....................................28. **Harris's Sparrow**.

L. Breast with yellow; throat with more or less black; bend of wing yellow ....52. **Dickcissel**.

L. Tail double-rounded; [6] middle and under pair of feathers about equal in length, the others gradually longer. (**P.**)

L. Tail rounded; [6] the middle pair of feathers about the longest; the under feathers gradually shorter. (**M.**)

M. Culmen, ¾ or more long; bend of wing yellow. (**O.**)

M. Culmen, less than ½ long. (**N.**)

N. Tail and wings almost exactly equal in length; back feathers black, bordered by buffy; no yellow in front of eye or on bend of wing....
....................................23. **Leconte's Sparrow**.

N. Tail measurably shorter than wing; back brown streaked with black; bend of wing pale yellow..........22. **Henslow's Sparrow**.

O. Back almost without streaks; breast slightly streaked with dusky; yellow in front of eye....................25. **Seaside Sparrow**.

O. Back somewhat streaked; breast broadly streaked with black; yellow in front of eye..............26. **Dusky Seaside Sparrow**.

O. Center of crown with a distinct stripe of ashy; breast and sides distinctly streaked with blackish.......24. **Sharp-tailed Sparrow**.

   1   2   3   4   5

P. Wing, $2\frac{3}{4}$–$3\frac{1}{4}$; back ashy, somewhat streaked with brownish; a white streak over the eye..18. **Ipswich Sparrow.**

P. Wing, $2\frac{3}{4}$–$2\frac{7}{8}$; pale yellow in front of eye and on bend of wing; back sharply streaked with black.................19. **Savanna Sparrow.**

P. Wing, $2\frac{1}{4}$–$2\frac{5}{8}$; bend of wing yellow; spot in front of eye orange....
....21. **Grasshopper Sparrow.**

P. Wing, $2\frac{5}{8}$–3; western species with *very* narrow and acute tail feathers; head buffy on crown and white on chin and throat............
....20. **Baird's Sparrow.**

 Q. Hind toe nail but little longer than that of middle toe; bend of wing chestnut; breast without yellow but streaked with black; under tail feathers almost entirely white; tail double-rounded[7]...
....17. **Vesper Sparrow.**

 Q. Hind toe nail but little longer than that of the middle toe; breast with more or less of yellow; under tail feathers not white........
....52. **Dickcissel.**

 Q. Hind toe nail about as long as the hind toe and nearly twice as long as that of the middle toe and but little curved.[8] (**R.**)

R. Bill stout, nearly as high at base as the culmen is long; under tail feathers almost entirely white; others, except the middle pair, tipped with black; bend of wing chestnut (western).........................
....16. **McCown's Longspur.**

R. Bill much more slender.[9] (**S.**)

 S. Two under tail feathers mostly white; under parts buffy.........
....14. **Smith's Longspur.**

 S. Second under tail feather but little white; breast with much black; belly whitish; legs and feet black........13. **Lapland Longspur.**

 S. Under tail feathers mostly or entirely white; all others with much white at base; legs pale.........15. **Chestnut-collared Longspur.**

T. Tail as long as or longer than the wings. (**W.**)

T. Tail shorter than the wings; wing, 3 or more long. (**V.**)

T. Tail shorter than the wings; wing, 2–3 long. (**U.**)

 U. Body yellow, with wings and tail black (*male*), or back brown, with more or less yellow below (*female*); bill very sharp and small...
....8. **American Goldfinch.**

 U. Plumage blue (*male*) or grayish-brown, with some tinge of blue on the outer web of the quills (*female*); the under side of the bill with a blackish stripe.....................49. **Indigo Bunting.**

 U. Head blue; back golden green; rump and under parts red (*male*), or above olive-green; below greenish-yellow (*female*)..........
....50. **Painted Bunting.**

6   7   8   9

    U. Head blue and red, belly reddish-purple (*male*), or brownish with whitish lower parts and no wing bars (*female*)..................
....50. **Varied Bunting.**

    U. Because of lack of distinct streaks, one of the small sparrows with narrow, acute-pointed tail feathers might be sought for here......
....25. **Seaside Sparrow.**

V. Blue, with chestnut on wings (*male*), or plain brown (*female*); tail even.[1]......................................48. **Blue Grosbeak.**

V. Bird with crimson, black, yellow, and white in its plumage.........
.... 10. **European Goldfinch.**

V. Brownish above and below, with rosy edgings to the quills; black or clear ash on head; tail slightly notched. The **Gray-crowned Leucosticte** (524. *Leucosticte tephrocotis*) of the Rocky Mountain region might be found east of those mountains.

V. Because of their finch-like bills, the bobolink and cowbird (Nos. 1 and 2 of the next family, page 141) might be looked for here.

    W. Under parts pure white or somewhat irregularly variegated with rusty; nail of hind toe twice as long as that of the middle toe and much curved;[2] wing, 4 or more long.............12. **Snowflake.**

    W. Belly white; sides chestnut-brown; under tail feathers tipped with white; back black (*male*), or grayish-brown (*female*) ......
....43. **Towhee.**

    W. Upper parts, head, and breast slate-color; belly and outer tail feathers white.................................35. **The Juncos.**

    W. Upper parts olive-green; under parts white, with pure white on the middle of the belly; head somewhat striped; edge of wings and under coverts of wings bright yellow ....42. **Texas Sparrow.**

X. Conspicuously crested,[3] with more or less of distinct red in the plumage.................. 44. **Cardinal** and 45. Texas Pyrrhuloxia.

X. No crest; small, southwestern birds, with wings less than 2¼ long...
.... 51. **Sharp's Seed-eater** and Grassquit (51).

X. No crest; large birds, with wings, 3¼ or more long. (Z.)

X. No distinct crest; smaller; wings, 2¼-3¼ long. (Y.)

    Y. Plumage with much red (*male*) or streaky olive-brown (*female*); no wing bars; crown with erectile feathers, slightly imitating a crest..3. **Purple Finch** and in the Rocky Mountains House Finch (3).

    Y. Streaky sparrow without yellow; wings, 3 or less long and with white or buffy wing bar[4].......... 11. **European House Sparrow.**

    Y. Wing with a large, conspicuous white or whitish patch; general color black (*male*), or brown streaked (*female*) ....
....53. **Lark Bunting.**

1   2   3   4

Y. Plumage blue (*male*) or brownish or tawny (*female*) ; wing bars chestnut or buffy ; tail, ½ inch shorter than the wings............
....48. **Blue Grosbeak.**

Y. On account of the stout bill, 16. **McCown's Longspur** might be looked for here. It has the nail of the hind toe very long and nearly straight.

Z. General colors rosy-red (*male*), or ashy-gray, with brownish-yellow on head and rump (*female*).......................2. **Pine Grosbeak.**

Z. General colors black and white, with rich red on breast and under wing coverts (*male*), or brownish streaked, with the under wing coverts rosy or orange (*female* and *young*) ; tail with white blotches...
....46. **Rose-breasted Grosbeak.**

Z. General colors black and white, with neck and under parts orange or yellow (*male*). or brownish streaked, with the under wing coverts sulphur-yellow (*female*) ....47. **Black-headed Grosbeak.**

Z. Upper tail coverts yellow ; inner secondaries and wing coverts white ; bill greenish-yellow ; wing over an inch longer than the tail........
....1. **Evening Grosbeak.**

1. Evening Grosbeak (514. *Corrothraustes (thréstes) vesper-tinus).* — A heavy-billed, olive-brown bird, with black and white wings, black crown and tail, and yellow forehead and rump. The *female* lacks the black crown and yellow forehead and rump, and has both wings and tail blotched black and white. A grosbeak of western North

Evening Grosbeak

America, which, rather irregularly in flocks, has been found as far east as Massachusetts.

Length, 8; wing, 4¾ (4-4½); tail, 3; culmen, ¾. Western British Provinces east to Lake Superior, and casually to the New England States.

2. Pine Grosbeak (515. *Pinicola enucleator).* — A large, winter, uncrested grosbeak, with a rosy tint over most of the body, but brightest on the head, breast, and rump, and blackest on the wings and tail. *Female* slate-gray, with much olive-yellow on head, breast, and upper tail coverts. On its somewhat rare winter visits to the northern United States, it comes in flocks, and can usually be found on the sumachs and mountain ashes, eating the berries.

Length, 8¼; wing, 4½ (4¼-5); tail, 3½; tarsus, ⅞; culmen, ⅗. Northern parts of the northern hemisphere; breeding from northern New England northward, and wintering irregularly southward into the northeastern states.

3. **Purple Finch** (517. *Carpódacus purpúreus*). — A common, small, rosy-red-bodied bird, with brownish wings and tail, and whitish belly. The rosy red is brighter on the head,

Pine Grosbeak

breast, and rump. The *female* is very much like a streaky, grayish-brown sparrow, having white under parts marked with many spots and streaks of dark brown. The *female* is somewhat difficult to determine, but the forked tail an inch shorter than the wings, and the tufts of feathers over the nostrils of the stout bill, distinguish it from all other birds.

Length, 6; wing, 3¼ (3-3½), tail, 2¼; culmen, ½. North America from the Plains eastward; breeding from New England northward (farther south in the mountains), and wintering in the Middle and Southern States. The House **Finch** (519. *Carpódacus mexicánus frontális*) of Colorado, western Texas to California, is similar in size and coloring to the purple finch but the tail is about square at tip. Both of these are excellent singers. The house finch is as common in the southern towns west of the Rocky Mountains as the English sparrow is in the towns east of them.

4. **American Crossbill** (521. *Lóxia curriróstra mínor*). — A climbing, dull-red-bodied, small bird with blackish wings and

tail, and no white on the wings. The back is brownish, the rump bright red, and the tail short and deeply notched. The *female* has the red replaced by olive-green, with the rump yellowish. These birds are very irregular in their appearance at any locality, but always come in flocks and are usually found among the cone-bearing trees, extracting the seeds by

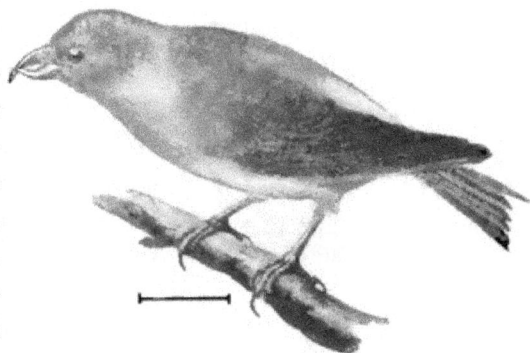

American Crossbill

their peculiar bills, the lower mandible of which curves upwards, its point crossing that of the upper one. In climbing about the trees, they act much like parrots. They fly in close ranks in a peculiarly undulating manner. (Red Crossbill.)

Length, 6; wing, 3¾ (3½–3¾); tail. 2½; tarsus, ⅝; culmen, ½. Northern North America; breeding in northern United States (south in the mountains to Georgia), and wintering irregularly south to Louisiana.

5. **White-winged Crossbill** (522. *Lóxia leucóptera*). — A bird similar to the last in action and coloring, but the pinkish red

White-winged Crossbill

of the body is much brighter, and the wings and tail blacker, and it has large white blotches on the wings. The female has a dull olive-green body, yellow rump, and white-blotched black wings. This bird is rare, but can be easily recognized by the white of the wing coverts and the greater noise it makes while feeding.

Length, 6; wing, 3½; tail, 2½; culmen, ⅜. Northern North America; breeding from northern New England northward, and wintering south in the United States to Pennsylvania.

6. Hoary Redpoll (527ª. *Acánthis hornemánnii exílipes*). — A bird similar to the next, but differing in having the rump nearly white (pinkish white in the *male*), without streaks, the feathers of back and wings with whitish edges, and the belly white without streaks.

Length, 5; wing, 3 (2½–3½); tail, 2½; tarsus, ½; culmen, ₁₆⁵. Arctic America and northern Asia; rarely wintering as far south as the northern United States.

7. Redpoll (528. *Acánthis linária*). — A small, winter, red-capped, black-chinned, streaky, brownish sparrow with (in the *male*) pink washings on the rump and breast. The very *young* lack the red cap. This bird comes into the northern United States rather irregularly in flocks, in the winter, and is usually found search-

Redpoll

ing for seeds on the grasses and low weeds which project above the snow in pastures. (Redpoll Linnet.)

Length, 5; wing 2⅞ (2½–3); tail, 2½; tarsus, ½; culmen, ₁₆⁵. North-ern portions of the northern hemisphere; breeding north of the United States, and wintering very irregularly south to Virginia and Kansas. **Holböll's Redpoll** (528ª. *A. l. holbœllii*) is a larger bird with a longer bill; it has been once recorded from the United States (Massachusetts). Wing, 3–3¼; culmen, fully ⅜. **Greater Redpoll** (528ᵇ. *A. l. rostráta*) is a larger bird with darker feathers on the back. This bird has been seen irregularly as far south as southern New York and northern Illinois. Wing, 3–3½; tail, 2½; tarsus, nearly ¾; culmen, ⅜.

8. American Goldfinch (529. *Spínus tristis*). — A very common, small, yellow-bodied bird with black cap, wings, and tail. It

flies through the air in a wave-like track, singing *per-chic-o-ree*
on the downward slopes of its passage. The *female* all the
year, and the *male* in winter, have only yellow washings on a brownish body; the black cap is
also lacking; the tail is deeply notched.
Except in early summer when nesting, these birds are found in small
flocks. (Thistle-bird; Yellow-
bird.)

Length. 5; wing, 2¾ (2½–
2⅞); tail, 2; culmen, ½.
Temperate North Amer-
ica; breeding from Vir-
ginia and Kentucky
northward, and winter-
ing throughout most if
not the whole of the
United States. The
Arkansas Goldfinch

American Goldfinch

(530. *Spinus psaltria*) differs in having the back dark olive-green to
black in color, the lower parts only being lemon-yellow. The black
wings have a large (or sometimes small) white patch on the base of the
quills. The *female* is grayish-olive-green on the back, and greenish-
yellow below; the white patch on the wings is smaller. This is a western bird found from the
Plains to the Pacific.

9. **Pine Siskin** (533.
Spinus pinus). — A small,
very streaky, dark-colored
winter sparrow with much
yellow on the wings and tail,
which is decidedly notched. It
has much the habits of the goldfinch, but is found more frequently upon the cone-bearing trees,
whence its name. (Pine Finch.)

Pine Siskin

Length, 5; wing, 2¾; tail, 1⅞; culmen, ½. North America; breeding
from the northern United States northward, and wintering very irregularly south to the Gulf States.

10. European Goldfinch (*Carduèlis carduèlis*). — A peculiarly bright-colored bird which has been introduced into this country from Europe, and has seemingly become naturalized in the vicinity of Boston and New York. It is a bright, brown-backed bird, with bright red around the base of the bill, black wings with a yellow band, black tail with white blotches on the under feathers, black crown with a black stripe on the side of the neck, and white belly. Its habits are somewhat like those of the American goldfinch.

European Goldfinch

Length, 5¼; wing, 3; tail, 2; tarsus, ½; culmen, ½.

11. European House Sparrow (*Pásser domésticus*). — An altogether too common, streaky-backed, gray-crowned sparrow, with whitish lower parts, having a black or blackish patch on the throat and breast. The back colors are black and chestnut; the rump ashy. There is chestnut on the sides and back of the gray crown, and a white bar on the middle coverts of the wings. The *female* has the head and rump grayish-brown, and the breast and sides washed with the same color, though lighter. The back has buff instead of chestnut, and the wing bar is not so distinctly white. The nearly universal conclusion is that the introduction of this bird was a great mistake, and a mistake which cannot be remedied. It has already spread over nearly the whole area covered by this book, and has driven out many of our best American birds. (English Sparrow.) See illustration on p. 382.

Length, 6; wing, 3; tail, 2¾; culmen, ½ nearly. The **European Tree Sparrow** (*Pásser montánus*) is a bird very similar in appearance to the last, which has become naturalized in the section near St. Louis, Missouri.

It can be recognized by the liver-brown color of the crown, and the wing rarely over 2⅓ long.

12. Snowflake (534. *Plectróphenax nivális*). — A sparrow-like, ground-living, winter bird, with much white on head, tail, wings, and under parts; up-per parts with much rusty-brown, streaked with black. This, our snow-colored snow-bunting, is to be found in the United States only when snow is on the ground. Like most of our small, winter birds, it is a seed-eater. It comes usually in large flocks. (Snow Bunt-ing; White Snow-bird.)

Snowflake

Length, 6⅓; wing, 4¼ (4–4½); tail, 2⅓; tarsus, ⅔; culmen, ⅓. Northern regions; breeding north of the United States, and wintering regularly in northern states, and irregularly to Georgia and southern Illinois.

13. Lapland Longspur (536. *Calcárius lappónicus*). — A streaky-backed, white-bellied, sparrow-like, ground-feeding, winter bird of the United States, with the nail of the hind toe longer than its toe. The length of the hind toe and its nail can often be determined by its tracks. The back has streaks of black, brown, and

Lapland Longspur

buff. This bird is found among flocks of shorelarks and snow-flakes, and has the habit of squatting back of some clod, where it will remain till almost trodden upon; then it will run a little distance and again attempt to hide.

Length, $6\frac{1}{4}$; wing, $3\frac{3}{4}$ ($3\frac{1}{2}$–$3\frac{7}{8}$); tail, $2\frac{5}{8}$; tarsus, $\frac{3}{4}$; culmen, $\frac{3}{8}$. Northern regions; breeding far north, and wintering in the northern United States, irregularly farther south even to South Carolina.

14. **Smith's Longspur** (537. *Calcarius pictus*). — A rare winter bird of the western plains, of size and habits similar to that of the Lapland longspur, but with much more buffy color to its plumage, and the head and back with much black. A line over the eye and the ear coverts white; a broad, white, wing bar, and the two under tail feathers mostly white.

Length, $6\frac{1}{4}$; wing, $3\frac{5}{8}$ ($3\frac{1}{2}$–$3\frac{3}{4}$); tail, $2\frac{1}{2}$; tarsus, $\frac{3}{4}$; culmen, $\frac{3}{8}$. Interior of North America; breeding far north, and wintering south to Illinois and Texas.

15. **Chestnut-collared Longspur** (538. *Calcarius ornatus*). — A beautiful western longspur of bright colors, with a chestnut collar, black breast and crown, and much white on head and tail. The bird may be distinguished by the great amount of white on the tail feathers (the under mostly white, the others with much white at base). *Female* usually without black.

Length, 6; wing, $3\frac{3}{8}$ ($3\frac{1}{4}$–$3\frac{1}{2}$); tail, $2\frac{1}{4}$; culmen, $\frac{3}{8}$. Interior of North America; breeding from western Minnesota west and north, and wintering south to Texas; accidental in Massachusetts.

16. **McCown's Longspur** (539. *Rhynchophanes mccownii*). — A heavy-billed, grayish-brown, mottled, western longspur, with black crown and crescent-shaped mark on breast and white-blotched tail feathers. The *female* lacks the black of head and breast, but both sexes have the under tail feathers white, and the others, except the middle pair, white at base with square, dark tips. (Black-breasted Longspur.)

McCown's Longspur

Length, 6; wing, 3½ (3⅛–3⅞); tail, 2½; tarsus, ⅔; culmen, ½. Interior North America; breeding from northern Kansas north to the Saskatchewan, and wintering south to Texas and northern Mexico.

17. **Vesper Sparrow** (540. *Poocætes gramineus*). A ground-living, streaky sparrow, with the bend of the wing chestnut and the outer tail feathers white. The back is mainly brownish-gray, and the under parts white, streaked with black and buffy. This, though mainly a field sparrow, will occasionally perch on fences and trees. It is one of the sweetest singers of

Vesper Sparrow

the morning and evening, the evening song giving it the name of vesper sparrow. Its notes are much like those of the song sparrow, but more plaintive. (Grass Finch; Bay-winged Bunting.)

Length, 6¼; wing, 3½ (2⅞–3¾); tail, 2½; tarsus, ⅞; culmen, ⅜. North America from the Plains eastward; breeding from Virginia and Missouri north to Nova Scotia, and wintering from south New Jersey southward. The **Western Vesper Sparrow** (540ª. *P. g. confinis*) averages slightly larger, is grayer in color, and is found from the Plains to the Pacific.

18. **Ipswich Sparrow** (541. *Ammodramus princeps*). — A rare, seacoast, brownish, much-streaked sparrow, with a white line over the eye, two buffy wing bars, and sometimes a spot of sulphur-yellow in front of the eye and on the bend of the wing. The upper parts are streaked with

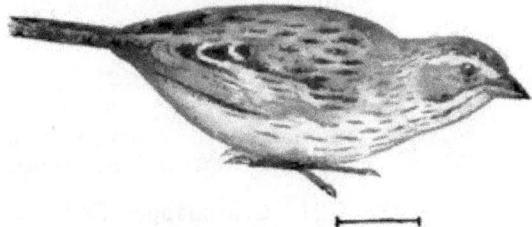

Ipswich Sparrow

brownish, black, and ashy: the lower parts are white, with streaks of blackish and buff on the breast and sides.

Length, 6⅛; wing, 3(2⅞–3⅛); tail, 2⅜; tarsus, ⅞; culmen, ⅜. Atlantic coast; breeding in Nova Scotia, and wintering as far south as Georgia.

19. **Savanna Sparrow** (542ª. *Ammódramus sandwichénsis savánna*).—A common, very streaky, ground sparrow, with some yellow in front of the eyes and on the bend of wing; in habits, size, and coloring much like the vesper sparrow. The streaky under parts and the method of flying are especially similar, but it lacks the chestnut bend of wing and the distinct white under tail feathers of that species, only the outer edge being whitish.

Length, 5¼; wing, 2¾ (2½–2⅞); tail, 2; tarsus, ⅞; culmen, ⅜. Eastern North America; breeding from northern New Jersey and Missouri to Hudson Bay, and wintering from North Carolina southward. The **Western Savanna Sparrow** (542ᵇ. *A. s. alaudinus*) has a smaller and more slender bill and is paler and more grayish in color. It is found from the Plains westward.

20. **Baird's Sparrow** (545. *Ammódramus bàirdii*).—A western, ground-living, pale-yellowish-brown sparrow, with a streaky, grayish-brown back and many sharp, small, dark streaks on its head and breast. From the breast the under parts are a dull white. Its notes have been written by Dr. Coues, "*zip-zip-zip-zr-r-r-r.*" This species is much like the last, but its tail feathers are more narrow and acute.

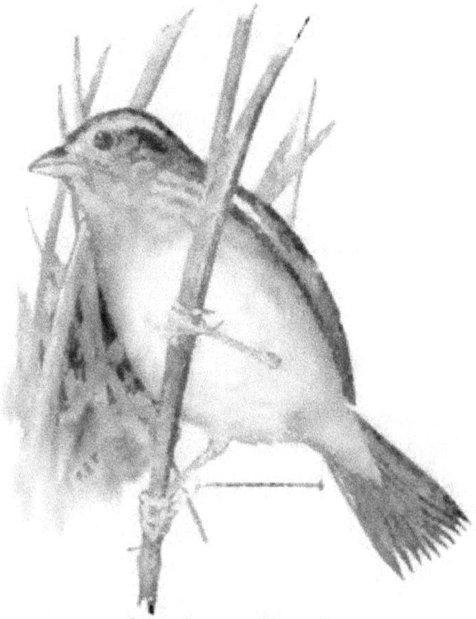

Grasshopper Sparrow

Length, 5½; wing, 2¾ (2½–3); tail, 2⅛; tarsus, ¾; culmen, ⅜. Interior North America from the Plains westward to Arizona.

21. **Grasshopper Sparrow** (546. *Ammódramus savannárum passerìnus*).—A common, streaky-backed, buffy-breasted, ground sparrow, with the sides much like the breast, but the

belly whitish and bend of wing yellow. The upper parts are streaked with black, brown, ashy, and buff, and the blackish crown has a buffy line through the center. There is an orange dot in front of the eye. The tail feathers are very acute, and their edges are decidedly lighter than the brown centers. This bird is one of the quietest and most easily overlooked of our common birds of the open fields. It takes its name from its voice, which is much like that of some grasshoppers. It rarely takes a higher position than that of the fences, and from such a perch it usually does its singing. (Yellow-winged Sparrow.)

Length, 5; wing, $2\frac{3}{8}$ ($2\frac{1}{4}$–$2\frac{1}{2}$); tail, $1\frac{7}{8}$; tarsus, $\frac{3}{4}$; culmen, $\frac{7}{16}$. United States from Plains eastward; breeding from the Gulf States to Canada, and wintering from Florida to Central America. The **Western Grasshopper Sparrow** (546[a]. *A. s. perpállidus*) has larger wings and tail, a more slender bill, and is paler in color. Wing, $2\frac{1}{2}$; tail, 2. It is found from the Plains westward.

22. **Henslow's Sparrow** (547. *Ammódramus henslówii*). — A ground-living, sharp-tailed, brownish sparrow, with the back, breast, and sides very much streaked with black, brown, and buffy. The bright brown on the back, wings, and tail and the olive tints of the head are the plainest characteristics of this rare, secretive, weed-inhabiting bird of the meadows or dry fields.

Henslow's Sparrow

Length, 5; wing, $2\frac{1}{8}$ (2–$2\frac{1}{4}$); tail, 2; tarsus, $\frac{5}{8}$; culmen, $\frac{3}{8}$ or a little more. United States from the Plains eastward; breeding from the Gulf States northward to southern New England and Ontario, and wintering in the Gulf States.

23. **Leconte's Sparrow** (548. *Ammódramus lecónteii*). — A sharp-tailed, streaky-backed, buffy-colored sparrow, with a cream-colored streak along the center of the blackish crown; the breast is practically without streaks, but there are some

streaks along the sides; the belly is white; the under tail feathers are nearly a half inch shorter than the middle pair. This is a western species of fields and marshes.

Length, 5; wing, 2 (1¾–2¼); tail, 2¼; tarsus, ⅝; culmen, ⅜. Interior United States; breeding from Minnesota north to Manitoba, and wintering from Iowa to Florida and Texas.

24. **Sharp-tailed Sparrow** (549. *Ammódramus caudácutus*). — A common, salt-marsh, sharp-tailed, streaky, olive-gray sparrow, with distinct orange-brown bands on the head, above and below the eye. The buffy lower parts are darkly streaked on the sides and breast, but the throat and belly are nearly white. This bird prefers to escape from a person by running and hiding among the grasses and reeds of the salt marshes (where it dwells), rather than to use its wings in flight.

Sharp-tailed Sparrow

Length, 5¼; wing, 2¼ (2⅛–2⅜); tail, 2; tarsus, ⅞; culmen, ⅝. Marshes of the Atlantic coast; breeding from North Carolina to Maine, and wintering along the south Atlantic and Gulf States. **Nelson's Sparrow** (549a. *A. c. nélsoni*) is a slightly smaller variety with the feathers of the back darker in the center and with wider whitish edges, and the sides, breast, and throat darker in tint, but much less streaked. Fresh marshes of the interior; breeding from Illinois northward, and wintering from Texas to South Carolina; accidental in New England. The **Acadian Sharp-tailed Sparrow** (549b. *A. c. subvirgàtus*) differs in having the sides, breast, and throat more creamy in tint and faintly streaked with gray rather than black. A salt marsh form; Nova Scotia to South Carolina.

25. **Seaside Sparrow** (550. *Ammódramus marítimus*). — A common, salt-marsh, sharp-tailed, slightly streaked, grayish-brown sparrow, with a little yellow at bend of wing and in front of eye. The white throat and middle of the belly and

the dusky breast indistinctly streaked with whitish are good distinguishing marks of this species. but a dusky and white stripe at each side of the white throat and the absence of any tint of reddish brown are still more characteristic. It is a ground bird, found nearly always among the reeds and grasses within both sight and sound of the sea.

Length, 6; wing, 2⅜ (2¼-2½); tail, 2¼; tarsus, ⅞; culmen, ½. Atlantic

Seaside Sparrow

coast; breeding from Georgia to Massachusetts, and wintering from Virginia southward. **Scott's Seaside Sparrow** (550ᵃ. *A. m. peninsulæ*) is a south Atlantic and Gulf coast variety of a very much darker color.

26. Dusky Seaside Sparrow (551. *Ammodramus nigrescens*).— A Florida species more nearly like Scott's seaside sparrow than any other form, but differing in having the feathers of the upper parts black, with grayish edges, and the under parts sharply streaked with about equal amounts of black and white.

Length, 6; wing, 2⅜; tail, 2⅜; tarsus, ⅞; culmen, ½. Eastern Florida.

Lark Sparrow

27. Lark Sparrow (552. *Chondestes grammacus*).— A common, western, ground-living, lark-like, streaky sparrow, with black and white tail, ashy-brown back, and a striped white and chestnut-colored head. The central tail feathers and the bases of the others are dark in color, forming a decided contrast to the ashy-brown back and white tips to the under tail feathers. There are black streaks on the side of throat and in the center of the breast.

This is a fine song bird of the middle west, with notes which somewhat resemble those of the song sparrow. When singing. it usually takes some elevated position on fence or tree.

Length, 6½; wing, 3¼ (3¼–3½); tail, 2¾; culmen, ₇/₁₆. The Mississippi Valley region to the Plains; breeding from Texas to Manitoba; accidental on the Atlantic coast.

28. **Harris's Sparrow** (553. *Zonotrichia querula*). — A large, beautiful, western, streaked, reddish-colored sparrow, with heavy brownish markings on the white of the breast and sides. The *male* when breeding has the head jet black excepting the cheeks, which are ash-colored; the throat and breast patch are also black. The *female* (also the male

Harris's Sparrow

out of season) has the head not especially marked and the breast patch brownish. There is no yellow anywhere, and the two white wing bars are distinct. This is the largest sparrow of the genus (*Zonotrichia*), and has been found from Illinois westward, mainly on the prairies and bushy bottom lands. (Black-hooded Sparrow.)

Length, 7¼; wing, 3¾ (3¼–3¾); tail, 3½; tarsus, 1; culmen, nearly ½. Interior United States from Illinois to Kansas, and Texas to Manitoba.

29. **White-crowned Sparrow** (554. *Zonotrichia leucophrys*). — A rare, beautiful, large, brownish sparrow, with the head striped black and white (three white and four black stripes), and the lower parts gray with some buff on the sides. There are two

white wing bars and no yellow on head or wings. It is found in the eastern United States only during the colder months; its singing is remarkable, resembling that of the white-throated sparrow.

White-crowned Sparrow

Length, $6\frac{3}{4}$; wing, $3\frac{1}{5}$ (3–$3\frac{1}{4}$); tail, 3; tarsus, nearly 1; culmen, $\frac{3}{8}$. North America; breeding north of the United States, and wintering from Virginia to Mexico.

30. **White-throated Sparrow** (558. *Zonotríchia albicóllis*). — A common, social, large, streaky, brownish sparrow, with a distinctly striped head and a square white patch on the throat, distinct from the grayish under parts. The head has two black and three white stripes, two of the white stripes yellow in front, and there are two distinct white wing bars. This beautiful sparrow is especially abundant in small flocks, in the autumn and winter, in the undergrowth of the woods and along the bushy fence rows. It is a good singer and says very distinctly *peabody, peabody*, whence it derives one of its names. (Peabody Bird.)

White-throated Sparrow

Length, $6\frac{3}{4}$; wing, $2\frac{7}{8}$ ($2\frac{3}{4}$–$3\frac{1}{8}$); tail, $3\frac{1}{8}$; tarsus, $\frac{7}{8}$; culmen, $\frac{3}{8}$. North America, from the Plains eastward; breeding along the northern border of the United States northward, and wintering from southern New England southward to the Gulf. The **Golden-crowned Sparrow** (557. *Zonotríchia coronáta*), a Pacific coast species with a back like the white-throated sparrow, and lower parts like the white-crowned sparrow, but with the central crown stripe yellow in front, has been seen in Wisconsin.

31. **Tree Sparrow** (559. *Spizélla montícola*). — A chestnut-crowned, streaky-backed, winter chippy, with whitish under parts unmarked except by a blackish dot on the center of the breast and some brownish washings on the sides. There are two distinct white wing bars. This common winter bird of the fields and bushes has a deeply notched

Tree Sparrow

tail like all of the genus (*Spizélla*), and is decidedly larger than the summer chippy (No. 32). (Winter Chippy.)

Length, 6¼; wing, 3 (2¾–3¼); tail, 2¾; tarsus, ¾; culmen, ¼. North America, from the Plains eastward; breeding north of the United States, and wintering through most of the eastern United States.

32. **Chipping Sparrow** (560. *Spizélla sociális*). — A common, small, chestnut-crowned, streaky-backed sparrow, with whitish under parts and line over the eye. The notched tail, black bill and forehead, and (even in the young) slaty-gray rump are all

Chipping Sparrow

points of importance in the determination of this species, as, in the autumn, the crown loses its bright chestnut color and becomes more or less streaked. This bird is one of the most quiet, familiar, and trustful frequenters of our dooryards and can readily be induced to eat out of our hands. Such song as it has is an insect-like repetition of its common name of *chippy*. (Chippy; Hair-bird.)

Length, $5\frac{3}{8}$; wing, $2\frac{3}{4}$ ($2\frac{5}{8}$-$2\frac{7}{8}$); tail, $2\frac{3}{8}$; tarsus, $\frac{5}{8}$; culmen, $\frac{3}{8}$. Eastern United States; breeding from the Gulf States to Great Slave Lake, and wintering from the Gulf States to Mexico.

33. **Clay-colored Sparrow** (561. *Spizélla pállida*). — A small, western, pale-colored sparrow, with much gray in its plumage, giving it its common name. Back brownish-gray; under parts white soiled with gray; sides of head brown, with irregular black and whitish markings; crown with a pale medium stripe; rump brownish-gray and not slaty-gray. In habits it is like the chipping sparrow but not so confiding and trustful.

Clay-colored Sparrow

Length, $5\frac{1}{4}$; wing, $2\frac{3}{8}$ ($2\frac{1}{4}$-$2\frac{1}{2}$); tail, $2\frac{3}{8}$; culmen, nearly $\frac{3}{8}$. Interior North America from the Rocky Mountains to Illinois; breeding from Iowa northward, and wintering from southern Texas into Mexico.

Field Sparrow

34. **Field Sparrow** (563. *Spizélla pusílla*). A buffy-breasted, reddish-billed, streaky-backed sparrow, with a dull-chestnut crown and gray line over the eye. The back is brightly marked with black, reddish-brown, and ashy, the breast unspotted buff; wings with two white bars. This bird is not an inhabitant of the open fields, but seeks bushy pastures. It is a fine

singer, especially of the early evening, and its notes have great variety.

Length. 5¾; wing, 2½ (2½–2¾); tail, 2½; culmen, ⅜. Eastern United States and southern Canada; breeding from South Carolina northward, and wintering from Illinois and Virginia southward.

35. **Slate-colored Junco** (567. *Júnco hyemális*). — A small, slate-colored, winter bird, with white belly and under tail feathers, and flesh-colored bill. The slate color of the breast abruptly changes to the white of the belly. This very common and easily recognized bird of the snowy season is usually found in flocks of twenty to thirty in the fields and among the bushes. When it flies, the white of the under tail feathers is readily seen. (Junco: Snow-bird; Black Snow-bird.)

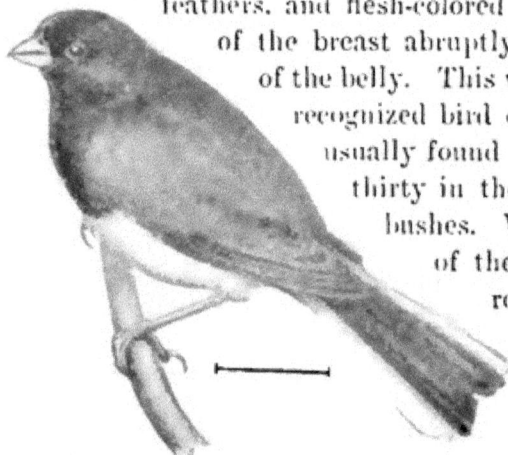

Length, 6; wing, 3 (2⅞–3⅛); tail, 2¾; tarsus, ⅞; culmen, ½ nearly.

Slate-colored Junco

North America mainly east of the Rocky Mountains; breeding among the higher parts of the Alleghanies and other mountains northward, and wintering throughout. The **Carolina Junco** (567e. *J. h. carolinénsis*) differs in having a darker colored bill and the back without any show of brownish, which can always be noticed on the common Junco. It is a common resident variety of the mountains of Virginia, North and South Carolina. A much browner variety than even the common one is **Shufeldt's Junco** (567b. *J. h. connectens*). In this the sides are almost a wine-brown. This belongs to the region from the Rocky Mountains westward, but has been seen in a number of the Eastern States, Massachusetts, Maryland, etc. The **White-winged Junco** (566. *Junco aikeni*) has two very distinct white wing bars. It breeds in the Black Hills, and in winter is found south to Colorado.

36. **Bachman's Sparrow** (575a. *Peucæa æstivalis bachmánii*). — A streaky, brownish-red-backed sparrow with the lower parts grayish-buff, deepest on the breast and almost white on the

belly; the bend of the wing is yellow and the tail much rounded; the under feathers are a half inch shorter than the upper ones, and all the tail feathers are very narrow though not acute-pointed. The bright rusty back is usually without black streaks. This is a wonderfully sweet and somewhat loud singer; found in its northern range in the open woods.

Length, 6; wing, 2½ (2¾-2½); tail, 2½; tarsus, ¾; culmen, ½ or more. The Carolinas west to Texas, north to southern Illinois, wintering in Florida. The **Pine-woods Sparrow** (575. *Peucæa æstiralis*) differs in having the back a light chestnut, streaked with black and margined with gray, and the breast sometimes spotted with black. It is an inhabitant of pine woods with an undergrowth of scrub palmetto. According to Frank M. Chapman it is the best singer among all our sparrows, and compares well in the exquisite tenderness and pathos of its melody with the hermit thrush. Florida and southern Georgia.

37. **Cassin's Sparrow** (578. *Peucæa cassini*). — This western species is similar to Bachman's sparrow, but differs in having the upper tail feathers decidedly barred with somewhat crescent-shaped dusky spots and the flanks broadly streaked with brownish.

Bachman's Sparrow

Length, 6¼; wing, 2¾; tail, 2⅞; tarsus. ⅞; culmen, ½. Central and western Kansas, south and west to Texas and Nevada.

38. **Song Sparrow** (581. *Melospiza fasciata*). — A very common, streaky, grayish-brown sparrow, with the sides of the breast especially marked with a cluster of stripes forming a blotch of brownish. It is abundant among shrubbery near water, and throughout the year of four seasons, and even throughout the day of twenty-four hours, it shows its wonderful powers of song. Of course the morning and evening of spring days are its especial times for singing. Though it

usually sings while perched on a twig, it occasionally sings a new and varied song while on the wing.

Song Sparrow

Length, $6\frac{1}{4}$; wing, $2\frac{5}{8}$ ($2\frac{3}{8}$-$2\frac{7}{8}$); tail, $2\frac{3}{4}$; tarsus, $\frac{7}{8}$; culmen, $\frac{1}{2}$. North America from the Plains eastward; breeding from Virginia northward, and wintering throughout the eastern United States.

39. **Lincoln's Sparrow** (583. *Melospiza lincólnii*).—A buffy-breasted, white-bellied, streaky, brownish-backed sparrow, with the throat white like the belly. The whole bird, except the middle of the belly, is sharply streaked. The creamy buff on the breast forms a band across it. This is a western bird of shy habits, rarely seen east of the Alleghanies. It is a singer of no great power, and of rather strange notes for a sparrow.

Length, $5\frac{3}{4}$; wing, $2\frac{1}{2}$ ($2\frac{3}{8}$-$2\frac{5}{8}$); tail, $2\frac{1}{8}$; tarsus, $\frac{3}{4}$; culmen, $\frac{3}{8}$. North America; breeding chiefly north of the United States, and wintering south of it.

40. **Swamp Sparrow** (584. *Melospiza georgiána*). — A common swamp or meadow-

Lincoln's Sparrow

dwelling, streaky-brown bird, similar in coloring to the song
sparrow, excepting that the breast is unstreaked. It is a little
smaller in size, and
has no such power of
song. Its notes con-
sist of a repetition
of *tweet-tweet*, with but
little if any change of
pitch.

Length, 5¾; wing, 2¾
(2¼-2½); tail, 2½; cul-
men, nearly ½. North
America from the Plains
eastward; breeding from
Virginia northward, and
wintering from Massa-
chusetts to the Gulf
States.

Swamp Sparrow

41. **Fox Sparrow** (585. *Passerélla ilíaca*). — A large, spotted-
breasted, rusty-red sparrow, with much bright chestnut on wings,
tail, and cheeks. The middle of the belly is unspotted white.
In the autumn and early spring, during migrations, this bird is
found among shrubbery in flocks, and at those times, but more
especially in spring, it shows more than the usual power of
song of sparrows.

Length, 7; wing, 3¾
(3¼-3⅗); tail, 2⅞; tarsus,
1; culmen, ½. North
America from the Plains
eastward; breeding north
of the United States, and
wintering mainly south
of the Potomac and Ohio
rivers.

Fox Sparrow

42. **Texas Sparrow**
(586. *Arrémonops ruficirgìta*). — A southwestern olive-green-
backed, brownish-white-breasted sparrow, with a striped head
and bright-yellow edge to the wing. The crown has two chest-

nut-brown stripes beside the central grayish one, and a brown stripe back of the eye below a side stripe of ashy. These stripes are not very sharply defined. (Green Finch.)

Length, $6\frac{1}{2}$; wing, $2\frac{5}{8}$ ($2\frac{1}{2}$–$2\frac{3}{4}$); tail, $2\frac{5}{8}$; tarsus, $\frac{7}{8}$; culmen, $\frac{1}{2}$. Western Texas and eastern Mexico, rarely to southern Louisiana.

43. **Towhee** (587. *Pipilo erythrophthálmus*).—A large, brightly marked bird, with black upper parts and breast, white belly and tips of under tail feathers, and chestnut sides. In the *female* the black is replaced by bright, grayish brown, excepting that the tail feathers are blackish, with similar white tips which can be distinctly seen while on the wing. This is a common, restless, ground bird of the bushy woods. The notes are clear, *chewink, towhee*, forming two of its common names. Ernest Thompson writes the full notes "*chuck-burr, pill-a-will-a-will-a.*" (Chewink; Marsh "Robin"; Ground "Robin"; Joree.)

Towhee

Length, $8\frac{1}{4}$; wing, $3\frac{1}{2}$ ($3\frac{1}{4}$–$3\frac{3}{4}$); tail, $3\frac{3}{4}$; tarsus, 1; culmen, $\frac{9}{16}$. United States east of the Plains, and southern Canada; breeding from Georgia northward, and wintering from Virginia southward. The **White-eyed Towhee** (587a. *P. e. álleni*) is a similar bird, but smaller, and with less white on the wings, and only two of the under tail feathers white at tip; eyes very light-colored, almost white. This is a shyer bird than the last, and is found among heavier growths. Wing, $3\frac{1}{4}$; tail, $3\frac{5}{8}$. Florida north to South Carolina. The **Arctic Towhee** (588. *Pipilo maculàtus árcticus*) is similar to the common towhee, but has white spots on the wing coverts and shoulders, the white on the shoulders lengthened into streaks, and that of the coverts forming two bars. The *female* is a dark brown, with the white markings as in the male. Wing, $3\frac{3}{8}$; tail, 4. The Rocky Mountains eastward to Kansas, and northward to the Saskatchewan River, wintering from Kansas to Texas.

44. Cardinal (593. *Cardinàlis cardinàlis*). — A large, distinctly crested, red bird, with black around the red bill, most extensive on the throat. The *female* has less bright red anywhere, the under parts are buffy and the throat blackish. The crest, wings, and tail are dull red. These birds are fine songsters and are frequently kept in cages; both sexes sing. They are resident birds wherever they are found, and as they are more easily and frequently seen when the foliage is off the trees they are often called winter red birds. (Cardinal Grosbeak.)

Length, 8½; wing, 3¾ (3½-4); tail, 4½; tarsus, 1; culmen, ¾. United States from the Plains eastward, north to the Great Lakes and central New England; wintering about as far north as its full range, and breeding throughout.

Cardinal

45. Texas Pyrrhuloxia (594ᵃ. *Pyrrhúloxia sinuàta texàna*). — A bird similar to the last, but lacking the black around the bill, and the very short and convex bill is yellow or slightly horn-color instead of red. Size practically the same as that of the cardinal except the shorter culmen. Southern Texas and Mexico, rarely to southern Louisiana.

46. Rose-breasted Grosbeak (595. *Zamelòdia ludoviciàna*). — A common, heavy-billed, beautifully marked black, white, and rose-colored bird. The head, back, wings, and tail are mainly black; breast and under the wings rose color; blotches on wings, rump, tips of under tail feathers, and belly white. The *female* is very different. Upper parts streaky grayish-brown, lower parts streaky buff. A broad conspicuous whitish line over the eye and orange under the wing. Its warbling notes are somewhat like those of the robin, but more melodious and very frequently given in the evening. It is one of our most beautiful birds and sings an exquisite song.

Length, 8; wing, 4 (3¼–4½); tail, 3¼; tarsus, ⅞; culmen, ⅝. Eastern United States, from the eastern border of the

Rose-breasted Grosbeak

Plains; breeding from the mountains of the Carolinas and Kansas northward to southern Canada, and wintering in Mexico to northern South America.

47. **Black-headed Grosbeak** (596. *Zamelòdia melanocéphala*). — An orange-bodied grosbeak with black head, wings, and tail. The wings are much blotched with white, and the belly and under wing coverts are bright yellow. The *female* is very different, a streaky-brown bird much like the female of the last species, but with the under wing coverts clear lemon-yellow instead of the salmon- or orange-yellow of that species. The dimensions of parts are practically the same as those of the rose-breasted grosbeak. Western United States from middle Kansas to the Pacific.

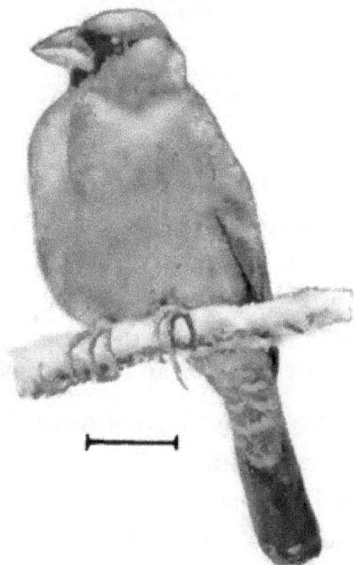

48. **Blue Grosbeak** (597. *Guiraca cærùlea*). — A southern, uncrested, dull-blue grosbeak, with a large

Blue Grosbeak

chestnut-colored blotch on the wings; wings, tail, chin, and lores mainly black. The *female* is grayish-brown above and creamy-buff below, sometimes with dull-blue on head and tail; the tail and wings are blackish-brown, with the wing coverts tipped with reddish-buff. This is a quiet, retiring bird of the thickets and weeds, of meadows and old clearings. The warbling notes of its song are said to be very beautiful, though weak.

Length, 7; wing, $3\frac{1}{2}$ ($3\frac{1}{4}$–$3\frac{5}{8}$); tail, $2\frac{3}{4}$; tarsus, $\frac{3}{4}$; culmen, $\frac{5}{8}$. United States from central Nebraska eastward; breeding from the Gulf to southern New Jersey, and wintering from southern Mexico to Central America. Casual to New England.

49. **Indigo Bunting** (598. *Passerina cyanea*). — A small, bright, blue bird, with no plain show of any other color anywhere, though the wing and tail feathers have much black on the hidden portions. The *female* is unstreaked grayish-brown, lighter below, with blackish wings and tail, having a gloss of bluish. The under side of the bill almost invariably has a stripe of blackish. This beautiful bird is a common inhabitant of old bushy pastures. (Indigo Bird.)

Indigo Bunting

Length, $5\frac{1}{2}$; wing, $2\frac{3}{4}$ ($2\frac{5}{8}$-$2\frac{7}{8}$); tail, $2\frac{1}{4}$; tarsus, $\frac{5}{8}$; culmen, $\frac{3}{8}$. United States from Kansas eastward; breeding from the Gulf to southern Canada, and wintering in Central America. The **Lázuli Bunting** (599. *Passerina amœna*), found from the Plains to the Pacific, is a similar bird, but the *male* has a white belly, white wing bars, and brown breast, and the *female* has brownish fore parts, and the rump and tail with much blue, and two whitish wing bars. Size a little greater.

50. **Painted Bunting** (601. *Passerina ciris*). — A brilliantly colored small bird of southern states, with blue head, green back, red rump, brown wings and tail, and bright red under

parts. The *female* has the upper parts bright olive-green and the under parts yellowish. This, though the most conspicuously colored of our birds, is, on account of its shy and retiring habits, seldom seen. Its home is among the densest and most thorny undergrowth of the wooded regions. (Nonpareil.)

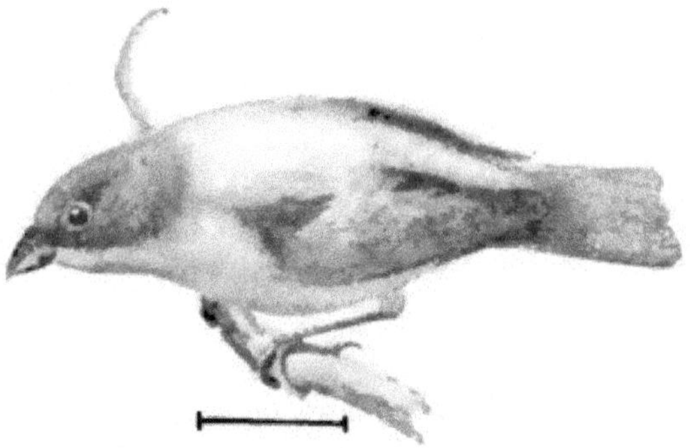

Painted Bunting

Length, $5\frac{1}{8}$; wing, $2\frac{3}{4}$ ($2\frac{5}{8}$–$2\frac{7}{8}$); tail, $2\frac{1}{4}$; tarsus, $\frac{5}{8}$; culmen, $\frac{3}{8}$. Eastern United States; breeding from southern Illinois and North Carolina southward, and wintering south of the United States to Central America. The **Varied Bunting** (600. *Passerina versicolor*) of southern Texas to Central America (accidental in Michigan) differs in having no green on the *male*. The forehead, hind neck, bend of wing, and rump, blue; wings and tail glossed with blue; throat and hind head, dull red; belly, reddish-purple. The *female* has brownish back and breast, whitish lower parts, and no whitish wing bars.

51. **Sharp's Seed-eater** (602. *Sporóphila morélleti shárpei*).— A very small, heavy-billed, southern Texas bird, with the upper parts black and the lower parts, including collar around neck, white or buffy. The rump is brownish, and the lower part of the collar is black. The *female* is plain olive above and olive-buffy below, and there are two whitish bars on the wings.

Length, 4; wing, 2; tail, 2; tarsus, $\frac{5}{8}$; culmen, $\frac{3}{8}$ nearly. Southern Texas and northwestern Mexico. The **Grassquit** (603. *Euethèia bicolor*) of the West Indies has been seen in southern Florida. It is a very small, stout-billed, olive-green bird with black head and lower parts becoming grayish on the belly. The *female* is olive-green with grayish head and lower parts. Length, 4; wing, 2; tail, $1\frac{3}{4}$. The **Melodious Grassquit** (603.1. *Euethèia canòra*), a similar small bird, also seen once in southern Florida, has a yellow, crescent-shaped band across the lower throat, extending back of the eyes. The chin and upper throat of the *female* is marked with chestnut-red. This species belongs to Cuba and has been seen only on one of the Florida Keys.

52. **Dickcissel** (604. *Spiza americàna*). — A yellow-breasted, black-throated, streaky-backed, sparrow-like bird with a reddish spot on the bend of the wing. The chin above the black throat is white, with more or less of yellow on the cheeks, and the crown and side of head are gray, with a yellow line over the eye. The back is brownish, streaked with black, and the belly whitish. *Female* duller and with but little yellow except on the breast, and the black throat patch almost wanting, sometimes slightly in-

Dickcissel

dicated by dusky spots. In summer it is abundant west of the Alleghanies in weedy fields, and the common notes it utters are expressed by its name. (Black-throated Bunting.)

Length, 6½; wing, 3¼ (2¾–3¾); tail, 2½; tarsus, ⅞; culmen, ½ or more. Eastern United States to the Rocky Mountains; breeding from Texas to southern Ontario, and wintering in northern South America. Very rare east of the Alleghanies.

Lark Bunting

53. **Lark Bunting** (605. *Calamospiza melanocòrys*). — A western, square-tailed, stout-billed, black bird, with a large white patch on the wings. The *female* (also the male in autumn and

winter) is a streaky, brownish, sparrow-like bird, with a distinct
whitish patch on the wings, in the position of the wing coverts.
This is a common bird of the Plains east of the Rocky Mountains.

Length, 6¾; wing, 3? (3¼-3¾); tail, 3; tarsus, 1; culmen, ½. Middle
Kansas to Manitoba, common east of the Rockies, but to be found all the
way to the Pacific, and south to Lower California. Accidental in some
eastern states (Mass., N. Y., S. C.).

FAMILY XIV. BLACKBIRDS, ORIOLES, ETC. (ICTÉRIDÆ)

A family (100 species) of American walking birds, which
vary greatly in sizes, habits, and colors. Our species are quite
naturally and easily separated into four subfamilies, under
which the characteristics will be given. (1) **Marsh Blackbirds.**
Medium-sized, generally black-colored, conical-billed birds, living mainly on the ground of marsh, meadow, or prairie. These
birds congregate together in great numbers, different species
in the same flock. Nos. 1–4. (2) **Meadow Starlings.** These
are long-conical-billed, short-tailed birds, with their plumage
consisting mainly of browns and yellows. Their name indicates their meadow-living habits. No. 5. (3) **Orioles.** This
group comprises brightly colored, tree-loving, song birds, with
very sharp-pointed, elongated bills.[1] The orioles build wonderfully woven hanging nests of fibrous materials. The plumage
of the males is mainly black, strikingly relieved with other
colors, among which are orange, chestnut, yellow, and white.
All the species are pleasing singers, and some have peculiarly
rich and flexible voices. Nos. 6 and 7. (4) **Grackles and Crow
Blackbirds.** A group of large to medium, ground-running,
black-plumaged birds, with long, sharp-pointed, somewhat
curved bills.[2] The black colors are often richly bronzed with
green, blue, and other tints. Their nests are rude and bulky,
and their notes, in most cases, far from musical. Many of the
species are to be found in flocks, throughout the year. Nos.
8–12. Many species of the family vary much in the sizes of
the males and females. In the dimensions given under the
different species the smaller numbers refer to the females.

Key to the Species

* Culmen, 1 or more long ; a black crescent-shaped spot on the breast ;
lower parts with much yellow ; tail with acute feathers and over an inch
shorter than the wings ; under tail feathers white...5. **Meadowlark.**

* Culmen less than ⅔ long ; tail feathers acute at tip, and the under ones
without white ; tail about an inch shorter than the wings. 1. **Bobolink.**

* Tail feathers usually rounded at tip, never very acute. (A.)

A. Bill very stout, decidedly more than half as high at base as long ;
head and neck seal-brown, rest of plumage black (*male*) or general
plumage brownish (*female*) ....2. **Cowbird.**

A. Bill less stout, usually less than half as high at base as long ; never
much more than half. (B.)

B. Culmen, 1 or more long ; bill with its tip conspicuously decurved ;²
tail with its under feathers ¾-4 inches shorter than the
middle ones ; no bright yellow or orange in the plumage. (E.)

B. Culmen, ⅝-1 long ; tail with the under feathers not over ⅜
inch shorter than the middle ones. (C.)

B. In the extreme south there may be found birds with culmen ¾-1½ long
and with the under tail feathers over ⅜ shorter than the middle ones.
These all have yellow or orange in their plumage and are described
under No. 7..............7. **Hooded Oriole and Audubon's Oriole.**

C. Black, with red or reddish shoulder patch (*male*), or rusty and
black streaked, with the under parts conspicuously black and white
streaked (*female*)....................4. **Red-winged Blackbird.**

C. Black, with yellow head and breast and large white wing patch
(*male*), or brownish-black with yellowish head and breast and small
white wing patch (*female*)..........3. **Yellow-headed Blackbird.**

C. Black throughout with more or less of rusty tips to the feathers
(*male*), or slate-color with feathers sometimes rusty tipped (*female*).
..................8. **Rusty Blackbird.** or 9. **Brewer's Blackbird.**

C. Not as above ; bill slender, less than half as high as long ;¹ plumage with some distinct yellow, orange, or chestnut.
(D.)

D. Under tail feathers about a half inch shorter than the
middle ones....................6. **Orchard Oriole.**

D. Under tail feathers nearly as long as the middle ones .....
....7. **Baltimore Oriole.**

E. Tail with the under feathers less than 1¾ inches shorter than the
middle ones ; wings and tail of about equal length...............
.............................10. **Purple Grackle.**

E. Tail with the under feathers over 1¾ inches shorter than the middle
ones ; wings and tail of about equal length. 12. **Boat-tailed Grackle.**

E. Tail decidedly longer than the wings and with its under feathers
2½-3½ inches shorter (Texas)..........11. **Great-tailed Grackle.**

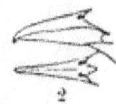

1. Bobolink (494. *Dolichonyx oryzivorus*). — *Male* in spring. A common meadow blackbird with white rump and shoulders; golden brown on the back of the head, and acute-tipped tail feathers. *Female* (also *male* in the autumn) much like an olive-colored streaky sparrow, with buffy belly, but with pointed tips to its tail feathers; wings and tail blackish-brown, the back streaked black and buffy-olive; crown blackish with a lighter central stripe, and all under parts buffy, slightly streaked on the sides. This is a wonderful singer throughout May and June, but during the rest of the year the notes are confined to merely a call of *chink*. The remarkable change of plumage in the *male* is accomplished by two complete molts each year. (Reedbird; Ricebird.)

Bobolink

Length, 7¼; wing, 3½ (3½–4); tail, 2¾; tarsus, 1; culmen, ⅝ or less. North America west to Utah; breeding from southern New Jersey north to Ontario, and wintering south of the United States.

2. Cowbird (495. *Mólothrus áter*). — A common, small, brown-headed blackbird, with feathers having a metallic gloss. The *female* is brownish-gray throughout, but lighter below and much smaller

Bobolink

in size. This bird, like the cuckoo of Europe, builds no nest of
its own, but deposits its eggs one at a time, in the nests of other,
generally   smaller,
species.   Many of
these   hatch   and
rear the young cow-
birds; though some
abandon the nests
into which the eggs
are placed, others
throw out the eggs,
and still others build

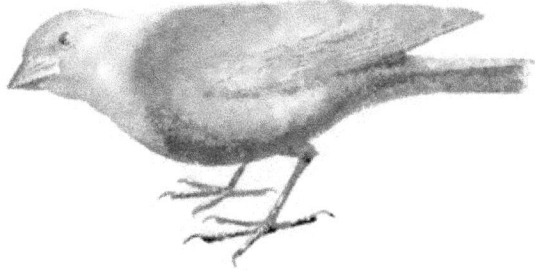

Cowbird

new nests over the one containing the parasite's egg. (Cow
Blackbird.)

Length, 7¾; wing, 4¼ (3½-4¾); tail, 3; tarsus, 1; culmen, ¾. Whole
United States and southern Canada; breeding throughout, and wintering
in the Gulf States and Mexico. The **Dwarf Cowbird** (495ª. *M. a obscúrus*)
of Texas to Lower California is very similar, though smaller. Wing of
female, 3¼; tail, 2¼. The **Red-eyed Cowbird** (496. *Callóthrus robústus*),
found in southern Texas to Central America, is larger. The *male* is black
with much bronzy luster, and the *female* brownish-gray, somewhat glossy
on the back. Length, 9; wing, 4½ (4-4¼); tarsus, 3¼-3½; culmen, ¾.

3. **Yellow-headed Blackbird** (497. *Xanthocéphalus xanthocéphalus*). — A western, ground-living, orange-yellow-headed

blackbird, with a
blotch of white near
the bend of the wing.
The breast and neck
are also yellow, but
the lores and chin are
black. The *female* is
a grayish-brown bird,
with most of the head
and breast a dirty

Yellow-headed Blackbird   yellow or yellowish

white. These birds gather together in companies and associate
with cowbirds, and like them are often found on the ground

among cattle and horses. Their notes are harsh and not in the least musical.

Length, 9–11; wing, 5¼ (4½–5⅞); tail, 4¼; culmen, ⅞. Western North America, from Wisconsin, Illinois, and Texas to the Pacific; wintering in the Southern States and southward, accidental in some of the Atlantic States.

4. **Red-winged Blackbird** (498. *Agelaius* (*lè-us*) *phœníceus*). — A very common, middle-sized blackbird, with the bend of the wing bright red. The red of the wing shades off to a buff. The *female* is a speckled or streaky brown. The back is made up of rusty, buffy, and black, and the under parts are of black and white. These birds are usually seen in flocks in reedy marshes and meadows. In the early spring, the *males* and *females* are found in separate companies. The notes are a rich and clear *con-qua-ree-e*. In July, after the short nesting season, these birds again gather in flocks which usually contain several of the different species of blackbirds. (Swamp Blackbird.)

Red-winged Blackbird

Length, 7½–10; wing, 3⅝–5; tail, 2¾–4; tarsus, 1; culmen, ⅞–1. North America north to Great Slave Lake; breeding nearly throughout, and wintering mainly in the Southern States. The **Sonoran Redwing** (498ª. *A. p. longiróstris*) of southern Texas, California, and northern Mexico averages a little larger, has a smaller bill, and the *female* is lighter colored, — especially the lower parts, which are mainly white, with fewer dusky markings. The **Bahaman Redwing** (498ᵇ. *A. p. brýanti*) of southern Florida and the Bahamas averages a little smaller and has a larger bill. In this variety the culmen of the *male* is a full inch in length, and the *female* has the crown marked with a pale medium stripe.

5. **Meadowlark** (501. *Sturnélla mágna*). — A somewhat large, common, ground-living, speckled-brown-backed, yellow-bellied bird, with a crescent-shaped black spot on the breast, and white under tail feathers. This is an abundant bird of the fields, meadows, and marshes. Its notes vary much for locality and season, as well as individually, and many attempts have been made to write out, in syllables and musical notes, its song. One of the attempts is, "*Spring-o'-the-year.*" (Field-lark.)

Length, 8-11; wing, 4-5; tail, $2\frac{1}{4}$-$3\frac{3}{4}$; tarsus, $1\frac{1}{2}$; culmen, 1-$1\frac{1}{2}$. United States from the Plains eastward; breeding from the Gulf of Mexico to Canada, and wintering coastwise and along rivers from New Jersey southward.

Meadowlark

The **Mexican Meadowlark** (501ª. *S. m. mexicána*) of southern Texas to Central America is a smaller bird, with proportionally larger feet; wing, $3\frac{7}{8}$-$4\frac{3}{4}$; culmen, 1-$1\frac{1}{4}$; tarsus, $1\frac{5}{8}$. The **Western Meadowlark** (501ᵇ. *S. m. neglécta*) of the western United States from Wisconsin to Texas, and west to the Pacific, is a duller and paler bird, with a generally grayish appearance; the yellow of the throat spreads over the cheeks. The size averages larger. Wing, $4\frac{1}{4}$-$5\frac{1}{4}$.

6. **Orchard Oriole** (506. *Ícterus spúrius*). — A bright-chestnut-bodied bird, with black head, upper

Orchard Oriole

back, wings, and tail. The black of the head extends to the

breast. The *female* is olive-green above, dull-yellow below, and has blackish wings, with two whitish wing bars. The olive-green is very bright on the head and rump. The *young male* begins to get the black on the throat during his second year; the under parts have a few spots of chestnut in the yellow. As its name indicates, it is generally to be found among orchard trees, but any separated trees of our lawns and parks suit it as well. It is an active, frolicsome bird, and a wonderfully sweet singer of short, rich, and flexible notes. Like the next species, it weaves its nest of fibrous material, and suspends it near the extremity of a limb, but makes it of a more globular form.

Length, 5¼–7¼ ; wing, 2¾–3¼ ; tail, 2¼–3¼ : tarsus, ⅞ ; culmen, ⅔. United States from the Plains eastward ; breeding from the Gulf of Mexico to Ontario, and wintering south of the United States to northern South America.

7. **Baltimore Oriole** (507. *Icterus gálbula*). — A beautiful orange-red-bodied bird, with black head, upper back, and wings.

Baltimore Oriole

The tail is orange, with some black near the base, and the wings have some white on the coverts and quills. The *female* is dull-orange below, mottled brown on the upper part including wings and head; rump and tail yellowish and wing bars white. This bird weaves a wonderful hanging nest of fibers, which it usually places near the ends of limbs twenty to forty feet from the ground. The song is an agreeable one of five to ten rich, mellow, though rather shrill notes. (Firebird; Hangnest; Golden "Robin.")

Length, 7-8¼; wing, 3⅜-3⅞; tail, 2⅞-3⅜; tarsus, ⅞; culmen, ¾ nearly. United States from about the Rocky Mountains eastward; breeding from the Gulf of Mexico to Ontario, and wintering south of the United States to Central America. The **Hooded Oriole** (505. *Íсterus cucullatus*) of southern Texas to Central America is an orange-colored bird with black wings, black tail, and a peculiar black hood covering the face and throat; the wings have white blotches on coverts and quills. The *female* lacks the black mask, but both sexes can be separated from all of our other orioles except the next, by the fact that the tail is longer than the wings. Wing, 3¼-3⅞; tail, 3½-4¼. It can be separated from Audubon's by the size. **Audubon's Oriole** (503. *Íсterus audubònii*) is found from southern Texas to Central Mexico. It is a very large, black-headed, orange-bodied oriole with black wings, tail, breast, etc. Length, 8¾-10½; wing, 3¾-4¼; tail, 4-4½; culmen, 1.

8. **Rusty Blackbird** (509. *Scolecóphagus carolinus*). — A common, medium-sized, glossy, bluish-black bird (in spring) with all the tail feathers of nearly equal length. In the autumn and winter the black is much hidden by the rusty-brown tips to the feathers. The *female* in spring is glossy slate-colored, but in the autumn and winter she, like the male, is rusty. This is a quiet, ground-living, swamp-loving species. (Rusty Grackle.)

Rusty Blackbird

Length, 8¼-9¾; wing, 4¼-5; tail, 3½-4¼; tarsus, 1¼; culmen, ¾. North America from the Plains eastward; breeding from northern New York northward, and wintering from New Jersey southward.

9. **Brewer's Blackbird** (510. *Scolecóphagus cyanocéphalus*). — A western blackbird similar to the last, but larger and with a conspicuous violet-purple iridescence to the head. The bill is stouter, and there are less rusty tips to the feathers at all seasons. *Female*, glossy slate-colored with a decided brownish tint near the head. (Blue-headed Blackbird.)

Length, 8¾–10¼ ; wing, 4½–5¼ ; tail, 3¾–4½ ; tarsus, 1¼ ; culmen, ¾. Western North America from the Plains to the Pacific. Accidental in Illinois.

10. **Purple Grackle** (511. *Quiscalus quiscula*). — A common, large, iridescent blackbird, with brilliant metallic reflections of greens and blues, arranged in bars on the back, rump, and belly. The *female* is much duller, but still a blackbird and somewhat iridescent. This is a gloomy bird with crackling notes which can hardly be called a song. (Crow Blackbird.)

Purple Grackle

Length, 11–13½ ; wing, 4½–6 ; tail, 4½–6, graduated, 1¼ ; tarsus, 1⅜ ; culmen, 1¼. Mainly east of the Alleghanies; breeding north to Massachusetts, and wintering from New Jersey south. The **Florida Grackle** (511a. *Q. q. aglæus*) of the southern portion of Gulf States, from Florida to Texas, is smaller and the head is decidedly violet-purple by reflections, and the back a rich green. The iridescent bars are not so distinct, though readily recognized. The *female* differs from the last only in being smaller. Length, 10–12 ; wing, 5–5⅜ ; tail, 4½–5¼ ; culmen, 1¼. The **Bronzed Grackle** (511b. *Q. q. æneus*) of the region east of the Rocky Mountains to the Alleghanies, north to Newfoundland and Great Slave Lake, and south to Texas, differs from the purple grackle more in the lack of iridescent bars on the bronze-colored back than in any other feature. The *female* is almost without metallic reflections and never has the iridescent bars.

11. **Great-tailed Grackle** (512. *Quiscalus macrourus*). — A very large, long-tailed, glossy-black bird with metallic-violet tints over the head, breast, back, and wing coverts, but without iridescent bars. *Female* a dark brown with metallic-greenish gloss on the back ; the head almost without gloss.

Length, 11½–18½ ; wing, 5¾–8 ; tail, 5⅜–9⅜ ; culmen, 1¼–1¾. Eastern Texas to Central America.

12. **Boat-tailed Grackle** (513. *Quíscalus májor*). — A very large, southern, beautifully metallic, bluish-black bird, with the head and breast more purplish, and the wings and tail less so. The female is much smaller in size; and brown in color, the lower parts being lighter. These birds are found in flocks in marshy places or near the water.

Length, $11\frac{1}{2}$-$17\frac{1}{2}$; wing, $5\frac{1}{2}$-$7\frac{1}{2}$; tail, 5-$7\frac{1}{2}$; culmen, $1\frac{1}{4}$-$1\frac{3}{4}$. The smaller numbers are the dimensions of females. Regions along the Atlantic and Gulf coasts from Virginia to Texas.

FAMILY XV. STARLINGS (STÚRNIDÆ)

A family (200 species) of Old World birds one species of which has apparently been successfully introduced into this country.

1. **Starling** (493. *Stúrnus vulgàris*). — A yellow-billed, metallic-purplish, walking bird, much spotted with buff; the wings, tail, and under tail coverts are dark brownish-gray more or less edged with buff. In *winter* the bill changes to dark brown and the entire under parts become heavily spotted with white. The bill is long and conical; [1] the tail only about half as long as the wings; and the first primary quill less than one inch long. [4]

Length, $8\frac{1}{4}$; wing, 5; tail, $2\frac{3}{4}$; tarsus, $1\frac{1}{4}$; culmen, 1. Europe and northern Asia. Introduced and apparently established in the vicinity of New York City.

FAMILY XVI. JAYS, CROWS, MAGPIES, ETC. (CÓRVIDÆ)

A large family (200 species) of rather large, heavy-billed, peculiarly intelligent birds, with the nostrils generally well covered with bristly feathers. [2] They have been divided into five subfamilies two of which are found almost everywhere. (1) The **Jays** are large, brightly marked, and usually brightly colored, saucy, noisy birds, with short, rounded wings and long, graduated tails. [3] With hardly any exceptions their voices are

1

2

3

4

harsh and discordant. Nos. 1–4. (2) The **Crows** are large, dark-colored (ours are black), walking birds, with long, pointed wings, short, nearly square tails, and unmusical cawing voices. As they can eat almost all vegetable and animal foods in almost any condition, they can be found nearly everywhere at all seasons. They usually associate together in large numbers. Their nests are rude and bulky. Nos. 5–8.

Key to the Species

* Tail only ⅔ as long as the pointed wings; plumage black. **(B.)**
* Tail, 1–3 inches shorter than the wings; plumage not black; extreme western. **(C.)**
* Tail, as long or longer than the rounded wings.[1] **(A.)**
> **A.** Tail, 2 or more inches longer than the wings; no crest; colors black and white; wings over 7 long............
> .... 1. **American Magpie.**
> **A.** Head crested; general color blue; tail tipped with white; a black breast patch.....................................2. **Blue Jay.**
> **A.** No crest; general colors blue and gray; tail without white tip....
> ....3. Florida **Jay.**
> **A.** No crest; general color gray; no blue or green in the plumage....
> .... 4. Canada **Jay.**
> **A.** Slightly crested; blue and black-headed, green-backed, yellow-bellied bird. The **Green Jay** (483. *Xanthoura luxuosa*) of eastern Mexico has been seen in southern Texas.
> **B.** Wings, 15 or more long; culmen, 2¼–3¼ long; neck feathers narrow and pointed.....................................5. Northern **Raven.**
> **B.** Wings, 13–15; culmen, 2–2⅞; neck feathers narrow and pointed, and those of the back neck peculiarly white at base. Extreme western ....6. White-necked **Raven.**
> **B.** Wings, 11½–14; culmen, 1⅜–2; neck feathers not sharp pointed....
> ....7. American **Crow.**
> **B.** Wings, 10–11½; culmen, 1¼–1⅜; neck feathers not sharp pointed...
> ... ... 8. Fish Crow.
>> **C.** Plumage, mainly gray; wings, glossy black........... ....
>> ....Clark's Nutcracker (8).
>> **C.** Plumage, mainly dull blue ....Piñon Jay (8).

1. **American Magpie** (475. *Pica pica hudsonica*). — A large, western, long-tailed, brightly iridescent black bird, with white shoulders, tip of wings, and belly. It is generally a ground-living, noisy bird, with an infinite variety of notes, harsh and pleasant, discordant and musical, squeaky and gurgling.

Length, 15–22 ; wing, 7½–8½ ; tail, 9½–12 ; tarsus, 1¼; culmen, 1¼. Northern and western North America from the Plains to the Cascade Mountains, and from Alaska to New Mexico. Casually or accidentally in Michigan and northern Illinois.

2. **Blue Jay** (477. *Cyanocitta cristàta*). — A very common, large, noisy, crested, brightly marked, blue bird, with white throat, belly, and tips of outer tail feathers. The wings and tail are barred with black, and the neck has a black collar. This very beautiful species has more bad traits than can here be mentioned ; among them are its fondness for eggs and nestlings. It has many notes of its own, and is a mimic, imitating the notes of a number of other birds.

American Magpie

Length, 11½ ; wing, 5½ (5–6) ; tail, 5½ ; tarsus, 1¾ ; culmen, 1. North America from the Plains eastward and from Florida to the fur countries. About resident throughout. The **Florida Blue Jay** (477ᵃ. *C. C. florincola*) of the Gulf coast region, from Florida to Texas, is smaller, grayer in color, and has the under tail feathers less tipped with white (usually under 1 inch). Wing, 4¾–5¼.

3. **Florida Jay** (479. *Aphelócoma floridàna*). — A large, very noisy, brownish-bodied, gray-

Blue Jay

ish-blue bird, with whitish under parts. There is a tint of blue on the sides of the breast and across the breast.

This jay is generally to be found on the ground, except when disturbed.

Length, 11½; wing, 4½ (4-4¾); tail, 5½; tarsus, 1½; culmen, 1. Florida, north of the center near the coasts.

4. **Canada Jay** (484. *Perisòreus canadénsis*). — A large, northern, loose-plumaged, gray bird, with the head mainly white, except the nape, which is blackish. The throat and sides of the neck are white, and the gray quills of the wings and tail are somewhat tipped with white. This, like all the jays,

Canada Jay

is a noisy bird, making many harsh and shrieking calls and uttering a few musical notes. (Whisky Jack; Gray Jay.)

Length, 11½; wing, 5¾; tail, 5¾; tarsus, 1½; culmen 1. Northern Michigan to northern New England northward. Not at all migratory; straggling from its home but very rarely. Has been seen as far south as central Pennsylvania.

5. **Northern Raven** (486ª. *Còrvus còrax principàlis*). — A somewhat rare, very large, crow-like black bird, twice the size of the common crow. The black everywhere shows

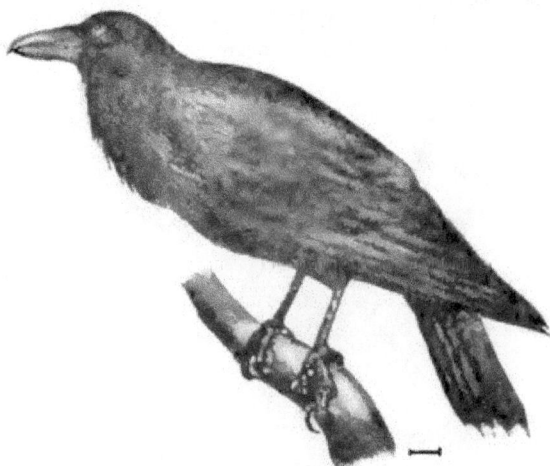

Northern Raven

bluish, metallic reflections. The feathers of the throat are narrow, pointed, and peculiarly independent of each other, not blended as is usually the case. When seen with crows, ravens can readily be recognized by the great difference in size, but when seen alone there is difficulty in distinguishing them, as the distance they are away is not easily determined. (Raven.)

Length, 20-27; wing, 15-19; tail, 10; tarsus, 2½; culmen, 2½-3½. Northern North America, not migratory, south to Michigan, New Jersey, and along the Alleghanies to North Carolina.

6. **White-necked Raven** (487. *Còrvus cryptoleùcus*). — A bird similar to the last, but smaller and with the feathers of the back neck white at base.

Length, 20; wing, 14; tail, 8; tarsus, 2½; culmen, 2¼. Texas to southern California and northern Mexico.

7. **American Crow** (488. *Còrvus americànus*). — A very abundant, large, black bird which is found in flocks everywhere, and can be recognized by its call notes of "*caw-w, caw-w.*" Though the crow has few friends and, were it not for his remarkable ability to escape the gunner, would soon become extinct, it seems to thrive under all circumstances and in all situations. Acting as though afraid of nothing, it always knows how to secure its own safety. In the winter the crows roost in immense colonies in particular places; during the day

American Crow

they forage for food over a great extent of country, but at night all return to the regular place for rest.

Length, 17-21; wing, 12½ (11¾-13½); tail, 7½; tarsus, 2¼; culmen, 1⅞. North America, south to Mexico; wintering from the northern United States southward. The **Florida Crow** (488ª. *C. a. floridanus*) has the bill and feet comparatively larger. Tarsus, 2½; culmen, 2¼. Florida.

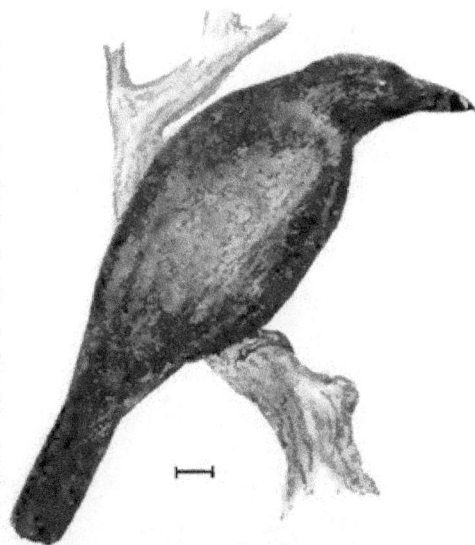

8. **Fish Crow** (490. *Còrvus ossífragus*). — Almost exactly like
the last, but smaller. This is a brighter, cleaner, smoother-
plumaged bird, with more metallic-purplish reflections on the
back and somewhat greenish ones below. The voice is different, but the difference cannot be readily described.

Length, 15–17 ; wing, 10–11½ ; tail, 6½ ; tarsus, 1¾ ; culmen, 1⅝.   Atlantic and Gulf coasts from Connecticut to Louisiana, mainly resident.
Clark's Nutcracker (491. *Nucifraga columbiàna*). — A large, gray, crow-
like bird with wings glossy black, except the white tips of the secondaries ;
and the tail white, except the black middle feathers.   Length, 12½ ; wing,
7½ ; tail, 5 ; tarsus, 1¾ ; culmen, 1⅝.   Western North America from the
Rocky Mountains to the Pacific, mainly in evergreen forests.   Accidental
in Kansas, Missouri, and Arkansas.   The **Piñon Jay** (492. *Cyanocéphalus
cyanocéphalus*) of the high lands between the Rocky Mountains and the
Sierra Nevada is a large, dull-blue bird, with the head somewhat brighter.
It has been seen both in eastern Kansas and eastern Nebraska.   It is
peculiar for this family in that its nostrils are not covered with bristly
feathers.   Length, 11 ; wing, 5¾ ; tail, 4¾ ; culmen, 1¼.

FAMILY XVII.   LARKS (ALAÚDIDÆ)

A family (100 species) of almost exclusively Old World
ground birds with the nail of the hind toe very long
and nearly straight,[1] and the inner secondaries much
lengthened.[2]   It is represented in North America
by but one species, if we except the noted singer,
the European skylark, which has been brought into
this country and allowed to escape several times, and
is thought to be established on Long Island.

Key to the Species

* First **primary** about as long as the longest ; tail nearly **even** at tip ;
   sides of head with a peculiar tuft of elongated **black** feathers.......
   ....1. Horned Lark.
* First **primary short** ; tail decidedly notched at tip........Skylark (1).

1. **Horned Lark** (474. *Otócoris alpéstris*). — A chocolate-
backed, ground-running, mainly winter bird, with distinct black
and yellow marks on the head and breast.   The under parts
are whitish, and the black tail feathers are somewhat margined

with white. The black mark from the bill to the eye and then downward along the side neck, and the black breast patch, can generally be seen, though in winter there is a veiling of the black by whitish tips to the feathers. The horned larks are usually found in flocks along the seacoast and in the open tracts of the interior. They sing both when at rest and when on the wing; they usually whistle a short note when taking wing, and frequently after a short flight return to the same spot from which they started. (Shore Lark.)

Length, 7½; wing, 4⅛ (4-4½); tail, 2⅜; tarsus, ⅞; culmen, ½. There are ten named varieties of this species in North America, three of which are to be found in the region covered

Horned Lark

by this book. The form above described ranges through northeastern North America, around Hudson Bay, and winters south to Illinois, the Carolinas, etc. The **Prairie Horned Lark** (474ᵇ. *O. a. praticola*) is slightly paler in color, somewhat smaller in size, and has less yellow about the head and breast, sometimes almost none, the line over the eye being white. It is distributed over the upper Mississippi Valley, around the Great Lakes, and New England, breeding mainly in the northeastern portion of New England, and wintering south to Texas and South Carolina. Wing, 3⅞ - 4⅛. The **Desert Horned Lark** (474ᵉ. *O. a. arenicola*) of the Plains and westward, and southward in winter to Mexico, is a paler-colored bird, but

Skylark

with the breast distinctly, often brightly, primrose-yellow. Wing, 4-4⅛. The **Texan Horned Lark** (474ᵈ. *O. a. giraudi*) of eastern Texas has the back a decided grayish color and the breast pale yellow, more or less dotted with grayish brown. The throat and line over eye are also a light yellow. Wing, 3½–3⅜. The **Skylark** (473. *Alaúda arvénsis*) of Europe and Asia may have been successfully started breeding on Long Island. It can be known by the short first primary, the notched tail, and the lack of the black, elongated feathers above the eyes. Its general color is a dull brown, much streaked both above and below. Length, 7½ ; wing, 3¾–4¼ ; tail, 2½ ; tarsus, 1 ; culmen, ½.

FAMILY XVIII. FLYCATCHERS (TYRNÁNIDÆ)

A large family (350 species) of American perching birds, with a broad, depressed, notched bill, slightly hooked at tip;[1] and almost no vocal powers. The flycatchers can be distinguished from most other birds by their peculiar method of feeding. They perch on some outlying twig or other support, watching for their prey ; when a passing insect is seen, they dart out, seize it with a characteristic *click* of the bill, and in an instant return to their old station, ready for another victim. Other birds (warblers, etc.) catch insects on the wing, but have not this deliberate plan ; they chase their prey. These solitary birds are to be found wherever there are trees and bushes, but are most abundant in the tropics. The sexual and seasonal differences are but slight, but in nesting habits there are great variations.

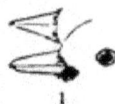

Key to the Species

* **Bird over 11 long,** with tail over 6 long, and deeply forked.
  — Cap ashy .... 1. Scissor-tailed **Flycatcher.**
  — Cap black........................... Fork-tailed Flycatcher (1).
* **Bird, 8–11** long ; wing, 4½–5½ long ; crown of adult with a partially concealed patch of bright yellow or red, which can be seen by displacing the feathers. **(G.)**
* Head somewhat crested ; wing and tail feathers with much chestnut on their edges ; no concealed crown patch of bright color...........
  ....5. **Crested Flycatcher.**
* Not as above. **(A.)**
  A. Wing, 2–3¼ long. **(E.)**
  A. Wing, 3–3¼ long. **(C.)**
  A. Wing, 3½–4½ long. **(B.)**

B. Wing about an inch longer than the tail; a tuft of fluffy, cotton-like feathers on the flanks at base of tail; sides, breast, and back olive-brown; throat and belly about white......9. **Olive-sided Flycatcher.**

B. Grayish-brown above, paler below, and with cinnamon color on the lower belly; tail, bill, and feet black............7. **Say's Phœbe.**

B. Blackish-brown, darkest on head and breast; belly abruptly white; wings less than ½ inch longer than the tail.........8. **Black Phœbe.**

C. Wings and tail nearly equal in length; entire bill black; under tail coverts yellowish-white ....6. **Phœbe.**

C. Wings ½ inch or more longer than the tail. (**D.**)

D. Olive-brown above, darker on the head; olive-gray on the sides and nearly across the breast; throat and belly nearly white............
....................10. **Wood Pewee** or 11. **Western Wood Pewee.**

D. Slightly crested bird with crest and under parts red (*male*), or upper parts including crest dull brown, and under parts white, tinged with red or orange (*female*). The **Vermilion Flycatcher** (471. *Pyrocéphalus rubineus mexicanus*) of Mexico has been seen in southern Texas, Arizona, and Utah.

E. Belly sulphur-yellow; throat and breast greenish-yellow; back olive-green without a shade of brown ....
.... 12. **Yellow-bellied Flycatcher.**

E. Under parts but slightly tinged with yellow. (**F.**)

F. Back olive-green without tint of brownish; head slightly crested; throat white; lower mandible light yellow........................
.... 13. **Green-crested Flycatcher.**

F. Back olive with more or less of brownish tint.
— Wing, 2½–3 long...........14. **Traill's** and **Alder Flycatcher.**
— Wing, 2¼–2⅝ long.....................15. **Least Flycatcher.**
— Wing, 2–2⅛ long (south Texas).....**Beardless Flycatcher** (15).

G. No distinct yellow below; all tail feathers abruptly tipped with white ....2. **Kingbird.**

G. No distinct yellow below; no abrupt white tips to the tail feathers; 5 or 6 of the outer primaries abruptly narrowed near tips.........
....3. **Gray Kingbird.**

G. Belly with distinct yellow; tarsus and culmen each an inch or more long...............................**Derby Flycatcher** (4).

G. Belly with distinct yellow; tarsus and culmen each under one inch long. (**H.**)

H. Outer web of outer tail feathers abruptly white. 4. **Arkansas Kingbird.**

H. Outer web but slightly pale edged; breast and belly bright yellow; throat and chin white; tail notched a half inch. (Texas) ....
.....................................**Couch's Kingbird** (3).

1. **Scissor-tailed Flycatcher** (443. *Milvulus forficatus*). — A beautiful, ashy-colored, small bird, having a deeply forked black

tail, and rosy or white edgings and scarlet patches on the sides of the body. This graceful bird can be recognized for a great distance by its forked tail, and especially so, as it has the habit of opening and closing the parts like a pair of scissors.

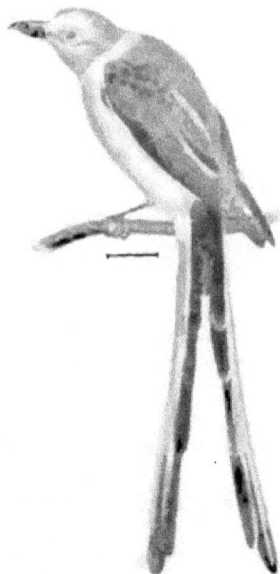

Length, 12–15; wing, 4¾ (4¾–5¼); tail, 7–12; tarsus, ⅝; culmen, ⅖. Southwestern Missouri, through Texas, eastern Mexico to Central America. Accidental in New England, New Jersey, Florida, etc. The **Fork-tailed Flycatcher** (442. *Milvulus tyrannus*) is a beautiful, tropical American bird, which has accidentally wandered to various parts of the United States (New Jersey, Kentucky, Mississippi, etc.). It is a black-capped, ashy-backed, black-tailed bird, with the lower parts white, and the outer webs of the outer tail feathers white for half their length. The tail is forked from 4–8 inches.

Scissor-tailed Flycatcher

2. **Kingbird** (444. *Tyrannus tyrannus*). — A pale, slate-colored kingbird, with nearly white under parts, and black tail tipped with a broad band of white. The under parts have a wash of gray on the breast. The *adult* has a concealed orange-red patch on the crown; this is lacking in the *young*. This bird has the habit of perching in an exposed position, with the body upright, like a hawk, watching for its insect prey. If food is seen, it suddenly darts into the air, seizes it, and returns to the same perch. (Bee Martin.)

Kingbird

Length, 8½ ; wing, 4½ (4½-4¾) ; tail, 3½ ; tarsus, ⅜ ; culmen, ⅜. North America, north to New Brunswick and Manitoba ; breeding from the Gulf northward, and wintering in Mexico and northern South America. Rare west of the Rocky Mountains.

3. **Gray Kingbird** (445. *Tyrannus dominicénsis*). — An ashy-gray-backed kingbird, with white or whitish under parts ; the tail is blackish, slightly notched, and without the white terminal band. The under wing coverts are pale yellow. Habits much the same as the last, but its notes very different.

Length, 9 ; wing, 4½ ; tail, 3¾ ; culmen, ⅝. South Atlantic States (South Carolina to Florida), West Indies to northern South America. **Couch's Kingbird** (446. *Tyrannus melanchólicus coûchii*), of Texas to Central America, has the breast and belly bright yellow, the throat white, the tail decidedly notched, and its outer feathers not white-edged. Length, 9½ ; wing, 4¾ ; tail, 4 ; culmen, ¾.

4. **Arkansas Kingbird** (447. *Tyrannus verticàlis*). — An olive-backed, yellow-bellied, ashy-headed kingbird, with a blackish square tail, without whitish tip. The outer web of the outer tail feathers entirely white, and the under wing coverts yellow like the belly.

Length, 9 ; wing, 5 (4¾-5¼) ; tail, 4 ; tarsus, ¾ ; culmen, ¾. Western United States from the Plains to the Pacific. Accidental in Maine, New Jersey, Maryland, etc. The **Derby Flycatcher** (449. *Pitángus derbiànus*), of southern Texas to northern South America, is an olive-brown-backed, yellow-bellied bird, with the top and sides of head black, a white line over the eye to the nape, and a yellow crown patch. The chin and throat are white, and the under wing coverts lemon-yellow like the belly. The culmen and tarsus are each an inch long, and the wing and tail feathers extensively bordered with chestnut.

5. **Crested Flycatcher** (452. *Myiárchus crinìtus*). — A crested, greenish-olive bird, with brown on head and wings and chestnut on tail ; lower parts yellow, except the

Crested Flycatcher

ashy throat and breast; two yellowish wing bars. This is a common woodland bird of quarrelsome nature, with a loud, piercing voice. It is noted for the habit of using snake skins in the structure of its nest. (Great-crested Flycatcher.)

Length, 8¾; wing, 4 (3⅞-4¼); tail, 3⅞; tarsus, ¾; culmen, ¾. Eastern United States and southern Canada west to the Plains; breeding from Florida northward, and wintering south of the United States from Mexico to Central America. The **Mexican Crested Flycatcher** (453. *Myiarchus mexicanus*), of southern Texas to Central America, differs from the last in having a broad dusky stripe on the inner web along the shaft of the outer tail feathers.

6. **Phœbe** (456. *Sayornis phœbe*). — A common, crested, dull olive-brown bird, having the lower parts dull white, with the breast tinged with brownish-gray. The head is darker, almost blackish, the belly has a very slight tint of yellow, and the eye has a whitish ring around it. The name is derived from the sound of its note, which is uttered in a harsh and abrupt manner. (Pewee.)

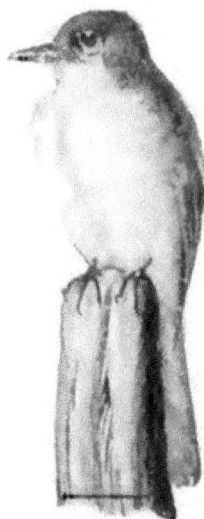

Length, 7; wing, 3¾ (3¼-3½); tail, 3; tarsus, ¾; culmen, ½. Eastern North America from Colorado and Texas eastward; breeding from South Carolina northward, and wintering from the South Atlantic and Gulf States to Cuba and eastern Mexico.

7. **Say's Phœbe** (457. *Sayornis saya*). — A western, grayish-brown bird, with the lower parts cinnamon-brown, darker on the throat. The tail, bill, and feet are black, and the wing bars whitish. This is a flycatcher of weedy and shrubby places rather than of wooded regions.

Phœbe

Length, 7½; wing, 4 (3¾-4¼); tail, 3½; tarsus, ¾; culmen, ½. Western United States from the Plains to the Pacific, north to the Arctic Circle, south to Central America. Accidental in Massachusetts.

8. **Black Phœbe** (458. *Sayornis nigricans*). — A very dark, almost black, Texas bird, with the belly abruptly pure white. The head and breast are the blackest portions; the bill and

feet are also black. The coloring is almost like that of the
juncos. This is a flycatcher of wooded regions, especially the
borders of rocky streams.

Length, 6½; wing, 3½ (3½–3½); tail, 3½; tarsus, ¾; culmen, ½. Southwestern United States from Texas to Washington, south to Lower California and central Mexico.

9. **Olive-sided Flycatcher** (459. *Contòpus boreàlis*). — A dusky,
olive-brown bird, with the lower parts white, except the olive-
brown sides, which give it its
name; the wing bars are very inconspicuous, and there are curious
tufts of fluffy feathers on the flanks,
nearly white in color. This is a
woodland bird found usually in
the tree tops.

Length, 7½; wing, 4½ (3½–4½); tail,
3; tarsus, ½; culmen, ½. North America;
breeding from New England northward,
and wintering from Mexico to northern
South America.

10. **Wood Pewee** (461. *Contòpus
vìrens*). — A dusky, olive-brown-
backed, whitish-bellied bird, with
the head, wings, and tail blackish.
The middle of the belly is yellowish, a ring around the eye

Olive-sided Flycatcher

white, and the two wing bars whitish. This is a very dark,
almost fuscous-backed bird, while the yellowish-white under
parts have some gray on the sides of throat and breast. It is
necessary to note the lengths of wings and tail in order to
separate this species from some other flycatchers. It is a
common wood-living, retiring bird, with sweet, pensive notes
sounding much like its name.

Length, 6½; wing, 3½ (3–3½); tail, 2½–2¾; tarsus, ½; culmen, ½. Eastern
North America from the Plains; breeding from Florida to southern Can-
ada, and wintering south of the United States in eastern Mexico to
Central America.

11. Western Wood Pewee (462. *Contópus richardsònii*). — This is a bird similar to the last, but darker and less olive-tinted on the back, with more of olive-gray across the breast, and less of yellowish on the belly.

Length. 6½ ; wing, 3¾ 3¼-3⅜) ; tail, 2¼ ; tarsus, ½ ; culmen, ½. Western United States from the Plains to the Pacific, north to British Columbia, and south in winter to northern South America.

12. Yellow-bellied Flycatcher (463. *Empidònax flavicéntris*). — A small, dark olive-green-backed, yellow-bellied flycatcher, with the yellow breast, sides, and throat washed with much olive-green. The wings and tail are blackish, and the two wing bars whitish. This has more sulphur-yellow on the lower parts than any other of our fly-catchers. No other eastern species has yellow of any shade on the throat. It is practically only a mi-grant in the United States, and dur-ing migrations is almost voiceless. In its summer home in the northern evergreen forests its rather plain-tive call as well as its harsh, abrupt "psĕ-ĕk" can be heard.

Yellow-bellied Flycatcher

Length, 5½ ; wing, 2¾ (2¾-2⅞) ; tail, 2¼ ; tarsus, ⅝ ; culmen, ½. North America from the Plains eastward; breed-ing from the northern border of the United States to Labrador, and wintering from eastern Mexico to the Isthmus. The **Western Flycatcher** (464. *Empidònax difficilis*) of the western United States from the Plains to the Pacific is very much like the last, but less yellow below, and with buffy wing bars. The tail averages 2¾.

13. Green-crested Flycatcher (465. *Empidònax viréscens*). — A slightly crested, dull-greenish-olive flycatcher, with the lower parts yellowish-white, and the distinct wing bars also yellowish-white. The breast has a slight tinge of green, the throat is white, and the wings and tail are blackish. The upper mandible is black, and the lower one flesh-colored. It is an inhabitant of wet woodlands, and is usually to be found on the lower branches. Its notes are very distinct from those of other fly-catchers, but difficult to express in words. (Acadian Flycatcher.)

Length, 5¾; wing, 2⅞ (2⅝–3⅛); tail, 2½; tarsus, ⅝; culmen, ¾. Eastern United States; breeding from Florida north to southern New England

Green-crested Flycatcher

and southern Michigan, and wintering south to Central America.

14. Traill's Flycatcher (466. *Empidònax tràillii*). — A small, western, slightly crested, olive-brown flycatcher, with ash-gray breast and sides, pale yellow belly, and two whitish wing bars. The wings and tail are blackish, the throat pure white, and the under mandible flesh color or whitish. It has more of a brownish shade than any other of our flycatchers. It is a silent, restless, retiring bird, frequenting bushy tracts instead of forests.

Length, 6; wing, 2¾ (2⅝–3); tail, 2½; tarsus, ⅝; culmen, ½. Western North America from Ohio westward to the Pacific; breeding from the northern border of the United States northward, and wintering in Mexico to northern South America. The **Alder Flycatcher** (466ª. *E. t. alnòrum*) of eastern North America is a variety with less of brown in the plumage; breeding range from the mountain region of New Jersey and Pennsylvania northward, and wintering south to Central America (west to Michigan).

15. Least Flycatcher (467. *Empidònax mínimus*). — A very small, common, olive-backed, whitish-bellied flycatcher, with grayish breast and sides, and whitish wing bars. The lower mandible is brown, and there is almost no yellow on the belly. This is the smallest of the flycatchers. The second common name it has is derived from an attempt to write the sound of its notes. It lives generally in our orchards

and among our shade trees, rather than in the wild woods.
(Chebec.)

Length, 5¼ ; wing. 2½ (2¼-2⅜) ; tail, 2¼ ; tarsus, ⅝ ; culmen, ¹⅕. Chiefly
eastern North America, west to Colorado ; breeding from Pennsylvania to
Quebec, and wintering in Central America.

ORDER II. HUMMINGBIRDS, GOATSUCKERS, AND SWIFTS (MACROCHÌRES)

An order containing one large American family, the **Hummingbirds**, and two smaller ones of general distribution, the
Goatsuckers and **Swifts**.

FAMILY XIX. HUMMINGBIRDS (TROCHÍLIDÆ)

A family (400 species) of small, brightly colored, American,
swiftly flying birds, living mainly on the insects and nectar
found in flowers, which they obtain while on the wing. Their
movements through the air are most swift and insect-like, the
wings vibrating so rapidly as to be lost to the eye in a hazy
mist. These birds are represented by many species in Mexico
and South America, and several species are found west of the
Rocky Mountains in the United States, but only one (if southern Texas is excluded) is to be found in the great region
covered by this book. Nearly all, except some tropical forms,
have weak, chippering, or squeaking voices.

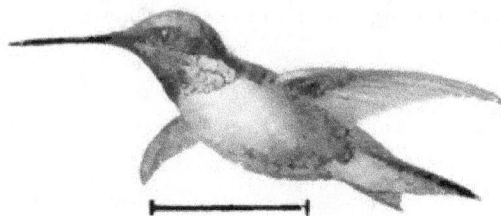

1. **Ruby-throated Hummingbird** (428. *Tróchilus cólubris*). — A
minute, long-billed, narrow-winged, greenish-colored bird, seen
hovering suspended
over flowers or flitting
rapidly from plant to
plant. The *male* has
a gorget, of brilliant,
metallic, ruby-red ; this
is wanting in the *fe-
male*. The *male* has

Ruby-throated Hummingbird

a notched tail of narrow feathers, and the *female* one not

notched, of rounded, white-tipped feathers. This bird is seen by most people only when hovering over flowers in search of food, and is hence thought to be always on the wing. If carefully watched, it will often be seen at rest on the twigs of tree tops. The only living forms which may be mistaken for hummingbirds are insects called hawk or sphinx moths. Most of these are found hovering over flowers in the evening; the hummingbirds visit the flowers in the daytime.

Length, 3; wing, 1½ (1½-1¾); tail, 1¼; culmen, ⅝. North America from the Plains eastward; breeding from Florida to Labrador, and wintering in Cuba, eastern Mexico to Central America. **Rieffer's Hummingbird** (438. *Amazilia fuscicaudata*) of southern Texas to northern South America is a brilliant green hummingbird, with dark purplish wings and deep chestnut tail; the belly is gray. Length, 4; wing, 2¼; tail, 1⅝; culmen, ¾. The **Buff-bellied Hummingbird** (439. *Amazilia cerviniventris*) of southern Texas to Central America is a similar green bird, with the belly pale cinnamon color. Length, 4¼; wing, 2¼; tail, 1½; culmen, ⅞.

FAMILY XX. SWIFTS (MICROPÓDIDÆ)

A family (75 species) of long-winged, close-feathered, small-bodied birds, with large, swallow-like or nighthawk-like mouths, and almost unrivaled power of flight. Almost the whole day is spent on the wing, catching enormous numbers of insects. These birds are found in immense flocks, especially when nesting or roosting.

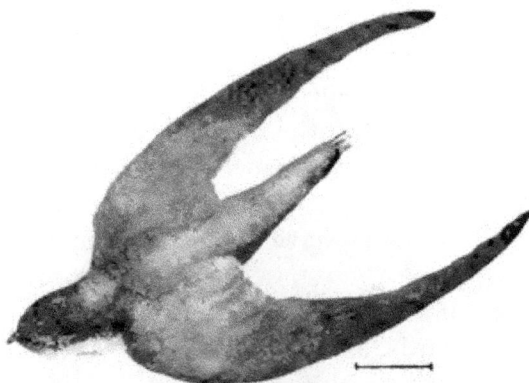

Chimney Swift

1. **Chimney Swift** (423. *Chætùra pelágica*). — An ashy-black bird resembling the swallow, with very long wings and short, rounded, spiny-tipped tail.[1] In certain places where large, unused chimneys are found, great flocks

of these birds will be seen in the early morning. flying out from the chimney top, and starting on their day's work of ridding the air of flying insects. In the evening all will be found returning, a steady stream, into the same roosting place. When resting on the inner wall of the chimney, the spiny tail is used as a support, much as the woodpeckers use their tails against the bark of tree trunks. · (Chimney Swallow.)

Length, 5; wing, 5; tail, 2; culmen, ¼. North America from the Plains eastward; breeding from Virginia to Labrador, and wintering south of the United States in Mexico. The **White-throated Swift** (425 *Aëronautes melanoleucus*) of the western United States from southern Montana to the Pacific is a blackish-backed, swallow-like bird, with a short, stiff, but not spiny-tipped tail, and most of the lower parts white; the male has the tail deeply notched. Length, 6¼; wing, 6; tail, 2¼.

FAMILY XXI. GOATSUCKERS, ETC. (CAPRIMÚLGIDÆ).

A family (nearly 100 species) of large, dull, mottled gray and brown, loose-plumaged, insect-eating birds which have enormous mouths, though the culmen or upper ridge of the bill is remarkably short.[1] Their heads are peculiarly large, broad, and flat, the legs small and weak, and the wings are rather long and pointed. They capture their prey while on the wing and, excepting during migrations, are solitary in their habits. The chuck-will's-widow has the widest mouth of any of our species; the gape measuring two inches from side to side. This enables it to swallow the largest of insects, and even hummingbirds and small sparrows have been found in its stomach. The night-hawks lack the rictal bristles which are so conspicuous in the other birds of the family. Our species are practically nocturnal birds, as silent in their flight as owls. During the day they recline rather than perch on limb of tree, or ground, in such position as to be entirely unnoticed, except by the most experienced observer. No nests are built or even hollows made, but the eggs are laid on the bare ground or on tree stumps. Their cries are among the most peculiar and striking of bird notes and from them many of the species derive their common names.

Key to the Species

* Rictal bristles very small, ¼ inch or less long; tail notched at tip; wings with a conspicuous white spot (or tawny in the *female* of a Texas species). (**B.**)

* Rictal bristles, ½ inch or more long and branching with short lateral hairs; wing, 7½ or more long...............1. **Chuck-will's-widow.**

* Rictal bristles long, but not branching.[1] (**A.**)

   A. Tail less than 2 inches shorter than the wing; no white blotches on the wings, which are usually 6 or more long...2. **Whip-poor-will.**

   A. Tail fully 2 inches shorter than the wings; no white blotches on the wings, which are less than 6 long ....3. **Poor-will.**

   A. Tail about as long as the wings; a large white blotch on the wings, which are 6-8 long..................... **Merrill's Parauque** (3).

 B. Wing over 7½ long........4. **Nighthawk** and **Western Nighthawk.**

 B. Wing, 7¼ or less long. 5. **Texas Nighthawk. Florida Nighthawk** (4).

1. **Chuck-will's-widow** (416. *Antróstomus carolinénsis*). — A large, finely mottled, brownish bird resembling the whip-poor-will, without any pure white markings. The mouth is very large, and the rictal bristles long and with hair-like branches for half their length. The *male* has an indistinct whitish band across the throat, and the *female* a buffy one.

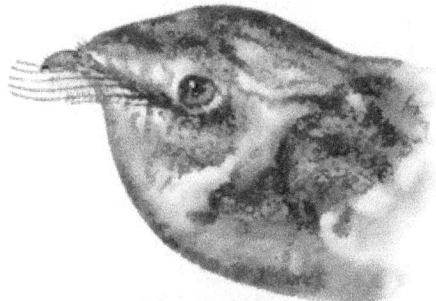

Chuck-will's-widow

Length, 11½; wing, 8½ (8-9); tail, 6; culmen, ¾. South Atlantic and Gulf States; breeding from Illinois and North Carolina southward, and wintering from our southern border to Central America. Accidental in Massachusetts.

2. **Whip-poor-will** (417. *Antróstomus vocíferus*). — A bird similar to the last, but smaller; the *male* is marked with a pure white collar, and the end half of each of the three outer tail feathers is white. The *female* has buff on neck and tail

Whip-poor-will

feathers instead of white. The rictal bristles are long, but not branching. This bird flies and makes its *whip-poor-will* notes after sunset and before sunrise. It is a common, low-flying bird, much more frequently heard than seen.

Length, 9½; wing, 6 (5¾–6¾); tail, 4¾; culmen, ⅜. North America from the Plains eastward; breeding from Virginia, north to New Brunswick and Manitoba, and wintering from Florida to Central America.

3. **Poor-will** (418 *Phalaenóptilus nuttállii*). — A western whip-poor-will, beautifully mottled with bronze-gray and silver-gray markings: both sexes have the white patch across the throat, and nearly white tips to the under tail feathers. In singing its notes the first syllable is dropped, and so this bird is called a *poor-will.*

Length, 8; wing, 5½ (5⅜–5¾); tail, 3½; tarsus, ⅝; culmen, ⅜. Western United States from Kansas and Montana, and southward and westward to Mexico. **Merrill's Parauque** (419. *Nyctidromus albicóllis mérrilli*). — A Texas species of very large size, long, rounded tail, and with a great white patch on the wings. The *male* has a broad white collar; the *female* a less distinct buff one. The under tail feathers are more or less white at tip. Length, 10½–13½; wing, 7 (6–7½); tail, 6½; tarsus, 1; culmen, ⅜. Southern Texas to northeastern Mexico.

4. **Nighthawk** (420. *Chordeiles (ďl-les) virginiànus*). — A bat-like, night and evening flying, dark-colored, finely mottled bird, with conspicuous white patches at about the middle of the wings, looking like holes when the bird is flying. It differs from the foregoing species in flying high in the air, and in having almost no rictal bristles. It varies its flight with occasional dives toward the ground with wings nearly closed; before reaching the earth there is a sudden check in the speed, and a slow upward movement again to the former elevation. (Bull-bat.)

Nighthawk

Length, 9½; wing, 7¾ (7⅜–8¼); tail, 4⅝; culmen, ¼. North America from the Plains eastward; breeding from the Gulf States to Labrador, and wintering in South America. The **Western Nighthawk** (420ᵃ. *C. v. hénryi*) of the western United States from the Plains westward is lighter colored and has the white spaces larger. The **Florida Nighthawk** (420ᵇ. *C. v. chápmani*) of Florida and the Gulf coast to South America is smaller and has more numerous white and buff markings. Wing, 7–7¼; tail, 4–4¼.

5. **Texan Nighthawk** (421. *Chordeiles acutipénnis texénsis*). — A southern, small, distinctly streaked and barred nighthawk with the white wing patch nearer the tips of the primaries than the bend of wing. The tail is blackish, crossed by grayish or tawny bars, with a complete white cross-bar near the tips of the feathers. The *female* has the wing patch tawny instead of white, and the white cross-bar of the tail is lacking.

Length, 8½; wing, 6⅜–7¼; tail, 4–4¾. Texas to southern California, south to Panama.

ORDER III. WOODPECKERS, WRYNECKS (PÌCI)

An order which with us includes only the following:

FAMILY XXII. WOODPECKERS (PÍCIDÆ)

A large family (350 species) of creeping or climbing birds with stiff, sharp-pointed tail feathers which are used as aids in supporting the body against the tree.[1][2] The toes are four in number; two directed forward and two backward[3] (in a few exceptional species there are only three, two in front and one behind[4]). These birds have stout, straight, chisel-pointed bills,[5] with which they are enabled to cut small holes in the wood for the purpose of securing insects, and large holes for nesting places. The tongue is peculiarly long, has a spear-like tip, and is so arranged that it can be thrust out to a wonderful distance. By its aid, the larvæ of insects are secured and brought from their retreats under the bark. Woodpeckers

1　　　　2　　　　3　　　　4　　　　5

have but poor vocal powers, and they make use of a tattoo with their bill for their love song. The eggs are in all cases white.

Key to the Species

* Head with a conspicuous crest ; large birds ; wings, 7¼ or more long.

  — Bill ivory-white.....................1. **Ivory-billed Woodpecker.**

  — Bill blackish.........................9. **Pileated Woodpecker.**

* Head not crested ; wings less than 7½ long. (**A.**)

  A. With only three toes ; two in front and one behind.[1] (**G.**)

  A. With four toes ; two in front and two behind.[2] (**B.**)

B. Back very distinctly barred crosswise with black and white. (**E.**)

B. Back not cross-barred, but with a broad central streak of white ; under parts white without spots or streaks. (**D.**)

B. Back black without cross-bars or lengthwise streaks ; the rump may be white. (**C.**)

B. Back olive-brown with numerous black bars ; breast with a broad black crescent-shaped band ; belly whitish, with numerous round black spots ; wing, 5¼ or more long ....14. **Flicker.**

B. Back irregularly variegated with black and yellowish ; belly with more or less of yellow...............8. Yellow-bellied Sapsucker.

  C. Secondary wing quills and rump wholly white (in the *young* nearly so) ; breast without broad black band ; head and neck of the adult red........................ ....10. **Red-headed** Woodpecker.

  C. Wing quills mainly black, but rump white ; a broad black band across breast, separating the white of throat and belly (extreme western). **Californian Woodpecker** (407. *Melanérpes formicivorus bairdi*).

  C. Western species with wings, rump, back, and tail a beautiful bronzy-black........................11. **Lewis's Woodpecker.**

D. Wing less than 4¼ long ; culmen less than 1 long ; under tail feathers white cross-barred with black............. 3. Downy Woodpecker.

D. Wing over 4¼ long ; culmen over 1 long ; under tail feathers white without black..................................2. Hairy Woodpecker.

  E. Culmen, 1 or more long ; belly with reddish tinge...............

  .. ..12. Red-bellied Woodpecker.

  E. Culmen, 1 or more long ; belly without reddish..................

  ....13. **Golden-fronted Woodpecker.**

  E. Culmen less than 1 long. (**F.**)

F. Wing, 4⅗–5¼ long ; a conspicuous white patch on the side of head and neck.................................4. Red-cockaded Woodpecker.

F. Wing, 3–4½ long ; the white of the side of the head and neck inclosing a curved black stripe..................5. Texan Woodpecker.

  G. Back uniformly black........ 6. **Arctic** Three-toed Woodpecker.

  G. Back barred crosswise......7. **American Three-toed** Woodpecker.

1. **Ivory-billed Woodpecker** (392. *Campéphilus principàlis*).—
A large, scarce, southern, white-billed, distinctly crested, black
woodpecker, with a white line on each side of neck and body
and a white blotch on tips of secondaries. The crest of the
male is scarlet, of the *female* black. This is a shy bird of the
dense, southern, cypress forests.

Length, 20 ; wing, 10 (9-10½) ; tail, 6½ ; tarsus. 2 ; culmen, 2¾. Now
found only locally in the lower Mississippi Valley and in the Gulf States.

2. **Hairy Woodpecker** (393. *Dryobàtes villòsus*).—A small,
white-spotted woodpecker, with much black on the upper parts

Hairy Woodpecker

and white below. The white spots of
the wings give them a barred appearance
when the bird is at rest, and the white of
the center of the back forms a longitudinal band; the head has streaks of black
and white. The *male* has a scarlet patch
on the nape. The under tail feathers are
white and not barred.

Length, 9½ ; wing, 4⅞ (4½-5) ; tail, 3½ ; tar-
sus, ⅚ ; culmen, 1⅛. Wandering but not migra-
tory. Northern and middle portion of the United
States from the Plains eastward. The **Southern
Hairy Woodpecker** (393 ᵇ. *D. v. audubònii*) of the
South Atlantic and Gulf States is a smaller vari-
ety and with less of white. Wing, 4⅜-4¾ ; tail, 3.
There is a northern variety which has a wing
5-5¼ long found in the northern portions of North
America.

3. **Downy Woodpecker** (394. *Dryobàtes pubéscens*).—A smaller,
more common woodpecker than the last, but with almost the
same arrangement of colors. This species has the white of the
under tail feathers cross-barred with black. After becoming
familiar with the notes of Nos. 2 and 3, one can distinguish them
by their voices. As Mr. Brewster says, the downy woodpecker
species "has a long unbroken roll," while the hairy woodpecker
has "a shorter and louder one with a greater interval between
each stroke." The downy woodpecker is much more abundant

in settled regions, where it can be found in woodland, orchards, and even in the shade trees along the streets of the towns.

Length, $6\frac{3}{4}$; wing, $3\frac{3}{4}$ ($3\frac{1}{2}$–4); tail, $2\frac{1}{2}$; tarsus, $\frac{5}{8}$; culmen, $\frac{3}{4}$. North America from the Plains eastward, and south to the Gulf of Mexico. Not migratory. (The northern form is 394c. *D. p. mediānus.*)

4. **Red-cockaded Woodpecker** (395. *Dryobātes boreālis*). — A small, southern "ladder-backed" woodpecker, with white sides to the head and a scarlet tuft of feathers on each side of the crown, back of the eyes and above the white cheeks. The *female* lacks the scarlet. This inhabitant of the pine woods of the Southern States has distinct black and white bands across the back, giving the appearance of a ladder. The crown and band between the white cheeks and throat are black.

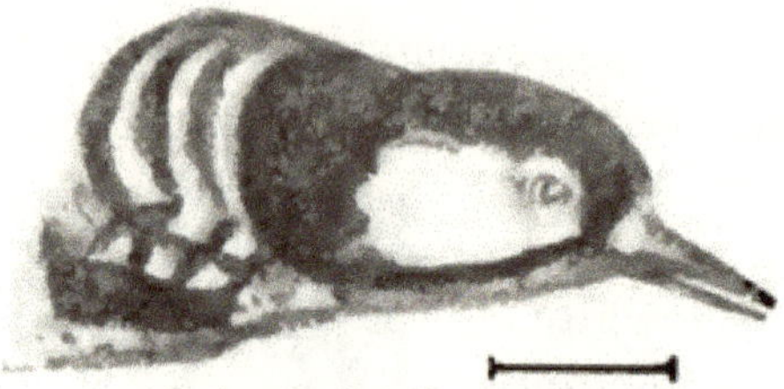

Red-cockaded Woodpecker

Length, $8\frac{1}{4}$; wing, $4\frac{3}{4}$ ($4\frac{1}{2}$–5); tail, $3\frac{1}{4}$; culmen, $\frac{3}{4}$. North Carolina to eastern Texas, south to the Gulf.

5. **Texan Woodpecker** (396. *Dryobātes scalāris bāirdi*). — A Texas "ladder-backed," gray-bellied woodpecker, with numerous small black spots on sides and crissum. The side of the head and neck is white, with a long, curved, black stripe extending from the eye downward and forward to the bill. The *male* has more or less of red on the crown; this is lacking in the *female*.

Length, $7\frac{1}{2}$; wing, $3\frac{3}{4}$ ($3\frac{1}{2}$–$4\frac{1}{4}$); tail, $2\frac{3}{4}$; culmen, $\frac{7}{8}$. Southern portion of the United States from Texas to California, and south to the table-lands of Mexico.

6. **Arctic Three-toed Woodpecker** (400. *Picoīdes ārcticus*). — An extreme northern, medium-sized, orange-crowned, black-backed, white-bellied woodpecker, with very small white spots on the otherwise black wings; outer tail feathers mainly white, and a line under the eye also white. The *female* has a black crown.

This active, restless bird takes long flights, with the characteristic undulating movements of woodpeckers in general, and at
every glide gives out its shrill note,
which sounds more like a mammal in
pain, than like a bird. (Black-backed
Woodpecker.)

Length, 9½; wing, 5½ (5–5½); tail, 3½; culmen, 1½. Northern North America, south to
New England, Michigan, and Idaho.

7. **American Three-toed Woodpecker**
(401. *Picoïdes americànus*). — A northern bird, similar to the last, but having the center of the back cross-barred
with black and white; the sides are
also barred. The orange spot is found
only in the crown of the *male*, the
female having a black and white spotted
crown.

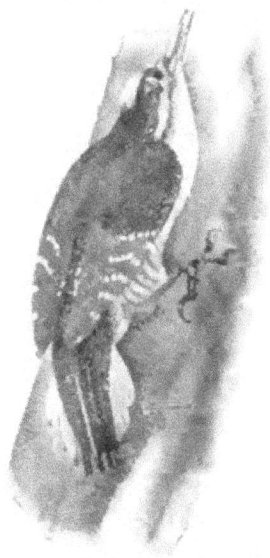

Arctic Three-toed Woodpecker

Length, 8¾; wing, 4½ (4¾–4¾); tail, 3½; culmen, 1½. Northern North
America, east of the Rocky Mountains, south to Massachusetts and New
York. Like most of the woodpeckers it is
not migratory.

8. **Yellow-bellied Sapsucker** (402.
Sphyrápicus várius). — A small,
rather common, scarlet-crowned, mottled-backed, yellowish-bellied woodpecker, with much white on the wings,
and black on the breast. The mottling of the back is of black and yellowish. The wings are black, with
many spots of white on the quills,
and the coverts are mainly white.
The tail is more or less barred with
black and white. The *male* has a red
throat, and the *female* a white one.

Yellow-bellied Sapsucker The crown of the *female* is sometimes black. This migratory woodpecker is a noisy bird during the breeding season in the north, but during the rest of the year is seldom seen; it lives in the densely foliaged trees and is seldom heard, as its notes are very weak. This bird, as its name indicates, feeds mainly on the juices of trees, and so probably does more harm than good.

Length, 8½; wing, 4⅞ (4⅞–5⅛); tail, 3⅛; culmen, 1. Eastern North America; breeding from Massachusetts northward, and wintering from Virginia to Central America.

9. **Pileated Woodpecker** (405. *Ceophlœus pileatus*). — A large, southern, red-crested, black-bodied, dark-billed woodpecker, with the sides of the head and the neck mainly white. When flying, much white can be seen on the wings, as the basal half of the feathers is white. The *female* lacks red on the fore part of the crown. While most woodpeckers have an undulating flight, this one moves in a direct course. This bird was formerly distributed generally over the wooded regions of North America, but is now becoming very rare except in the wilder sections. (Logcock.)

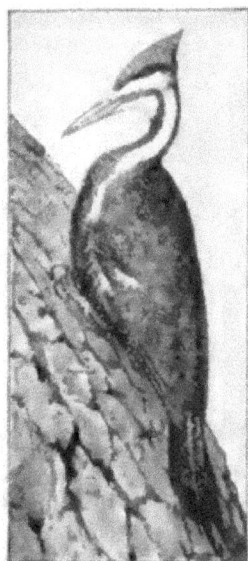

Pileated Woodpecker

Length, 13–19; wing, 9 (8–10); tail, 7; culmen, 1⅜–2½. North America; very rare in the settled portions of the Eastern States.

10. **Red-headed Woodpecker** (406. *Melanerpes erythrocéphalus*). — A common, medium-sized, black-backed, white-rumped, white-bellied woodpecker, with the whole head and neck bright red. The secondary quills are white, forming a large white wing patch. The *young* has a grayish-brown head and neck, more or less mixed with brownish, and the back and wings are somewhat barred. It is a noisy, active bird, with ability to resist the most extreme cold of the Northern States in winter, if food is abundant.

Length, 9½; wing, 5½ (5¼-5¾); tail, 3½; culmen, 1⅛. United States from the Rocky Mountains eastward; breeding throughout and north into Canada, and wintering irregularly throughout, but more abundantly in the Southern States.

11. Lewis's Woodpecker (408. *Melanérpes torquátus*). — An extreme western, red-bellied, bronze-black-backed, red-faced woodpecker, with a bluish-gray band around the neck. The crown and the neck above the gray band are black. The wings, tail, back, and crissum are a rich iridescent green-black.

Red-headed Woodpecker

Length, 11; wing, 6¾ (6½-7); tail, 4½; culmen, 1⅛. Western United States from the Black Hills to the Pacific; wintering in western Texas, and casual in Kansas.

Red-bellied Woodpecker

12. Red-bellied Woodpecker (409. *Melanérpes carolínus*). — A southern, medium-sized, "ladder-backed," whitish-bellied woodpecker, with the crown and back neck bright scarlet, and the breast and belly often tinged with red. The black and white bars of the back and wings are numerous and distinct. The *female* lacks the red on the center of the crown, this being replaced by an ashy color.   This is a common southern bird, peculiarly spasmodic in its movements along a tree trunk.

Length, 9½ ; wing, 5¼ (4⅞–5½) ; tail, 3½ ; culmen, 1¼.   Eastern and southern United States, north casually to Massachusetts, southern Michigan, and eastern Kansas.   Not migratory.

13. **Golden-fronted Woodpecker** (410. *Melanérpes aúrifrons*). — A southern Texas species, similar to the last, but with the belly yellowish instead of reddish, the *male* having the crown but not the back neck red.   In the *female* the red crown is wanting, but both sexes have the back neck more or less yellow or orange in the form of a band, and the forehead golden-yellow, giving the name to the species.   The head and under part are ashy-gray.

Length, 10 ; wing, 5¼ (5–5½); tail, 3½ ; culmen, 1¼.   Central Texas and south to the city of Mexico.

14. **Flicker** (412. *Colàptes auràtus*). — A common, brown-backed, white-rumped woodpecker, with a scarlet band across the back of the head, a golden lining to the wings and tail, a black crescent on the breast below the reddish throat, and a light-colored belly, thickly spotted with round black dots.   When at rest, this bird can easily be recognized by the red crescent on the back of the head and the black crescent on the breast ; when flying, by the white rump and the golden lining to the wings.   It often perches on limbs. Its peculiar habits, notes, and colors have given it nearly two-score names, the commonest of which are here given. (Golden-winged Woodpecker ; Yellow-hammer; Pigeon Woodpecker; High-hole; Tucker; Clape.)

Flicker

Length, 12½; wing, 6 (5½–6½); tail, 4½; culmen, 1¾. North America from the Plains eastward, breeding throughout and wintering mainly south of the Middle States.

ORDER IV. CUCKOOS, KINGFISHERS, ETC.
(COCCYGES)

An order of tropical, Old World birds containing families differing widely in their characteristics, and classified together in one miscellaneous group only because they belong under no other order, and it would be inconvenient to classify each family by itself. We have representatives belonging to three of these families.

FAMILY XXIII. KINGFISHERS (ALCEDÍNIDÆ)

A large family (nearly 200 species, mainly Malayan) of chiefly tropical birds. The American species are solitary and exclusively fish-eating birds, found only near the water. A few Old World species feed upon insects, snails, etc., and live in the forests, though most of them have the habits of our forms. They are heavy-straight-billed, large-headed, bright-colored birds, with small feet and short tails.

Key to the Species

* Wing, 5–7 long; culmen about 2 . . . . 1. **Belted Kingfisher.**
* Wing, 3–4 long; culmen less than 2 . . . . 2. **Texas Kingfisher.**
* Wing over 7 long; culmen over 3. **Ringed Kingfisher** (390–1. *Céryle torquáta*). A Mexican species casually found in southern Texas.

1. **Belted Kingfisher** (390. *Céryle álcyon*). — A noisy, short-tailed, large-straight-billed, crested, blue-backed bird, with white lower parts and bluish band across the breast. The wing quills and tail feathers are black, more or less blotched and barred with white. The *female* is similar, but has a brown band across the belly. A common inhabitant of the wooded shores of streams and lakes, where its harsh, rattling cry can

often be heard. When watching for fish, which form its only food, it sits on some support projecting over the water and can readily be recognized by the large, crested head and short tail.

Length, 11-14½ ; wing, 6¼ (6-6½) ; tail, 4; culmen, 2. Throughout North America; breeding from the southern border of the United States, and wintering from the Middle States to Panama.

2. Texas Kingfisher (391. *Céryle americána septentrionális*).

Belted Kingfisher

— A small, bronze-green kingfisher with the collar and belly white. The *female* has the green band across the breast replaced by a rufous one.

Length, 8; wing, 3½ ; tail, 2¾ ; culmen, 1¾. Southern Texas to Panama.

FAMILY XXIV. TROGONS (TROGONIDÆ)

A family (50 species) of brilliantly colored, tropical birds represented in southern Texas by the following :

1. Coppery-tailed Trogon (389. *Trógon ambíguus*). — A beautiful, long-tailed, red-bellied bird, with the back and breast golden-green, face black, and a white collar between the carmine belly and the golden-green of the throat. Bill serrated.[1] The middle tail feathers of coppery-green give the species its name.

Length, 11½ ; wing, 5½ ; tail, 7 ; culmen, ½. Southern and Central Mexico, north to southern Texas.

FAMILY XXV. CUCKOOS, ANIS, ETC. (CUCÙLIDÆ)

This large, tropical family (200 species) of birds includes species of various forms, colors, and habits, so that it has been separated into about ten subfamilies. All have two toes in front, and two behind,[2] more or less downwardly curved bills, and elongated, rounded, to strongly graduated tails. Our birds belong to three of these subfamilies. Only the Old World cuckoos use the nests of other birds in which to place their eggs. The anis are very peculiar in their nesting habits; several females join together and build a single nest for all their eggs and then take turns in the work of incubation.

Key to the Species

* Bill nearly as high as long, and much flattened sideways.[3] (**C.**)
* Bill elongated, only about a third as high at base as long. (**A.**)
    A. Bill nearly straight almost to the tip when it is abruptly decurved ; tail 10 or more long. Western ground bird . . . . . . 2. **Road-runner.**
    A. Bill regularly curved downward for nearly its full length;[4] tail, 8 or less long. (**B.**)
  B. Bill nearly black throughout ; wings with little or no cinnamon color. . . . . 5. **Black-billed Cuckoo.**
  B. Bill with much yellow below ; belly white ; wings with much cinnamon color. . . . . 4. **Yellow-billed Cuckoo.**
  B. Bill yellow below ; belly tawny or buffy . . . . . . 3. **Mangrove Cuckoo.**
    C. Upper mandible smooth or slightly wrinkled . . . . 1. **Ani.**
    C. Upper mandible with several distinct grooves parallel with the top of the bill . . . . **Groove-billed Ani** (1).

1. Ani (383. *Crotóphaga áni*). — A long-tailed. large, southern, bronze-black bird. with a large, much compressed bill. The back shows steel-blue reflections, the lower parts are a dull black, and the tail is much rounded. This is a ground-living bird.

Length, 12–15; wing, 5¾ (5½-6); tail, 8; tarsus, 1½; culmen, 1¼. West Indies and eastern South America, casual in Florida and Louisiana, and accidental near Philadelphia. The **Groove-billed Ani** (384. *Crotóphaga sulciróstris*). of Mexico and Texas. is similar to the last, but with a grooved bill. Length, 12–15; wing. 6; tail, 8.

1   2   3   4

2. **Road-runner** (385. *Geocóccyx califórniánus*). — A large, long-tailed, crested, coarse-plumaged, ground bird, with the body striped with buffy and bronze-brown, somewhat glossed with green. Skin around the eye naked. (Ground Cuckoo.)

Length, 20–24; wing, 6¾ (6¼–7); tail, 10–12; tarsus, 2; culmen, 2. Kansas and Colorado to California, and south to central Mexico.

3. **Mangrove Cuckoo** (386. *Coccýzus minor*). — This rare summer resident of the extreme south is similar to the next, but with the ear coverts black and the under parts a rich buff.

Length, 12¼; wing, 5¼ (5–6); tail, 7; culmen, 1. Florida, Louisiana, and West Indies to northern South America. **Maynard's Cuckoo** (386ᵃ. *C. m. maynárdi*) differs from the last in having the lower parts a pale buff instead of ochraceous buff. Dimensions a little less. Bahamas and the Florida Keys.

4. **Yellow-billed Cuckoo** (387. *Coccýzus americánus*). — A common, long, slender, long-tailed, brownish-gray bird, with a slender, curved bill and conspicuously white-tipped outer tail feathers. The under parts are whitish, the wings have much cinnamon color, and the under mandible is yellow at base. This bird destroys great numbers of that pest of our trees, — the tent caterpillar.

Yellow billed Cuckoo

Its notes are a harsh, grating *cl-uck, cl-uck* varied by *cow, cow.* (Rain "crow.")

Length, 12; wing, 5¾ (5¼–6); tail, 6; tarsus, 1; culmen, 1. Eastern North America; breeding from Florida to Canada and Minnesota, and wintering south of the United States to Central America.

5. **Black-billed Cuckoo** (388. *Coccýzus erythrophthálmus*). — A bird similar to the last in form, colors, and habits, but with less white and no black on the tail, the under mandible black, and no cinnamon on the wings. The voice is less harsh.

Length, 11¾; wing, 5½ (5!-5¾); tail, 6!; culmen, 1. North America from the Rocky Mountains eastward; breeding from the Gulf of Mexico to Labrador, and wintering south of the United States to northern South America.

ORDER V. PARROTS, MACAWS, ETC. (PSITTACI)

An order of about 500 species of almost exclusively tropical birds, here represented by only one. belonging to :

FAMILY XXVI. PARROTS AND PAROQUETS (PSITTÁCIDÆ)

A large family (400 species) of tropical, gaudily colored, harsh-voiced, hooked and cered-billed birds; having feet with two toes in front and two behind. which they use for walking, climbing, and as hands. Their discordant voices are, in most species, readily trained to utter the words of human speech. They are inhabitants of dense forests. When necessary, they fly well. They live upon fruits and seeds.

1. **Carolina Paroquet** (382. *Conùrus carolinénsis*). — A rare, southern, green paroquet, with a yellow head and neck, and bright orange fore-head and cheeks. The bend of the wing is also orange. The *young* have the head, neck, and bend of wing also green. This, our only repre-sentative of the par-

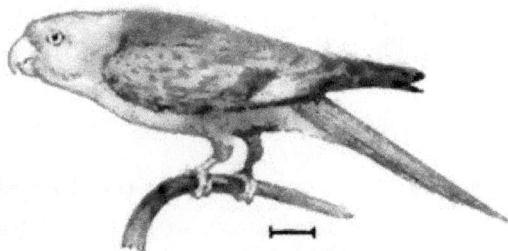

Carolina Paroquet

rots, was formerly found as far north as the Great Lakes, but is becoming every year more rare and local even in Florida, Arkansas, and Indian Territory, the only divisions of the United States where it has recently been found.

Length, 12½; wing, 7½ (7-8); tail, 6¼.

ORDER VI. BIRDS OF PREY (RAPTORES)

An order of usually large, rapacious, land birds, with hooked and cered bills; living exclusively upon animal food. They are found in all lands, and form several well-marked families. Some are night-flying (owls), some are carrion-eating (buzzards and vultures), some live mainly on mammals, fish, and birds (eagles and larger hawks), and some eat mice and insects (the smaller hawks).

FAMILY XXVII. HORNED OWLS, HOOT OWLS, ETC.
(BUBONIDÆ)

A large family (200 species) of owls, with rounded eye disks and toe nails, without saw-like teeth. It contains all our species except one, the barn owl. The owls differ from all other birds in having the face so broadened that both eyes look forward instead of sidewise, and they are so surrounded by radiating feathers as to make these features of the head seem larger. The eyes are immovable in their sockets, so that the whole head has to be turned when the bird wishes to look in a new direction. This gives a live specimen a very strange appearance. Many of the birds of this family have tufts of erectile feathers appearing like external ears and popularly called ear tufts.[1] These birds are regarded by many with superstitious awe because of their uncanny appearance, their strange actions, and their harsh, hooting, weird voices.

Key to the Species

* Wing, 5½–7½ long; tarsus partly bare of feathers and twice as long as the middle toe..........................11. **Burrowing Owl**.

* Wing, 3–4½ long; tarsus partly bare and but little longer than the middle toe..........................13. **Elf Owl**.

* Tarsus fully feathered. (A.)

  A. Head with conspicuous ear tufts or horns.[1] (F.)

  A. Head without ear tufts.[2] (B.)

 B. Wing, 15–19 long. (E.)

 B. Wing, 11–14 long. (D.)

 B. Wing, 8–10 long ....10. **American Hawk Owl**.

B. Wing, 3–8 long (**C.**)
  C. Wing, 6¦–8 long.........................5. **Richardson's Owl.**
  C. Wing, 5–6¦ long........... ....6. **Saw-whet Owl.**
  C. Wing, 3–5 long; tarsus densely feathered and not longer than
    middle toe......................12. **Ferruginous Pygmy Owl.**
  D. Belly with longitudinal stripes; back and breast with cross bars....
  ....3. **Barred Owl.**
  D. Belly and back with longitudinal stripes (there are small, possibly
    unnoticed ear tufts of few feathers.)...........2. **Short-eared Owl.**
  D. Belly and back dotted with black; the nail of the middle claw has a
    saw-like ridge on the inner side, so this species belongs to the next
    family; face heart shaped?............ **American Barn Owl**, p. 192.
    E. Plumage chiefly white; tail rounded.............9. **Snowy Owl.**
    E. Plumage mottled and barred with blackish and whitish..........
    ....4. **Great Gray Owl.**
  F. Wing, 14–18 long........................ .8. **Great Horned Owl.**
  F. Wing, 5–8 long.................................7. **Screech Owl.**
  F. Wing, 11–13 long. (**G.**)
    G. Ear tufts large, of 8–12 feathers.....1. **American Long-eared Owl.**
    G. Ear tufts small, of few feathers.......... ...2. **Short-eared Owl.**

1. **American Long-eared Owl** (366. *Asio wilsoniànus*). — A
large, common, night-flying, long-eared, brownish, mottled
owl, with the lower parts lighter, streaked on the breast
and barred on the belly. The ear
tufts are an inch or more long,
nearly black, with a light border.
During the daytime, this tame
bird is usually to be found in
deep, and, by preference, evergreen forests. Its food consists
mainly of mice and other small
mammals.

Length, 13–16; wing, 11¾ (11–12);
tail, 6; tarsus, 1¼; culmen, 1. Temperate North America south to central
Mexico; breeding throughout.

2. **Short-eared Owl** (367. *Asio
accipitrínus*). — A large, ochraceous, brown mottled, and streaked,
marsh-living owl, with ear tufts

Short-eared Owl

so small as often to be unnoticed. There is much of buffy
tints. especially on the lower parts, which are streaked on
both breast and belly. This inhabitant of wet, grassy places
is fearless and will allow itself to be almost stepped upon
before it will fly. Like the last, it feeds principally upon
mice.

Length, 14–17 ; wing, 12½ (12–13); tail, 6 ; tarsus, 1¾ ; culmen, 1¼.
Found in almost all lands; breeding in North America, locally from
Virginia northward.

3. **Barred Owl** (368. *Syrnium nebulòsum*). — A large, night-
flying. much-barred. brownish, hooting owl, without ear tufts.

The cross bars are fine and numerous
on all parts of this bird, except the
belly and sides, which are white,
broadly streaked with blackish. Most
owls have light. usually yellow eyes,
but the barred owl's eyes are nearly
black. This inhabitant of large, dense
woods is the one whose hooting call
can be heard nearly a mile — *whōō-
whōō-whōō-whōō-äh*.   Its notes are
more frequently heard soon after dark
and before sunrise. but during moon-
light nights it may be heard all night,
and occasionally even during the day.
Its food consists of small mammals,
insects, and birds. (Hoot Owl ; American Wood Owl.)

Barred Owl

Length, 17–25 ; wing, 13½ (12–14) ; tail, 9 : culmen, 1¼.   United
States from Nebraska and Texas eastward. north to Quebec : breeding
throughout. The **Florida Barred Owl** (368ᵃ. *S. n. alleni*) differs in having the toes almost bare of feathers ; a few bristly feathers are to be found
along the outer side of the middle toe only. South Carolina to Texas,
near the coast.

4. **Great Gray Owl** (370 *Scotiáptex cinèrea*). — A northern,
very large, yellow-eyed, ashy-brown, mottled owl, without ear

tufts. The under parts are pale gray, streaked on the breast, and barred on the belly and sides with dark brown. This immense owl of the Arctic regions is practically found only within the United States, very irregularly in winter. (Spectral Owl.)

Length, 24–30; wing, 17 (16–18); tail. 12; culmen, 1½. Arctic America, straggling southward in winter to southern New England, New Jersey, Illinois, and Idaho.

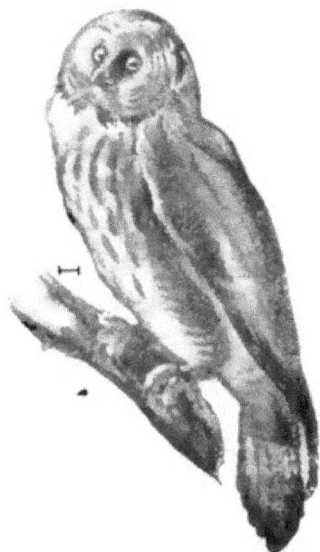

5. Richardson's Owl (371. *Nyctala téngmalmi richardsoni*). — A small, northern, night-flying, yellow-eyed, white-spotted, brown owl, without ear tufts. The under parts are white, thickly, but very irregularly, streaked with brown. This, like the last, is an inhabitant of northern regions, seldom seen in our Northern States, and only in winter. (Arctic American Saw-whet Owl.)

Great Gray Owl

Length, 9–12; wing, 7 (6½–7½); tail, 4½; tarsus, 1; culmen, 1. Arctic America; breeding from the Gulf of St. Lawrence northward, and wintering south to our northern range of states.

6. Saw-whet Owl (372. *Nyctala acàdica*). — A very small, yellow-eyed, night-flying, brownish-mottled owl, without ear tufts. The head is finely streaked, and the back spotted with white. The under parts are white, heavily streaked with light and dark

Saw-whet Owl

brown. This is a night-flying bird, passing its time during the day among dense trees. It gets its name from a resemblance of its notes to the noise made in filing a saw. Mice form its main food. (Acadian Owl.)

Length, 8; wing, 5½ (5¼-5½); tail, 2⅞; tarsus, ⅞; culmen, ½. North America; breeding from the Middle States northward (south to Mexico, in the mountains), and wandering irregularly southward in winter.

7. **Screech Owl** (373. *Mégascops ãsio*). — A very common, small, night-flying owl, with conspicuous ear tufts, and weird, whistling notes. It is found in varieties of two colors, distinctly reddish and distinctly grayish. The back is finely streaked and dappled; the under parts are white-streaked, blotched and barred with dark colors, giving them a variegated appearance. The *young* is more regularly barred than the adult. This owl prefers orchards near human habitations, to the wild woods, for its home. Its food consists of insects, mice, and birds.

Length, 7½-10; wing, 6½ (6-7½); tail, 3½; culmen, ⅞. North America from the Plains eastward, south to Georgia, and north to New Brunswick and Minnesota; practically resident throughout. The **Florida Screech Owl** (373ᵃ. *M. a. floridãnus*), of South Carolina to Louisiana, mainly near the coast, is a smaller bird, with the colors deeper, and the markings more distinct. Wing, 6; tail, 3. The **Texas Screech Owl** (373ᵇ. *M. a. trichópsis*), of southern Texas to Central America, has the small size of the Florida bird, and the two phases of color of the common screech owl, but is more regularly streaked and barred; thus it lacks the blotchy appearance which is so characteristic of the northern form.

Screech Owl

8. **Great Horned Owl** (375. *Bùbo virginiãnus*). — A very large, yellow-eyed, long-eared, finely mottled, brownish owl, usually marked with a white collar. The under parts are reddish-buff barred with black. This bird inhabits dense forests of the sparsely settled sections, and is probably the only owl that

kills poultry and game birds in any
great numbers, though even this one
is more apt to feed on the smaller
mammals. It can scream in a most
terrifying manner as well as *whōō-
whōōōō.* (Hoot Owl; Cat Owl.)

Length, 18–25; wing, 15½ (14½–17); tail,
8¾; tarsus, 2⅛; culmen, 1½. North America
from the Mississippi Valley eastward, north
to Labrador and south to Central America.
The **Western Horned Owl** (375ª. *B. v. sub-
árcticus*), of the western United States from
the Great Plains westward (and east casually
to Illinois), is lighter in color, having the buff
markings changed to gray or white. The
Arctic Horned Owl (375ᵇ. *B. v. árcticus*),
of arctic America south to South Dakota,
Wyoming, and Idaho, has much whiter
plumage, the under parts being pure white
with very restricted dark markings.

Great Horned Owl

9. **Snowy Owl** (376. *Nýctea nýctea*). — A very large, mottled,
white owl, with densely feathered
feet and no ear tufts. The *female*
is more heavily barred. It is a day-
flying, yellow-eyed owl, found in
the United States only in winter.
Though a day-flying owl, it is more
active in the early morning and
evening. In the United States it
is more apt to be found in marshy
flats bordering bays and rivers, and
along the seashore.

Length, 20–27; wing, 17 (15½–18¾);
tail, 9½; tarsus, 2; culmen, 1½. Arctic
regions of the northern hemisphere;
breeding north of the United States,
migrating in winter to the Middle States
and straggling to South Carolina and
Texas.

Snowy Owl

10. American Hawk Owl (377ᵃ. *Súrnia ùlula cáparoch*). — A medium-sized, day-flying, much-mottled, grayish-brown owl, with a long and rounded tail and no ear tufts. The breast and belly are regularly barred with reddish-black upon a white ground; across the upper breast there is a more or less perfect dark band. The upper parts of the back and head are marked with round white spots. This is as much a day-flying bird as any of

American Hawk Owl

the hawks, and its shrill cry is frequently uttered while on the wing. Its perch is usually on some dead-topped tree in an open place. (Day Owl.)

Length, 14½–17½; wing, 9; tail, 7; tarsus, 1; culmen, 1¼. Arctic America; breeding north of the United States, and wintering south to the northern border states.

11. Burrowing Owl (378. *Speótyto cunicu-lària hypogàa*). — A small, burrowing, day-flying, grayish-brown owl, without ear tufts, but with excessively long legs that are nearly bare of feathers. The brown of the back is both spot-

Burrowing Owl

ted and barred with whitish, and the buff-colored under parts are barred with grayish-brown except on the throat. This bird lives in the deserted holes or burrows of "prairie dogs" and other quadrupeds, including foxes and badgers.

Length, 10; wing, $6\frac{1}{2}$ ($5\frac{3}{4}$–$7\frac{1}{4}$); tail, $3\frac{1}{4}$; tarsus, $1\frac{3}{4}$; culmen, $\frac{5}{8}$. Western United States, including the Great Plains, north to southern British America and south to Central America. Accidental in New York and Massachusetts. The **Florida Burrowing Owl** (378[a]. *S. c. floridana*) of southern Florida and the Bahamas is much like the last, but averages slightly smaller; the legs are even more nearly free from feathers; the lower parts have a more purely white ground color, and the upper parts a sepia-brown with pure white dottings.

12. **Ferruginous Pygmy Owl** (380. *Glaucidium phalænoïdes.* —A very small, southwestern, olive-brown, or reddish owl, without ear tufts, and with a chestnut-red or white tail crossed by about eight blackish-brown bars. The top of the head is sharply streaked with whitish. The olive-brown or reddish of the back is mainly free from markings excepting the shoulders, which have large, round, white spots. This small species, like the screech owl, is found in some varieties,—a grayish-brown, a red one and some that are intermediate in color. Sometimes the red is so intense and uniform as to destroy the barring of the tail and the wings. (Ferruginous Gnome Owl.)

Ferruginous Pygmy Owl

Length, $6\frac{3}{4}$; wing, 4 ($3\frac{1}{2}$–$4\frac{1}{2}$); tail, 3; tarsus, $\frac{3}{4}$; culmen, $\frac{3}{8}$. Texas to Arizona and south to southern Brazil.

13. **Elf Owl** (381. *Micropállas whitneyi*).—A very small, western, mottled, and grayish or grayish-brown owl, without ear tufts and with a white or whitish, more or less interrupted, collar around the neck. The lower parts are white, with more

or less longitudinal blotches of brownish or rusty color. The tarsus is longer than the middle toe, very bristly with hairs in front, and naked behind.

Length, 6; wing, 4¼ (4–4½); tail, 2¼; tarsus, ⅞; culmen, ₁⁷₆. Southern Texas to southern California, and south to Mexico.

FAMILY XXVIII. BARN OWLS (STRÍGIDÆ)

A very small family (8 species) of owls, with triangular-shaped eye disks, a saw-toothed nail to the middle toe, and very downy plumage. Our only species is so nocturnal in its habits, and in the daytime so well able to hide from obser-

American Barn Owl

vation, that, though not rare, it is seldom seen. The peculiar form of face, due to the eye disks, gives it somewhat the appearance of a monkey.

1. **American Barn Owl** (365. *Strix pratincola*). — A large night-flying, monkey-faced, black-eyed, brownish owl, with fine mottlings of white and black and no ear tufts. It has been said to appear like a closely hooded, toothless old woman with a hooked nose. Its food consists almost entirely of mice and other small mammals.

Length, 15–21; wing, 13½ (12½–14); tail, 6½; tarsus, 2¾; culmen, 1¼. United States, more abundant south of New York; breeding from Pennsylvania southward, and very rare in southern New England. Not migratory.

FAMILY XXIX. HAWKS, EAGLES, VULTURES, ETC. (FALCÓNIDÆ)

This is the largest family (350 species) of the birds of prey (*Raptores*), and representatives are found in all lands. The American species can be naturally divided into seven groups,

under which divisions the peculiarities will here be given.
(1) **Kites.** Slender, graceful, small birds of prey with very
long, pointed wings, often forked tails, and slender, weak
bills.[1] They are particularly birds of the air, and in their
method of flying remind one of swallows by their grace and
ease of movement. Nos. 1–4. (2) **Harriers.** Hawks with long
legs, unnotched, lengthened bills,[2] long wings and tail, and
slender form, having the feathers radiating around the eyes,
ruff-like, somewhat imitating the owls. No. 5. (3) **Hawks.**
A large group of medium to small birds of prey with short,
stout bills,[3] long, nearly square tails, and long bare legs. They
have wonderful powers of flight, and rarely fail to capture
their prey, which consists chiefly of small quadrupeds and
birds. Nos. 6–8. (4) **Buzzards and Eagles.** A large group of
medium to large-sized birds of prey with heavy bodies and
mainly toothless bills. They are inferior in power of flight
to the hawks or falcons, and usually capture their prey by
stealth rather than by open fight like most other birds of
the family. In many species the tarsus is more or less feathered; in some, the feathering extends to the toes. Buzzards,
Nos. 9–20; Eagles, 21–23. (5) **Falcons.** A large group of medium to small, but strong, birds of prey with toothed bills,[4]
long, strong wings, short, stiff, rounded tails, short legs, and
stout nails (talons) to the strong toes. This is the most typical
group of the family, and includes the bravest and most daring
of birds. They capture their prey with the most sudden and
violent of movements. It includes our smallest species. Nos.
24–32. (6) **Caracaras.** A small group of sub-tropical, vulture-
like, sluggish, mainly terrestrial birds of prey, with short,
toothless bills, long necks, and fully feathered heads. No. 33.
(7) **Osprey.** This probably consists of but one species, of world-
wide distribution,—namely, the well-known fish hawk or osprey,

1   2   3   4

of which the peculiarities are given in the specific description.
No. 34.

Practically, in this whole family the female is the larger and
stronger bird. Where dimensions are given, the smaller numbers refer to the male and the larger to the female.

Key to the Species

* Nails (talons) all of the same length, narrowed and rounded on the
  lower side ; wing, 17–22 long ; scales of the tarsus small, rounded [1]..
  ....33. **American Osprey.**
* Nails of graduated length, the hind one longest, the outer shortest. (**A.**)
  A. Tarsus densely feathered all around and down to the toes ; wing,
  22–28 long ....21. **Golden Eagle.**
  A. Tarsus feathered to the toes in front but with a bare strip behind ;
  wing, 15–20 long. (**P.**)
  A. Tarsus bare for at least one third of its length. (**B.**)
  B. Wing over 19 long ; tail under 16 long ; head not crested...........
  ....23. **Bald Eagle.**
  B. Wing over 19 long ; tail over 16 long ; head conspicuously crested ;
  Texas...22. **Harpy Eagle.**
  B. Wing under 18 long. (**C.**)
  C. An extreme southern, ground bird, with the front of the tarsus
  covered with numerous rounded scales ; wing, $14\frac{1}{2}$–$16\frac{1}{4}$ long ;
  culmen, $1\frac{1}{4}$ or more long..............32. **Audubon's Caracara.**
  C. Tail deeply forked (6 inches or more) ; wing, 15–18 long........
  ....1. **Swallow-tailed Kite.**
  C. Wing, 7 or more times as long as the tarsus ; bill with no sharp
  teeth or notches ; nostril elongated and without inner bony tubercle ;
  tail at most but slightly notched. (**O.**)
  C. Wing about 7 times as long as the tarsus ; bill with a sharp notch
  and tooth back of the tip ; nostril circular and with an inner bony
  tubercle ; [2] wing over 11 long. (**N.**)
  C. Wing, 6 or less times as long as the tarsus. (**D.**)
  D. General plumage black, with almost no portions of lighter color
  except some bands on the tail. (**M.**)
  D. Upper tail coverts white ; base of tail white also in some of the
  species. (**L.**)
  D. With neither the general plumage black nor the upper tail coverts
  white. (**E.**)

E. Nostril circular and with a conspicuous central bony tubercle ; [2] upper mandible with a strong tooth and notch back of the hooked tip. (**J.**)

E. Nostril oval and the upper mandible without more than one lobe or tooth, and that not a strong one.[3] (**F.**)

F. Tail about ¾ as long as the wing. (**I.**)

F. Tail not over ¾ as long as the wing. (**G.**)

   G. Outer web of the primaries with white, buffy, or reddish spots ; four outer primaries notched on the inner web ;[4] wing, 7½–10 long.....
.....................................11. **Red-shouldered Hawk.**

   G. Under parts white, very slightly if at all streaked ; upper parts nearly black ; Florida...................16. **Short-tailed Hawk.**

   G. Not as above. (**H.**)

H. Four outer primaries notched on the inner web ;[4] tail generally quite red...................................10. **Red-tailed Hawk.**

H. Three outer primaries notched ;[5] wing, 14–18 long... ....
....14. **Swainson's Hawk.**

H. Three outer primaries notched ; wing, 9½–12 long..... ....
.....................................15. **Broad-winged Hawk.**

   I. Wing under 9 long ; tail square[6] ....6. **Sharp-skinned Hawk.**

   I. Wing, 8¾–11 long ; tail rounded[7] ....7. **Cooper's Hawk.**

   I. Wing, 11½–14½ long ....8. **American Goshawk.**

J. Wing, 11–17 long ; only one primary notched on the inner web.[8] (**N.**)

J. Wing, 9½–11½ long ; two primaries notched ; southern Texas........
.....................................30. **Alpomado Falcon.**

J. Wing, 5–9½ long ; two primaries notched. (**K.**)

   K. Back or belly with more or less of bright brownish-red..........
.........31. **American Sparrow Hawk or Cuban Sparrow Hawk.**

   K. Back bluish slate color, or blackish and without bright rufous....
.................28. **Pigeon Hawk or** 29. **Richardson's Merlin.**

L. Tail gray, barred with blackish ; wing, 13–16 long ; common.......
....5. **Marsh Harrier.**

L. Tail mainly white but much barred ; wing, 14–18 long ; Texas ......
.............................13. **Sennett's White-tailed Hawk.**

L. Tail zoned black and white ; extreme southwestern hawk with wing 9½–12 long...............................18. **Mexican Goshawk.**

   M. Base and tip of tail white ; shoulders and tibia chestnut ; wing, 12–15 long ; western ....9. **Harris's Hawk.**

   M. Three white bands across the tail at base, tip, and center ; Texas
.............12. **Zone-tailed Hawk or** 17. **Mexican Black Hawk.**

   M. Wing, 10–13 long ; Florida ; black phase of..16. **Short-tailed Hawk.**

5   6   7   8

M. Wing, 14½-17½; western; black phase of..14. **Swainson's Hawk.**
M. Wing, 13½-16½; northeastern; black variety under..............
  ....25. **Gray Gyrfalcon.**
N. Tarsus hardly at all feathered at the upper part....27. **Duck Hawk.**
N. Tarsus feathered less than half way down in front; back grayish-
  brown...................................26. **Prairie Falcon.**
N. Tarsus feathered over half way down in front and on the sides.....
  ....24. **White Gyrfalcon.** 25. **Gray Gyrfalcon.**
 O. Tail white without bars and square tipped; wing, 11-14 long.....
  ....2. **White-tailed Kite.**
 O. Tail white at base and whitish at tip; tail coverts above and below
  mainly white; upper mandible lengthened and hooked; culmen,
  1 or more long; wing, 12-16 long; Florida....4. **Everglade Kite.**
 O. Slaty-blue above, gray below; tail black, unbarred..............
  ....3. **Mississippi Kite.**
P. Legs bright brownish-red with black bars.20. **Ferruginous Rough-leg.**
P. Legs more or less buffy without brownish-red...................
  ....19. **American Rough-legged Hawk.**

1. Swallow-tailed Kite (327. *Elanoïdes forficatus*). — A beautiful, large, glossy, bluish-black kite, with the head, rump, and

under parts white,
and the tail deeply
forked. The neck and
under wing coverts
are also white. This
is a graceful bird,
generally seen on the
wing, where its move-
ments remind one of
those of a swallow.

Swallow-tailed Kite

It is remarkable in that it can drink as well as eat, while
coursing through the air. Common in the south.

Length, 20-25; wing, 16½ (15½-17¾); tail, 13½; tarsus, 1¼; culmen. 1.
Interior United States, west to the Great Plains, north to North Carolina
and Minnesota; casual to New England and Manitoba; breeding locally
throughout its regular range, and wintering in Central and South America.

2. White-tailed Kite (328. *Elanus leucurus*). — An ashy-
backed, white-headed, white-tailed, white-bellied kite, with the
wing coverts conspicuously black. The *young* have the whites

more or less marked with reddish-brown, and the tail with an ashy bar near the tip. The tarsus is feathered half way down in front, and the exposed portion is finely
reticulated. This is one
of the strongest of the
kites; its food consist-
ing of birds, quadru-
peds, reptiles, and in-
sects. Common in
marshy regions. west
of the Mississippi.
(Black-shouldered
Kite.)

White-tailed Kite

Length, 15-17; wing, 12½ (11½-13½); tail, 7; tarsus, 1¾; culmen, ⅞. Southern United States; breeding north to South Carolina and southern Illinois, and south throughout most of South America. Casual in Michigan. It winters south of our territory.

3. **Mississippi Kite** (329. *Ictinia mississippiénsis*). — A kite with slate-colored wings and back, light gray head. neck, and belly, and black, unbarred tail. The primaries are blotched with much chestnut. The *young* lacks the chestnut of the wings, has the head more or less streaked with black and white, and the tail marked with a few white, irregular bars.

Length, 13-15½; wing, 11¼ (10½-12½); tail, 6½; tarsus, 1½; culmen, ⅞. Southern United States east of the Rocky Mountains; breeding north to South Carolina, southern Illinois, and Kansas, and wintering in the tropics.

Mississippi Kite

4. **Everglade Kite** (330. *Rostrhámus sociábilis*). — A dark, slate-colored kite, with the upper tail coverts and the base of

the tail white; the tip of the tail is somewhat whitened and notched; the upper mandible is peculiarly lengthened and hooked. This bird acts much like a gull, flying over the shallow, fresh waters of southern Florida; it dives for snails, which form its main food. (Snail-hawk.)

Length, 17; wing, 14 (13-15); tail, 7½; tarsus, 2; culmen, 1½. Florida, Cuba, and eastern Mexico, south to the Argentine Republic.

5. **Marsh Hawk** (331. *Circus hudsònius*). — A large, common, ashy-colored or gray hawk, with white tail coverts, and white belly, spotted or barred with reddish. The primaries are blackish, and the tail is silvery-gray, irregularly barred with blackish. The *female* has a brownish back, head, and neck, darker primaries and tail, and the under parts more buffy, streaked on the belly with blackish. This is a common, low-flying hawk of the open country, easily determined by the white tail coverts. It may often be found perching on a low elevation, or even in the grass. (Marsh Harrier.)

Marsh Hawk

Length, 18-24; wing, 14½ (13-16); tail, 9½; tarsus, 3; culmen, 1 nearly. North America; breeding throughout, south to Panama.

6. **Sharp-shinned Hawk** (332. *Accipiter velox*). — A common, long, square-tailed, medium-sized, dark-brownish or slate-colored hawk, with much-barred, buffy under parts. The tail has blackish cross-bars and a white tip; the primaries are also barred with blackish. The *young* has brownish markings on

the back, and blackish streaks or spots on the whitish lower parts. This is a fearless, swift, low-flying hawk, living mainly on birds, including poultry.

Length, 10-14 ; wing, 7½ (6-9) ; tail, 5-8 ; tarsus, 2 ; culmen, ⅝. North America, south to Panama ; breeding throughout.

7. Cooper's Hawk (333. *Accipiter cooperii*). — A hawk similar to the last, but larger and with a decidedly rounded tail. It is a dark-brown hawk, with grayish- and brownish-spotted under parts. In habits and food, it is much like the sharp-shinned hawk.

Length, 14-20 ; wing, 9-11 ; tail, 7-10; tarsus, 2½ ; culmen, 1. North America, south to southern Mexico ; breeding throughout.

Sharp-shinned Hawk

8. American Goshawk (334. *Accipiter atricapillus*). — A large, dark, slate-colored hawk, with grayish, wavy bars on a white ground on all the lower parts. The head is blackish, and has a white line over the eye, and the throat and breast are s o m e w h a t streaked with blackish. This is one of the strongest and

Cooper's Hawk

American Goshawk

Harris's Hawk

most daring of all of the hawks, feeding upon birds and quadrupeds in about equal proportions.

Length, 20-26; wing, 12-14; tail, 9-13; tarsus, 3; culmen, 1¼. Northern and eastern North America; breeding north of the United States except in the higher mountains, and wintering south to the Middle States.

9. **Harris's Hawk** (335. *Parabuteo unicinctus harrisi*).—A large, southwestern, dark-brown hawk, with reddish shoulders and tibiæ. The tail has a white base and tip, the middle portions being unbarred. This is a sluggish, carrion-feeding bird, associating with buzzards, and having the loral region bare of feathers back to the eyes.

Length, 19-23; wing, 12½-14½; tail, 8½-11; tarsus, 3¼; culmen, 1¼. Mississippi and Texas to Lower California, south to Panama; breeding from southern Texas westward and southward.

10. **Red-tailed Hawk** (337. *Buteo borealis*). — A common, mottled, brownish hawk, with a bright, brick-red tail; the tail feathers are tipped with white, and have a dark bar near the tip; under parts nearly white, with many brownish streaks, especially on the upper breast. The *young* is similar, but has the tail crossed with many, more or less distinct, blackish

Red-tailed Hawk

bars. Four of the outer primaries are notched on the inner web, and the shoulders are not marked with reddish. Its food consists mainly of small quadrupeds, but it will not refuse birds, insects, or reptiles. (Hen Hawk; Chicken Hawk.)

Length, 19-25; wing, 13½-17½; tail, 8½-10½; tarsus, 3; culmen, 1½. North America from the Plains eastward, south to eastern Mexico; breeding about throughout. **Krider's Hawk** (337ª. *B. b. kridérii*) of Minnesota to Texas and westward (casual in Iowa and Illinois) is a light-colored form, pure white below and with the tail bar nearly lost. **Western Red-tail** (337ᵇ. *B. b. calùrus*) of North America, west of the Rocky Mountains (casual in Illinois), is a nearly evenly colored, dark chocolate-brown hawk, with the red tail crossed by several black bars. **Harlan's Hawk** (337ᵈ. *B. b. hárlani*) of the Gulf States (casually north to Pennsylvania, Iowa, and Kansas) is nearly uniform black, with the tail rather longitudinally mottled with dusky and white, and having more or less of the red tinge and the zone of black near the tip. (Black Warrior.)

11. Red-shouldered Hawk (339. *Búteo lineátus*). — A common hawk, with much brownish-red on head, shoulders, breast, and belly. The tail and primaries are black, with broad bars of white. The throat is streaked with blackish, and the breast and belly are much barred with white or whitish. The *young* is very different and hard to determine; above plain, dark brown, with little indication of the red shoulders; head, neck, and under parts are nearly white, fully streaked with dark brown; tail and wing quills brown, crossed with many indistinct, lighter and darker bars. Four primaries are notched on the inner web. This is a bird of well-watered woods, living on small quadrupeds, insects, and reptiles, in the order given. (Misapplied names: Hen Hawk; Chicken Hawk.)

Red-shouldered Hawk

Length, 17½-22; wing, 11¼-14¼; tail, 8-10; tarsus, 3; culmen, 1. North America from the Plains eastward, north to Manitoba and Nova Scotia, south to Mexico; breeding throughout. The **Florida Red-shouldered Hawk** (339ª. *B. l. álleni*) of South Carolina to Texas, mainly coastwise, is a smaller hawk, with a streaked, grayish-white head, grayish throat, indistinctly barred, buffy under parts and no red shoulders.

12. Zone-tailed Hawk (340. *Buteo abbreviatus*). — A south-western glossy-black to blackish-brown hawk; the tail has three slate-colored bands above, and three pure white ones below. The *young* has a grayish-brown tail crossed above with numerous oblique black bands, and showing below mainly the white inner webs. When disturbed, the feathers of this hawk show much white, especially on the head and breast, as much of the plumage is pure white at base.

Length, 18½–21½; wing, 15–17½; tail, 8½–10¾; tarsus, 2½; culmen, 1. Texas to southern California. south to northern South America.

13. Sennett's White-tailed Hawk (341. *Buteo albicaudatus sennetti*). — A Texas, ashy or lead-colored, short-tailed hawk, with the tail coverts. tail. and entire under parts white, and the wing coverts chestnut. The tail has numerous narrow, broken, zig-zag lines, and a broad black band near the tip. The *young* is a brownish-black bird, with a grayish tail, becoming darker near the tip. This hawk, like the last species, has white bases to many of the feathers, which show when the plumage is disturbed.

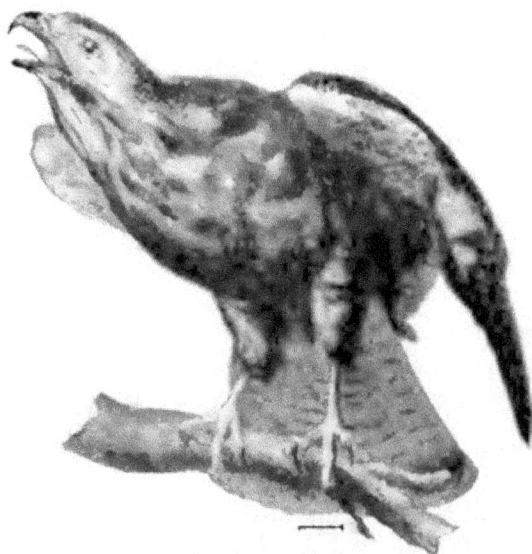

Swainson's Hawk

Length, 23; wing, 14½–18; tail, 7–10; tarsus, 3½; culmen, 1¾. Southern Texas and southward into Mexico.

14. Swainson's Hawk (342. *Buteo swainsoni*). — A western, dark-colored hawk, very variable in color. but usually with conspicuous dark patches on the sides of breast. and many (8–12) dark tail bars. The breast has a large, cinnamon-red patch; the primaries are unbarred, and the three outer ones are notched on the inner web; the belly is much blotched and barred with blackish, whitish, and buffy markings. There is a very dark (melanistic) form of this bird, in which the whole plumage is evenly blackish. Of course there are birds of intermediate colors. The size and the peculiar primaries as above given are distinct enough to fix the species.

Length, 19-22; wing, 14½-17¼; tail, 8-10; tarsus, 2½; culmen, ⅞. Western North America from Wisconsin, Arkansas, and Texas to the Pacific, north to the Arctic regions, and south to the Argentine Republic; breeding throughout its North American range. Casual to Massachusetts and Maine.

15. Broad-winged Hawk (343. *Buteo latissimus*). — A dark-colored hawk, with grayish tail, crossed by two broad, dark bars; under parts brownish, heavily barred. The primaries are without reddish markings, and the three outer ones are notched on the inner web. The *young* has a grayish-brown tail, crossed by three to five indistinct black bars, but has the narrow whitish tip of the adult. A sluggish, unsuspicious hawk, feeding on insects, small mammals, batrachians, and reptiles.

Broad-winged Hawk

Length, 13-17; wing, 10-11½; tail, 6½-8; tarsus, 2½; culmen, ¾. Eastern North America north to New Brunswick, south to northern South America; breeding throughout its United States range.

16. Short-tailed Hawk (344. *Buteo brachyurus*). — A rare, Florida, slaty-gray to grayish-brown hawk, with all under parts pure white, except some brownish markings on the sides of the breast. The grayish tail is barred with black and narrowly tipped with white. The *young* has the under parts washed with buffy. This species, like No. 14, is found in a very dark (melanistic) phase, in which nearly the whole plumage is a somewhat glossy black.

Length, 15–18 ; wing, 10½–13 ; tail, 6–7½, tarsus, 2¼ ; culmen, ¾. South America north to Florida, where it breeds.

17. Mexican Black Hawk (345. *Urubitinga anthracina*). — A southern Texas, coal-black hawk, with a central broad white band across the tail, and a white tip: the ends of the upper tail coverts are also white. The *young* is mottled blackish-brown above and streaked buffy below ; the tail is crossed with about seven bands of blackish and grayish.

Length, 21–23 ; wing, 13–16 ; tail, 8–11 ; tarsus, 3¼ ; culmen, 1½. Northern South America north to southern Texas.

18. Mexican Goshawk (346. *Asturina plagiata*). — A southwestern, ashy-backed, white-bellied hawk, with a black tail crossed by several somewhat broken white bands and a white or whitish tip. The white belly and breast are beautifully and finely barred with dark lines. The *young* is blackish-brown above and whitish below, much mottled with reddish above and blackish below ; tail, like the back, crossed with numerous blackish bars.

Mexican Goshawk

Length, 17 ; wing, 9½–11½ ; tail, 7–8 ; tarsus, 2¾ ; culmen, ¾. Southwestern border of the United States, south to Panama. Once seen in Illinois.

19. American Rough-legged Hawk (347ᵃ. *Archibuteo lagopus sancti-johannis*). — A large, dark-brownish hawk, with rough, feathered legs, and the under parts spotted with black and buffy. The basal half of the tail is almost white and the rest very dark. but usually showing two or three grayish bars. The spotted under parts form a dark band across the belly. This rather sluggish, low-flying. almost exclusively mouse-eating hawk. is more nocturnal in its habits than any other of our species.

Length, 19-23; wing, 16-18; tail, 9-11; tarsus, 2⅜; culmen, 1⅛. Northern North America; breeding north of the United States, and wintering south to Virginia.

20. Ferruginous Rough-leg (348. *Archibuteo ferrugineus*). — A large. western, somewhat mottled, brownish-red hawk, with the under parts white,

American Rough-legged Hawk

much barred with rufous across the belly. The tail is grayish-white tinged with rufous. The *young* is more grayish-brown, with the base of tail white. This is a hawk of the open prairies west of the Mississippi.

Length, 21-25; wing, 16-19; tail, 9-11; tarsus, 2⅞; culmen, 1⅛. Western North America from North Dakota to Texas, and west to the Pacific; breeding from Utah northward; casually east to Illinois.

21. Golden Eagle (349. *Aquila chrysaëtos*). — A very large blackish-brown bird. with lighter, almost golden, back head and back neck; base of the tail for more than half its length is white, and the tarsus is white-feathered to the toes. The *young* is blacker in general plumage, and the base of the tail is more or less banded with

Ferruginous Rough-leg

grayish bars. The food consists of the larger mammals and birds, though carrion also is eaten: rabbits, lambs, turkeys, and ducks are prey for this bird. Rare east of the Mississippi.

Length, 30–40; wing, 23–27; tail, 15; tarsus, 4; culmen, 2. Northern portions of Old and New Worlds, south in America to Mexico; breeding, practically, only in the mountains of sparsely settled regions.

22. **Harpy Eagle** (350. *Thrasaëtos harpyia*). — A rare Texas eagle, with the back ashy-gray, mottled with glossy black, and the belly white, more or less blotched with ashy. The head and neck are grayish, darker on the crown, and whiter on the throat. The tail is more or less irregularly banded with black and ashy. The *young* has the head, neck, and entire lower parts white, with ashy gray on crown and breast.

Golden Eagle

Length, 33–40; wing, 21–25; tail, 16–19; tarsus, 4¾; culmen, 2¾. South America, north to southern Texas and possibly Louisiana.

23. **Bald Eagle** (352. *Haliæetus* (ä-e-tus) *leucocéphalus*). — A very large, dark-colored eagle, with white head, neck, and tail. This adult condition is not reached till the third year; before this, the whole plumage is nearly black, but white mottlings gradually appear on the portions that finally become entirely white. The lower part of the tarsus is bare of feathers and is covered with numerous rounded scales. This eagle is seldom found far from water, as its food consists principally of fish and ducks; dead fish thrown on the shore, fish stolen from the fish hawk, or, if the need is very great, fish captured from the water by its own exertions.

Length, 30–43; wing, 20–27; tail, 11–15; tarsus, 3¼; culmen, 2¼. North America, south to Mexico; breeding locally throughout.

24. White Gyrfalcon (353. *Falco islándus*). — This arctic falcon, which has been found in northern Maine, is, in good plumage, a slightly mottled white all over. There is apt to be more or less of brownish and grayish in bars or streaks on the shoulders, central tail feathers, and head.

Length, 22; wing, 16; tail, 9½; tarsus, 2¼; culmen, 1. Arctic regions, wandering south to northern Maine.

25. Gray Gyrfalcon (354. *Falco rusticolus*). - A northern falcon, with the upper parts, including the tail, blackish-gray, barred with buffy-white, and the under parts white, blotched and streaked with blackish, but the under tail coverts are barred with brownish.

Length, 20-25; wing, 13½-16½; tail, 8-10; tarsus, 2½; culmen, 1. Arctic regions, straggling south in winter to the northern United States. The Gyrfalcon (354ª. *F. r. gyrfálco*) of the Arctic regions has been found as far south in winter as Rhode Island. It is similar to the gray gyrfalcon, but lacks the regular bars of the upper parts; sometimes there are no bars at all; the under parts are always heavily streaked with blackish. The **Black Gyrfalcon** (354ᵇ. *F. r. obsoletus*) of Labrador is casually found as far south as Long Island. This, as its common name indicates, is a very dark-colored gyrfalcon. The upper and lower parts are an unbarred slaty-black; even the tail is nearly unbarred. All the gyrfalcons are rare in the United States.

Gray Gyrfalcon

26. Prairie Falcon (355. *Falco mexicánus*). — A bold, graceful, low-flying, western, grayish-brown falcon, with the lower parts white, streaked and spotted with the color of the back. The primaries and the inner webs of all but the middle tail feathers are blotched or barred with buffy. The *young* has buffy margins to the feathers of the upper parts. This, as its common name indicates, is a bird of the plains west of the Mississippi.

Length, 16-20; wing, 12-14¼; tail, 6½-9; tarsus, 2; culmen, 1. Western United States from the eastern border of the Plains to the Pacific; breeding throughout. Casual east to Illinois.

27. Duck Hawk (356. *Falco peregrinus anatum*). — A dark, bluish-slate-colored hawk, with the under parts cream-buff, much spotted with black, except on the breast; tail indistinctly barred with blackish and tipped with a narrow, white band. The *young* has the blackish upper parts margined with orange-buffy, and the under side of the tail barred with the same. This is a beautiful, swift-flying, daring bird generally found near the water, as it feeds mainly on ducks and other water birds. No bird can fly swiftly enough to escape its talons. (Peregrine Falcon; Great-footed Hawk.)

Prairie Falcon

Length, 14-20; wing, 11½-15; tail, 6-9; tarsus, 2; culmen, 1. North America, and south to central South America; breeding locally over most of its United States range.

28. Pigeon Hawk (357. *Falco columbarius*). — A small, slate-blue hawk, with all the under parts light creamy or brownish, much streaked with dark;

Duck Hawk

tail with three or four broad, lighter-colored bars, and the neck usually with a rusty collar. The primaries are barred with white. The *young* has the upper parts blackish and the barring of the primaries reddish-yellow. The pigeon hawk has a resemblance to the wild pigeon both when perching and when in flight. It is to be found in the open country, near the edge of woods, especially where there are large bodies of water. Its food consists mainly of small birds and insects. (American Merlin.)

Length, 10-13; wing, 7½-8½; tail, 5¼; tarsus, 1¾; culmen, ½ or more. North America; breeding north of the United States, and wintering through most of the states and into northern South America.

29. Richardson's Merlin (358. *Falco richardsonii*). — A western, very small, bluish - gray - backed hawk, with the lower parts including the front of the head whitish, much streaked with brown to black, especially on the breast and sides. The chin and throat are about the only portions without any shaft streaks on the feathers; even the brown back is so marked with black. Tail with five blackish, five grayish, and one terminal white band. The *female* has the back more earthy-brown in color, and the outer webs of the quills marked with buffy spots (the *male* has these spots light-grayish.) (Richardson's Pigeon Hawk.)

Length, 10-13½; wing, 7½-9½; tail, 4½-6¼; tarsus, 1½; culmen, ½ or more. North America from the Mississippi to the Pacific, north into the British Possessions, and south to Texas and probably Mexico.

30. Aplomado Falcon (359. *Falco fusco-coeruléscens*). — A Texas, medium-sized, heavy-billed, lead-colored falcon, with the chin, throat, and breast unspotted white; sides and a

Pigeon Hawk

broad belly band blackish-barred, and the thighs and lower tail coverts rusty or reddish-brown; tail tipped with white and crossed by about eight narrow, white bands. The *young* is duller colored, with the back inclined to brownish.

Length, 15-18; wing, 9¼-11½; tail, 6¼-8¾; tarsus, 1⅜; culmen, 1. South America north to southern Texas and Arizona.

31. American Sparrow Hawk (360. *Falco sparvèrius*). — A common, beautiful, little hawk, with much chestnut on back and

American Sparrow Hawk

tail, and usually on crown also. The wings are slaty-blue, with black and white barred primaries, and the tail has a black band near the white tip. The white cheek has a black patch both in front and behind it. The under parts are buffy, very heavily streaked with darker in the *female*. The wing coverts are slaty-blue in the *male*, and chestnut, barred with black, in the *female*. With almost all other hawks the male is much the smaller bird, but in this species there is but little, if any, variation in size. Generally the sexes are colored alike, but in this case there is a decided difference in markings. This is an insect-eating hawk, though mice and small birds form part of its diet. (Rusty-crowned Falcon; Killy Hawk.)

Length, 8¾-12; wing, 6½-8; tail, 4½-6; tarsus, 1¼; culmen, ½. North America from the Rocky Mountains eastward; breeding from the Gulf States to Hudson Bay, and wintering from New Jersey southward. The **Cuban Sparrow Hawk** (361. *Falco dominicénsis*), which has been found casually in southern Florida, has the rufous coloring only on the breast and neighboring lower parts; the *female* and *young male* have a touch of the same tint on the back. The so-called "mustache" stripe, which is so plain on the cheek of the American sparrow hawk, is hardly to be noticed in the Cuban species. The Cuban bird has a conspicuous white line over the eye, wanting in the other. There is a color phase of the Cuban sparrow hawk, in which the usual rufous coloring of the under parts is lacking.

32. **Audubon's Caracara** (362. *Polyborus chériway*). — An extreme southern, large, dark-colored bird, strong in flight, with bare, red skin on face, buffy neck and breast, and white tail.

Audubon's Caracara

tipped and barred with black. These birds associate with the buzzards and vultures, feeding on carrion as they do, but in their flight there is no resemblance. Besides the carrion, they eat many kinds of reptiles, which they capture for themselves.

Length, 20–25; wing, 14½–16½; tail, 8–10; tarsus, 3¼; culmen, 1⅔. Florida, Texas, and Arizona, south to northern South America; breeding in all sections of the United States where found.

33. **American Osprey** (364. *Pandion haliaëtus carolinénsis*). — A large, blackish-backed, white-bellied bird, with much white on top of head and upper neck. Tail with six to eight obscure bands, more distinct below. It is seen flying slowly over the water of our coasts.

American Osprey

watching for the fish which form its only food. When its prey is seen, it closes its wings and drops with wonderful velocity into the water, and generally it secures the fish observed. Its food is usually eaten while the bird is perched on some favorite tree in the vicinity of its fishing grounds. These birds live in colonies of greater or less size, and return each year to their old nesting place. (Fish Hawk.)

Length, 20–25; wing, 17–21; tail, 7–10; tarsus, 2¼; culmen, 1¼. North America and northern South America; breeding throughout its North American range, and wintering along the South Atlantic States and southward.

FAMILY XXX. AMERICAN VULTURES (CATHÁRTIDÆ)

A small family (8 species) of New World vultures of large size, living upon decaying flesh, and having the head and much of the neck bare of feathers.[1] Our species are in size and appearance much like turkeys. The bill is more lengthened and weaker than in the other families of birds of prey (*Raptores*), and the feathers are very dark and dull colored. In all the southern states these birds can usually be seen sailing in great circles in the air.

1. **Turkey Vulture** (325. *Cathártes aúra*). —A very large black bird, with bare neck and head, seen abundantly in the Southern States, soaring in graceful circles with outstretched wings, throughout the day. During life the skin

Turkey Vulture

of the head and neck, and the base of the bill are bright red.
The tail rounded and the nostril large and broad. The edges of
the glossy-black feathers are brownish. This is a very useful
bird, as its only food is dead and decaying animal matter.
In southern towns this and the next species are depended upon
to keep the streets free from carrion. (Turkey Buzzard.)

Length, 26-32; wing, 20-24; tail, 10½-12; tarsus, 2¼; culmen, 2¼.
Temperate North America (and all of South America) from New Jersey,
Ohio, and British Columbia south to Patagonia; breeding and wintering
about throughout.

2. **Black Vulture** (326. *Catharista atrata*). — A bird similar
to the last, but smaller, stouter, and blacker; the bare skin
of head and neck and
base of bill is also
blackish. Its heavier
weight and shorter
wings make it more
labored in flight, so
the flapping of the
wings is more fre-
quent. This differ-
ence in flying. the rel-
atively short, square
tail, the silvery under
surface of the wing
quills and the small
and narrow nostril
will enable any one

Black Vulture

to distinguish this bird from the last. The black vulture is
much more common near the seacoast, and decidedly more
abundant in cities and towns.

Length, 22-27; wing, 16½-17½; tail, 7½-8½; tarsus, 3; culmen, 2¼.
South Atlantic and Gulf States, and southward throughout most of South
America; breeding in the United States from North Carolina to Texas,
northward in the Mississippi Valley to Illinois and Kansas, and straggling
to New England and South Dakota.

ORDER VII. PIGEONS, ETC. (COLÚMBÆ)

An order represented, in our region, only by the following:

FAMILY XXXI. PIGEONS (COLÚMBIDÆ)

This large family (300 species) of land birds, found in the warmer regions of all portions of the earth, is represented in the eastern United States by but few species, only four being found north of southern Texas and southern Florida. They cannot be said to frequent any particular kind of haunt; many live most of the time on the ground, some are tree birds, some seek open places, while others are to be found only in forests. They are short-billed, small, round-headed, plump-bodied, short-legged, smooth-plumaged birds, with a peculiar, more or less iridescent, grayish and brownish coloration. In one way they are very different in habit from other birds; they hold the bill in the water till they finish drinking, instead of raising the head at each mouthful. Most species produce a whistling sound of the wings while in flight.

Key to the Species

* Wings, 7-9 long. (D.)
* Wings, 5-7 long. (A.)
* Wings, 3-4 long; tail shorter than the wings, 2½-3 long..............
 ....8. Ground **Dove.**
* Wings, 3-4 long; tail longer than the wings, 3½-4½ long. 9. Inca Dove.
 A. Tail about the length of the wings, 5¼ or more. 4. **Mourning Dove.**
 A. Tail nearly two inches shorter than the wings; southern doves, mainly of Florida and Texas. (**B.**)
 B. Forehead white, changing to bluish-gray on the crown ....
 .... 6. **White-fronted Dove.**
 B. Forehead not white. (**C.**)
  C. A conspicuous white patch on the wing coverts....................
 ....7. **White-winged Dove.**
  C. No white wing patch or white stripe under the eye ....
 ....5. **Zenaida Dove.**
  C. No white wing patch, but a broad white band under the eye ......
 ....10 and 11. Quail Doves.
 D. Tail as long as the wings................ ......3. **Passenger Pigeon.**

D. Tail an inch and a half shorter than the wings. (**E.**)
  E. Neck all around of the same color and without metallic gloss......
    ....1. **Red-billed Pigeon.**
  E. Top of head white or pale buffy; hind neck with a cape of metallic
    bronze; each feather of the cape edged with velvety black.......
    ....2. **White-crowned Pigeon.**

1. **Red-billed Pigeon** (313. *Colúmba flaviróstris*). — A dark,
richly colored pigeon of Texas, with the head, neck, and breast
a purplish wine-color, and the back olive-brown with a bronzy
gloss. Other portions of the body more or less slate-colored.
Tail rounded and without white tips to its feathers. Base of
bill red in life.

Length, 14; wing, 7¾; tail, 5½; tarsus, ?; culmen, only ½, because of
the curious extension of the frontal feathers. Arizona to Texas, and
southward to Central America.

2. **White-crowned Pigeon** (314. *Colúmba leucocéphala*). — A
large, rare, southern, rich-slate-colored pigeon, with a white
crown (pale buffy on the *female*), and greenish, metallic reflections on the hind neck. The feathers of this "cape" are edged
with velvety black, and have a bronzy luster.

Length, 12–14; wing, 7½ (7–7¾); tail, 5½; culmen, ¾. Southern
Florida, West Indies, and coast of Honduras.

3. **Passenger Pigeon** (315. *Ectopístes migratórius*). — A large,
long-tailed, slate-blue-backed pigeon, with the lower parts
chestnut-colored to-
ward the chin, and
whitish toward the
tail. Tail pointed,
and the outer (un-
der) feathers with
much white; sides
of the neck with a
purplish iridescence.
The *female* has the

Passenger Pigeon

upper parts less iridescent, and the lower parts decidedly
grayish. Probably the largest number of birds of any kind

ever seen together were in the flocks of passenger pigeons early
in the nineteenth century; single flocks were carefully esti-
mated, and declared to contain more birds than there are
human inhabitants on the whole earth. Now at the close of
the century they are practically extinct. (Wild Pigeon.)

Length, 15-17; wing, 8¼ (7¼-8½); tail, 8¼; culmen ⅜. North America
from the Great Plains eastward and north to Hudson Bay; breeding now
only along the northern border of the United States and in Canada.
Stragglers have been found as far west as Washington.

4. **Mourning Dove** (316. *Zenaidùra macroùra*). — A very common, pointed-tailed, brownish-backed, ground dove, with brownish to yellow or buff
under parts. The
sides of neck are
brightly iridescent,
with a small, black
mark below the ear.
Tail feathers with a
black bar, and the
outer (under) ones
tipped with white.
This species resem-
bles the last in ap-
pearance, but is much smaller. During the breeding season,
these birds are usually in single pairs in open woodlands.
Later in the season they are to be found in grain fields in
flocks, sometimes of great size. The peculiarly sad *coo-o-coo-
o-oing* of the male has led to the application of the common
name. (Wild Dove; Turtle Dove.)

Mourning Dove

Length, 11-13; wing, 5¾; tail, 5¾; tarsus, ⅞; culmen, ½. Temperate
North America; breeding from southern Canada southward, and wintering
from southern Pennsylvania to Panama.

5. **Zenaida Dove** (317. *Zenaida zenaida*). — A rare, extreme
southern, short-square-tailed, olive-brown-backed, reddish-bel-
lied dove, with the secondary wing quills tipped with white,
and the outer tail feathers having a black band near the ashy

tips. The neck has a metallic iridescence, and a velvety black spot on the sides. Though the bird is often found on the branches of trees, it spends most of its time on the ground.

Length, $10\frac{1}{2}$; wing, $6\frac{1}{4}$; tail, $4\frac{1}{4}$; culmen, $\frac{5}{8}$. Florida Keys, Bahamas, West Indies, and coast of Yucatan.

6. **White-fronted Dove** (318. *Leptótila fulciventris brachýptera*). — An extreme southern, large, silky, brownish-olive-backed dove, with much of the head and neck iridescently coppery-purplish, but the forehead white, and the top of the head bluish with a "bloom." Belly and chin are pure white, fore breast wine-color, and other under parts more or less shaded with the tint of the back. The outer (under) tail feathers are slate-colored, tipped with white.

Length, 12; wing, $6\frac{1}{4}$; tail, $4\frac{3}{4}$; tarsus, $1\frac{1}{4}$; culmen, $\frac{5}{8}$. Mexico and Central America, north to southern Texas.

7. **White-winged Dove** (319. *Melopèlia leucóptera*). — An extreme southwestern, common, generally bluish-ashy dove, with a large white blotch on the wings, which are formed of the wing coverts and the tips of the secondary quills. The middle tail feathers are much like those of the back, but the outer (under) ones are slaty, with conspicuous white tips. The sides of the head and neck are iridescent with golden-green, and marked with a steel-blue spot. The wing quills are mainly black, but somewhat white-edged.

Length, 12; wing, $6\frac{1}{2}$ ($6\frac{1}{4}$–$6\frac{3}{4}$); tail, $4\frac{3}{4}$; tarsus, $\frac{3}{4}$; culmen, $\frac{7}{8}$. Southern border of the United States, Florida, Texas, Arizona, and southward to Central America and the West Indies; straggling north to Colorado.

8. **Ground Dove** (320. *Columbigallìna passerìna terréstris*). — A common, very small, southern, ground-living, grayish-olive-backed, purplish-red-bellied dove, with a gloss of blue on the head and neck. *Female* grayish below instead of purplish. This fearless bird can be found almost everywhere in the south, from city streets to dense pine growths, but is more common near the coast.

Length, 6¾; wing, 3½; tail, 2¾; tarsus, ⅝; culmen, ¼. South Atlantic and Gulf States, West Indies, and northern South America; breeding from South Carolina to Louisiana.

9. **Inca Dove** (321. *Scardafélla Inca*). — A Texas dove, with a peculiar scaled appearance due to the crescent-shaped black marks on most of the feathers, especially abundant on the belly. The upper parts are grayish-brown, and the lower parts ashy-lilac in front and ochraceous at the back. There is much rich chestnut on the wings; the middle tail feathers are like those of the back, but the outer (under) ones are blackish, with white tips. (Scaled Dove.)

Length, 8; wing, 3¾; tail, 4; tarsus, ½; culmen, nearly ¼. Mexico, north to Texas and Arizona, and south to Central America.

Ground Dove

10. **Key West Quail-Dove** (322. *Geotrygon chrysia*). — A rare, Florida, very iridescent, wine-red dove, with the under parts lighter and more creamy, and, toward the tail, white. A plain white band below the eyes. This is a ground dove found in wooded regions.

Length, 11; wing, 6¼; tail, 5; culmen, ¼. A West Indian dove, found on the Florida Keys in the summer.

11. **Blue-headed Quail-Dove** (323. *Starnönas cyanocéphala*). — A rare, Florida, blue-crowned, black-throated, chocolate-backed, cream-buff-bellied dove, with a white line beneath the eye. This quail-dove is much like the last, both in habits and

appearance, and they both get their common name from the fact that in form they resemble the quail. They have short, broad tails, without white tips to the under feathers.

Length, 11 ; wing, 5½ ; tail, 4¼ ; culmen, ½. Cuba and accidentally on the Florida Keys.

ORDER VIII. GALLINACEOUS BIRDS (GALLÌNÆ)

The birds of this order derive their name from their characteristic habit of scratching the ground in search of food, which trait is almost exclusively confined to them. Nearly all of our representatives belong to the Grouse Family.

FAMILY XXXII. CURASSOWS (CRÁCIDÆ)

This small family (15 species) of tropical American birds is represented, in southern Texas, by the following:

1. **Chachalaca** (311. *Órtalis cétula maccálli*). — A crested, long-tailed, large, slender, generally olive-green-colored, ground bird, with naked sides to the head, and naked stripes on the chin. The tail is a bright lustrous green, and the under parts are least bright and least green. The outer (under) tail feathers are tipped with whitish. A peculiar bird, easily domesticated, and very noisy in the breeding season, with notes which are expressed in its name.

Length, 20-24 ; wing, 8 (7½-9) ; tail, 9-11 ; tarsus, 2 ; culmen, ⅞. Mexico and Central America, north to southern Texas.

FAMILY XXXIII. PHEASANTS, TURKEYS, ETC. (PHASIÁNIDÆ)

This family (nearly 100 species) of Old World fowl includes all our birds of the barnyard, except the ducks, the geese, and the pigeons. They have one distinctive difference from the members of the next family in that the *males* have spurs on their legs. Our only native species is the following:

1. **Wild Turkey** (310. *Meleágris gallopávo*). — A very large, broad-tailed, lustrous-plumaged, game bird, with head and

upper neck bare of feathers, and with more or less of erectile processes on the bare portions. The breast is furnished with a tuft of hair-like feathers, and the tail feathers and upper tail coverts are tipped with chestnut. The common domestic turkey has white tips to the tail feathers and tail coverts. This noble game bird of wooded regions is the original of the domestic race, and is becoming each year more rare. This wild species is divided into four varieties, the form given and the three following.

Length, 40–50; wing, 18–22; tail, 16–19. United States from Chesapeake Bay southward to the Gulf of Mexico and westward to the Plains. The **Mexican Turkey** (310ª. *M. g. mexicana*) of the southwest, from Texas to Arizona, and southward into Mexico, has the upper tail coverts tipped with buffy white. The **Florida Wild Turkey** (310ᵇ. *M. g. osceola*) of southern Florida is a smaller, and darker bird. The primaries are much less regularly barred with white. The **Rio Grande Turkey** (310ᶜ. *M. g. ellioti*) of the lowlands of Texas and northeastern Mexico can be distinguished from all the others by the dark buff edgings on the tail, and upper and lower tail coverts, in contrast with the white on the same parts of the Mexican turkey, and the deep, dark, reddish-chestnut of the common wild turkey. The brilliantly colored **European** and **Ring-necked Pheasants** have been introduced and more or less acclimated in several of the states. The *males* are 36 long and the *females* 25 long. The *females* are plainly colored. As most of the specimens are hybrid forms descriptions would be of little value.

Wild Turkey

FAMILY XXXIV. GROUSE, PARTRIDGES, ETC. (TETRA-ÓNIDÆ)

A large family (100 or more species) of game birds of all countries, living almost entirely on the ground, and having mainly brown and gray colors. They have the habit of hiding rather than flying to escape the gunners, and, if it were not for the ability of dogs to detect their presence, they would generally escape the fowler's shot. When they fly, their flight is rapid, accompanied by a whirring noise caused by the beating of their small, concave wings. Like the hens of the barnyard they scratch the ground to obtain their food, which consists of worms, insects, seeds, etc. They are generally large birds with short bills, heavy bodies, short, more or less feathered legs, and, in many species, rather long tails. (For European Pheasants, see p. 220.)

Key to the Species

* Wing, 10 or more long; tail stiff and pointed and about as long as the wing...11. **Sage Grouse.**
* Wing, 8–10 long. (**C.**)
* Wing, 6–8 long. (**B.**)
* Wing, 4–6 long; tarsus bare of feathers. (**A.**)
  A. Tail less than an inch shorter than the wing; Texas...2. **Scaled Partridge.**
  A. Tail about 2 inches shorter than the wing; common.1. **Bob-white.**
  A. Tail nearly 3 inches shorter than the wing; Texas...3. **Massena Partridge.**
  B. Tarsus bare of feathers for half its length........6. **Ruffed Grouse.**
  B. Tarsus entirely feathered, but the toes bare......5. **Canada Grouse.**
  B. Tarsus and toes entirely feathered.................7. **Ptarmigans.**
    C. Tail about 2 inches shorter than the wings and square...4. **Dusky Grouse.**
    C. Tail, 3½–5 inches shorter than the wings. (**D.**)
  D. Tail pointed, wedge-shaped; tarsus full feathered...10. **Prairie Sharp-tailed Grouse.**
  D. Tail rounded. (**E.**)
    E. Tarsus full feathered, no bare stripe behind......8. **Prairie Hens.**
    E. Tarsus scantly feathered, exposing a bare stripe behind..9. **Lesser Prairie Hen.**

1. **Bob-white** (289. *Colinus virginiànus*). — A common grass-inhabiting, brownish-mottled, white-throated, game bird, with the belly much lighter than the back. The *female* has a buff throat patch instead of the white of the *male*. The notes *bob-white* so often heard in spring are given by this bird. In summer, the crown is blacker, and the buffy markings lighter than in winter. The tints of the back have much of reddish-brown and chestnut, and the lower breast and belly are white barred with black. (Quail; Virginia Partridge.)

Bob-white

Length, 10; wing, 4½ (4¼–4¾); tail, 2½; tarsus, 1¼; culmen, ⅔. United States from Kansas eastward and north to southern Ontario. It is also found locally in many places west of the Rocky Mountains, even to the Pacific. The **Florida Bob-white** (289ᵃ. *C. r. floridànus*) of Florida is a smaller bird, with darker plumage, especially with more black on the back. The regular northern bob-white will occasionally take to the trees when flushed, but the Florida bird is more apt to do so. The **Texan Bob-white** (289ᵇ. *C. r. texànus*) of Texas and Mexico is a small bird like the last, but paler, having much gray and tawny in the plumage.

2. **Scaled Partridge** (293. *Callipépla squamàta*). — A Texas, crested, bluish-lead-colored quail, with the neck and most under parts peculiarly "scaled" by crescent-shaped black tips to all the feathers. The crest is dark brown, ending in pure white, and the back belly orange-brown. (Blue Quail.)

Length, 10–12; wing, 4½ (4¼–5); tail, 4¼; tarsus, 1¼. Table-lands of Mexico, north to central Texas and southern Arizona.

3. **Massena Partridge** (296. *Cyrtonyx montezumæ*). — A Texas, crested, short-tailed, brownish and purplish quail, with the sides of the head and neck fantastically marked with black and white stripes, and the sides of the body crowded with numerous round white dots on a dark ground. The middle line of breast and belly is mahogany-colored, the under tail coverts are black, and the crest is brown. The *female* lacks the peculiar black and white stripes of the head, and the sides are mottled instead of dotted. The prevailing color of the female is pinkish-cinnamon.

Length, 9; wing, 4⅜; tail, 2; tarsus, 1¼. Table-lands of Mexico, north to western Texas and Arizona.

4. **Dusky Grouse** (297. *Dendrágapus obscúrus*). — A large, Rocky Mountain, dark brown to blackish grouse, with slate-colored belly, a rather short, broad tail, nearly white throat patch, and red, bare skin around the eyes. This bird is finely mottled with lighter tints everywhere, and the tail is tipped with a distinct gray band.

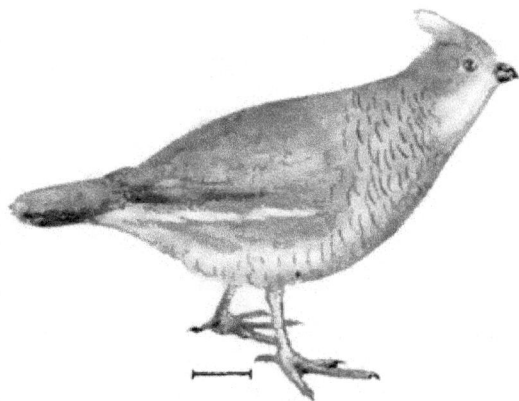

Length, 18-24; wing, 9½ (8½-10); tail, 7½; culmen, ¾. Rocky Mountains from central Montana east to the Black Hills of South Dakota and west to Nevada.

Scaled Partridge

5. **Canada Grouse** (298. *Dendrágapus canadénsis*). — A large, northern, forest-living, short-billed, dark-colored grouse, with much white mottling, especially on the under parts. There are brown tips to the tail feathers and a red patch of bare skin over the eyes. The *female* is much browner, especially on the head and neck. These birds have the upper parts much barred with blacks, grays, and browns. They are com-

mon in the evergreen forests of the north, and are usually resident where found.   (Spruce Partridge.)

Length, 14¾–17 ; wing, 7 (6¼–7¾) ; tail, 5 ; culmen, ⅓.   Northern portions of Minnesota, Michigan, New York, and New England, and northward and westward to Alaska.

6. **Ruffed Grouse** (300. *Bonása umbéllus*). — A large, common,
woodland-living, brown-mottled grouse, with a glossy black ruff
of feathers on each side of the neck, and a dark band near the
end of the broad, fan-
shaped tail. The *fe-
male* has the neck ruff
much smaller. The
male produces a loud
"drumming" noise by
rapidly beating the
air with his wings.
This bird is improper-
ly called "partridge"
in the New England
States, and just as im-
properly "pheasant"
in the Middle and
Southern States.

Ruffed Grouse

Length, 15¼–19 ; wing, 7½ (7–7¾) ; tail, 6¼ ; culmen, ⅘.   United States
from Minnesota eastward, north to southern Canada and south to Georgia,
Mississippi, and Arkansas. The **Canadian Ruffed Grouse** (300ᵃ. B. u.
togáta), of the spruce forests of the northern portions of New York and
New England, north to the southern portion of Hudson Bay, and westward to Oregon and British Columbia, differs in having the upper parts
gray rather than reddish-brown, and the lower parts, including the breast
and belly, fully barred.

7. **Willow Ptarmigan** (301. *Lagópus lagópus*). — An extreme
northern, large ptarmigan with blackish outer tail feathers, and
a coloration of body depending on the season. In winter the
whole body is white ; in summer the back, head, and neck are
mottled in browns or rufous. The *female* in summer has the
plumage more regularly and more fully barred with rufous.

This is an abundant bird in the Arctic regions, but does not nest farther south than central Labrador, though in winter it migrates southward, even into northern New York. The ptarmigans have the toes fully feathered.

Length, 14–17 ; wing, 7½ ; tail, 4½ ; culmen, ½. Northern portions of the northern hemisphere ; south in winter occasionally into the northern border of the United States. The **Rock Ptarmigan** (302. *Lagòpus rupéstris*), of Arctic America south to the Gulf of St. Lawrence, has in winter the outer tail feathers blackish, generally tipped with white, and the lores black, while the rest of the plumage is pure white. In

Willow Ptarmigan

summer it has mottled and barred grayish plumage with almost no rufous. **Welch's Ptarmigan** (303. *Lagòpus wélchi*), of Newfoundland, has in winter the whole tail blackish, except the white tips of the central feathers, and the lores black, while the rest of the plumage is white. In summer the upper parts are black with wavy lines of buff and white, and the belly white. Probably none but the **Willow Ptarmigan** has ever been found in the United States.

8. **Prairie Hen** (305. *Tympanùchus americànus*). — A large, ground-living, short-tailed, very much mottled, brownish, somewhat crested grouse, with a tuft of ten or more, narrow, stiffened, mottled, black feathers on the side of the neck, under which there is a patch of bare, inflatable, yellow skin. The peculiar neck feathers have their tips rounded, and

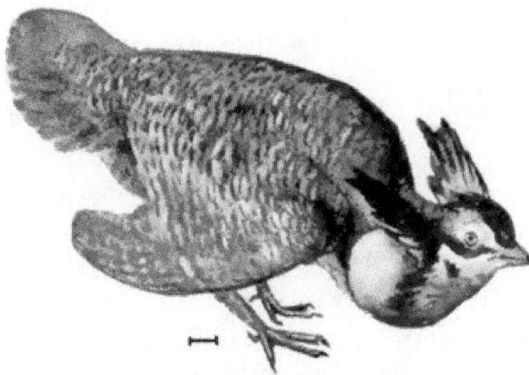

Prairie Hen

the rounded, blackish tail is white tipped. The *female* has the neck tufts much smaller. This is a bird of the open prairies, rarely found, except during severe storms, within timbered tracts. (Pinnated Grouse; Prairie Chicken.)

Length, 17-19 ; wing, 9 (8½-9½); tail, 4 ; tarsus, 2 ; culmen, ⅔. Prairies of the Mississippi Valley, south to Louisiana, east to Ohio, north to Ontario, and west to Nebraska. The **Heath Hen** (306. *Tympanuchus cupido*), of Martha's Vineyard (formerly New England and Middle States), differs from the last in that the neck tufts consist of less than ten pointed feathers. There are but few (less than 100) of these birds left on the island.

9. **Lesser Prairie Hen** (307. *Tympanuchus pallidicinctus*). — A southwestern bird similar to the common prairie hen in dimensions of parts, but paler and browner in color, and with the tarsus much less fully feathered. The darker bars of the back appear in sets of threes, there being a continuous broad bar inclosed between two narrower and darker ones in each set. From Texas to Kansas along the eastern edge of the Great Plains.

10. **Prairie Sharp-tailed Grouse** (308b. *Pediocætes [ped-i-è-se-tes] phasianéllus campéstris*). — A large, northwestern, sharp-tailed, very much mottled, brownish grouse, with the central tail feathers projecting and rounded at tip, and the outer ones sharp-pointed. There are no neck tufts of peculiar feathers, but the breast has many

Prairie Sharp-tailed Grouse

V-shaped, black marks. The middle of the belly is white. This is a somewhat migratory bird, living in the open prairies in summer, and in wooded tracts in winter.

Length, 15-19 ; wing, 8½ ; tail, 4¾ ; culmen, ⅔. Plains and prairies of the United States east of the Rocky Mountains, east to Illinois, and south to New Mexico.

11. Sage Grouse (309. *Centrocércus urophasiánus*). — A very large, western, much-mottled, dark-colored grouse, with long, sharp-pointed tail feathers, and having inflatable, bare skin on the sides of the breast. The *female* has a shorter tail.

Length, 24-30; wing, 12 (10½-13); tail, 8-13; culmen, 1¾. Sage-brush regions of the Rocky Mountains, east to North Dakota, Nebraska, and Colorado, south to New Mexico, and west to California.

ORDER IX. SHORE BIRDS (LÍMICOLÆ)

A large order of plover-like and snipe-like birds, usually found in open places, near the water. They are most of them small; they have slender and frequently long bills, small and, as a rule, fully feathered heads, long-pointed wings, short tails, and long legs, with more or less of the tibia exposed and bare of feathers. A few species have the legs short and the tibia fully feathered. The hind toe is short and elevated, or completely wanting (with one exception, the jacana of the first family). With us this order is represented by seven families.

FAMILY XXXV. JACANAS (JACÁNIDÆ)

A small family (10 species) of peculiar, somewhat plover-like, wading birds, with very long toes and long, straight claws, the hind claw fully as long as the toe.[1]

1. Mexican Jacana (288. *Jacána spinósa*). — A small, Texas, long-legged, long-toed, purplish-chestnut-colored, wading bird with a horny, yellow spur on the bend of the wing, and a peculiar, yellow, leaf-like lobe of skin extending on the forehead from the plover-like bill. The rich chestnut color is brightest on the wings and tail, and darkest on the back, breast, and sides. The *young* is grayish-brown above, buffy below, and has but little of the frontal lobe of skin.

Length, 8½; wing, 5 (4½-5½); tail very short and soft; tarsus, 2; middle toe and nail, 2½; culmen, 1¼. Southern Texas, Mexico, and Central America.

FAMILY XXXVI. OYSTER-CATCHERS (HÆMATOPÓDIDÆ)

The birds of this small family (10 species) are found only on the outer beaches of ocean shores, searching for the shellfish left by the receding tide. They are large birds, with stout, long, hard bills.[1] stout, rather short legs, and pointed tails. Our one species has but three toes.

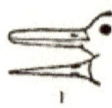

1. **American Oyster-catcher** (286. *Hæmátopus palliátus*). — A large, shy, rather solitary, long, red-billed, three-toed, seacoast bird, with black head, neck, and back, and white belly. There is a large, white patch on the center of the wing and also on the rump. When disturbed, it gives a shrill cry and flies to a great distance. It runs swiftly or walks in a stately manner, and feeds mainly on bivalves, which it opens with its long, strong bill.

Length, 17-21; wing, 10½ (10-12); tail, 4½; tarsus, 2¾; culmen, 3-4.

American Oyster-catcher

Seacoast of America, from New Jersey to Patagonia (occasionally north to Massachusetts); breeding along the Southern States, and wintering south of the United States.

FAMILY XXXVII. TURNSTONES, ETC. (APHRÍZIDÆ)

A small family (4 species) of seacoast birds of rather small size, short, hard bill, and (for shore birds) short legs.

1. **Turnstone** (283. *Arenária intérpres*). — A common, shore-living, stout-billed, brightly marked bird, with a back marked like calico, and a white belly with a black breast patch. The center of the back, as seen while flying with scapulars separated,

is white. In summer there is much rufous, black, and white on the upper parts; in winter the bright, reddish-brown is lacking, and the colors of the back are mainly blacks and grays. This bird is often seen turning over stones and shells along the outer shore for food. (Calicoback.)

Length, 9½; wing, 6; tail, 2½; tarsus, 1; culmen, ⅞. Along nearly all shores of lakes, rivers, and oceans. In the New

Turnstone

World, from Greenland to the southern part of South America. More or less common along the great rivers and lakes of the interior; breeding in the Arctic regions, and wintering mainly south of the equator.

FAMILY XXXVIII. PLOVERS (CHARADRIIDÆ)

This large family (100 species) of snipe-like birds with long wings, short, pigeon-shaped bills,[1] and (in most species) three toes, is represented throughout the world, though only eight species are found in North America. These are short-billed, round-headed, short-necked, plump-bodied, long-winged, short-tailed, wading birds with (in most species) rather short legs for waders, and but three toes.

Many species inhabit the shores of water, both salt and fresh, but some are found on the dryest plains. They move rapidly when running or flying, and their note is a mellow whistle.

Key to the Species

* A hind toe present about ⅛ long.
    — Head without crest.....................1. **Black-bellied Plover**.
    — Head crested; back metallic green. The **Lapwing** (269. *Vanéllus vanéllus*) of the Old World has once been seen on Long Island. Wing, 8½–9; culmen, 1.
* Toes only three; hind toe absent. (**A**.)
    A. Plumage speckled on the back with whitish or yellow ....
    ....2. **American Golden Plover**.

A. Plumage of the back about uniform in color. **(B.)**
B. Wing, 6–7 long; rump, orange-brown .... .3. **Killdeer.**
B. Wing, 5½–6 long; no black band across breast. .8. **Mountain Plover.**
B. Wing less than 5½ long. **(C.)**
   C. Culmen, ¾ or more long; a black or dark brown band across breast
   ....7. **Wilson's Plover.**
   C. Culmen about ½ long; no black band across breast ...
   ....6. **Snowy Plover.**
   C. Culmen less than ½ long. **(D.)**
D. All toes distinctly webbed at base; [1] feathers black between the eye and the bill.....4. **Semipalmated Plover.**
D. Inner toes without distinct webbing; no black from the eye to the bill . . ..5. **Piping Plover.**

1. **Black-bellied Plover** (270. *Squatàrola squatàrola*). — As seen in the autumn and winter in the United States: a short-billed, short-tailed, large (for a plover), mottled, grayish-brown, shore bird, with grayish or whitish under parts mottled with more or less of blackish on the breast. This is our only plover with a hind toe; it is minute, being only about ½ inch long. The bird derives its name from its very black under parts, in the breeding

Black-bellied Plover

season, in the far north. During its northward migration in the spring, it is found with a more or less complete black breast and fore belly. The axillary plumes [2] (long feathers growing from the armpit and seen underneath the wings) are black. (Black-breast; Bull-head Plover; Beetle-head.)

Length, 11½; wing, 7½ (7–7½); tail, 3; tarsus, 2; culmen, 1¼. Generally throughout the northern hemisphere, though not confined to it; breeding far north, and wintering in Florida, the West Indies, and northern South America.

2. American Golden Plover (272. *Charádrius domínicus*). — As
seen in the United States, a short-billed, three-toed shore
bird, with the entire upper parts blackish, brightly dotted and
marked with golden and whitish spots, and the lower parts
grayish-white, with brownish streakings on the sides. In late
spring, while migrating northward, some of these birds are
seen with the black bellies of the breeding season. These
graceful, quick-moving birds are found in marshes and old
fields as well as on the sand flats exposed by the tide. They
have the habit, common among plovers, of rapidly running a
few yards, then stopping, elevating the head and looking
around. (Greenback.)

Length, 9½-11; wing, 7 (6¾-7¾); tail, 3; tarsus, 1¾; culmen, ⅞.
America; breeding in the Arctic regions, and wintering from Florida to
Patagonia.

3. Killdeer (273. *Ægialítis vocífera*). — A common, noisy,
active, beautifully marked, short-billed, three-toed, brown-
backed, white-bellied plover, with two dark bands across the
breast, the upper one extending around the neck. The rump
is very brightly colored, often decidedly red, and the wings have
much black and white. This bird is very abundant, spending
most of the time on the ground, often far from water. Its
shrill notes give it its name, *kil-dee*. Though scattered while
feeding, it usually moves in flocks when on the wing.

Length, 10½; wing, 6½ (6-6¾); tail, 4; tarsus, 1½; culmen, ¾. United
States, north to Newfoundland and Manitoba; breeding throughout, and
wintering from Virginia to northern South America, including the West
Indies.

4. Semipalmated Plover (274. *Ægialítis semipalmáta*). — A
common, short-billed, ashy-brown-backed, white-bellied plover,
with a rather broad, complete ring of black around the neck,
and distinctly marked black, white, and brown head, including
a black band from the eye to the bill. The *female* has the
neck band and head markings brown instead of black. This
is an abundant seacoast plover, with the toes nearly half

webbed. In feeding, the small flocks of five to ten scatter, but on the wing form a compact bunch. (Ring-neck.)

Length, 7; wing, 4¼ (4½–5); tail, 2¼; tarsus, 1; culmen, ½. North America; breeding in the Arctic regions, and wintering from the Gulf States to Brazil.

5. **Piping Plover** (277. *Ægialitis meloda*). — A wary, coast-living, short-billed, ashy-backed, white-bellied plover, with a narrow, black collar on the sides. but not complete across the breast, and a narrow, black stripe from eye to eye above the forehead. In winter the black is replaced by brownish gray. Its notes are peculiarly sweet and musical, a *peep-peep-peep-o*. (Pale Ring-neck.)

Piping Plover

Length, 7; wing, 4¾ (4½–4¾); tail, 2¼; tarsus, ⅞; culmen, ½ nearly. Eastern North America; breeding from the coast of Virginia north to Newfoundland, and wintering from Florida southward. The **Belted Piping Plover** (277ᵃ. *Æ. m. circumcincta*) is much like the last, but has the black collar complete across the breast. The *young* lack this complete collar. Mississippi Valley; breeding from northern Illinois northward, and wintering from the Gulf southward. Occasionally eastward to the Atlantic coast.

6. **Snowy Plover** (278. *Ægialitis nivosa*). — An extreme western, grayish-brown-backed plover, with the forehead, line over eye, somewhat of a collar around the back neck, and all lower parts pure white. Above the white forehead there is a black patch on the crown, another on the ear coverts, and a third on the side of the breast. The *young* has the black markings replaced by ashy-brown.

Length, 6½; wing, 4¼; tail, 2; tarsus, 1; culmen, ¾. Western North America from California eastward to Kansas and Texas; wintering in Central America and western South America.

7. **Wilson's Plover** (280. *Ægialitis wilsònia*). — A southern, common, brownish-gray-backed, white-bellied plover, with a broad black band across the upper part of the breast and blackish wing quills. The forehead and line over the eye are white. lores blackish, and a black band across the front of the crown. There is a more or less complete white band across the back neck. The *female* has the breast band brownish-gray. This is a gen-

Wilson's Plover

tle, fearless bird, of the sandy marine beaches and mud flats.

Length, 7½; wing, 4¾ (4½–5); tail, 2; tarsus, 1¼; culmen, ⅞. Coasts of America from Long Island and Lower California to Brazil and Peru; breeding from Virginia southward, and wintering from Mexico southward.

8. **Mountain Plover** (281. *Ægialitis montàna*). — A tame, western, grayish-brown-backed, whitish-bellied plover, with blackish wing quills. The fore part of crown and a stripe from the eye to the bill are black; forehead and stripe over the eye white; the breast has an indistinct cross band of ochraceous, darkest on the sides. The feathers of the back are margined with rufous. The *young* has the head, neck, and upper breast like the back. This bird inhabits the dryest of the plains and grassy districts of the west in large flocks. It rises from the ground by several quick flaps of the wings, and, usually near the ground, circles through the air most gracefully.

Length, 9; wing, 5¾ (5½–6); tail, 2¾; tarsus. 1⅝; culmen. ⅞. Chiefly on the Plains; breeding from central Kansas to the British boundary, and wintering mainly southwestward to central California, and south into Mexico. Accidental in Florida.

FAMILY XXXIX. SNIPES, SANDPIPERS, ETC. (SCOLOPÁCIDÆ)

A large family (100 species) of generally long-legged, short-tailed, shore birds, divided into many groups which grade into one another, but which have characteristics distinct enough to give them different common names. The peculiarities of the main groups will be given. (1) **True Snipe and Woodcock.** Long-billed, mottled-brown birds of swampy meadows and woodlands, where the mud is soft. Their legs are relatively shorter than those of the other groups of the family, and the eyes are placed farther back on the head than in any other birds. Nos. 1 and 2 show these characteristics fully, while 3 and 4 are intermediate between this group and the next. (2) Sandpipers. This is a large group, and contains the smallest species of the family, as well as some of large size. They are short, straight-billed, long-legged, slender-bodied birds, of open, wet places, with a piping, resonant voice and unbarred, short tails. They have little, and, in some cases, no webbing to the toes. Their colors are grays, browns, and dull yellows. Their movements are quick and graceful in both running and flying. After a run, many of them have a way of teetering the body in a see-saw way, which is very characteristic. Nos. 5–16, 24–26 are somewhat intermediate between this group and the tattlers. (3) **Godwits.** These are large, snipe-like birds, with long legs and very long and slightly upwardly curved bills.[1] They are found abundantly on marshes and salt meadows, around bays and lakes. Nos. 17 and 18. (4) Tattlers. This is about as large a group as that of the sandpipers, and the different species vary so in their characteristics that but little can be said of them as a whole. The bill is generally about straight, and longer than the head; it is harder and less sensitive than the bills of the other groups. Their noisy and restless character has given them their name;

1   2   3   4

many have the habit of continually bobbing the head, as though they were confiding some wonderful news. Nos. 19–24. (5) **Curlews**. These might be called godwits with long, slender, downwardly curved bills.[2] Their habitat is mainly in the marshes, though some are found on dry plains. Nos. 27–29.

Key to the Species

* Bill very long and much decurved;[2] tarsus scutellate only in front, reticulate behind. (**P.**)

* Bill not strongly decurved; tarsus scutellate in front and behind.[5] (**A.**)

    A. Toes only three, the hind toe wanting.... ......16. **Sanderling**.

    A. Toes four, the hind toe present. (**B.**)

B. Eyes situated back of the middle of the head;[3] bill twice as long as the tarsus and with the upper mandible thickened at the tip; toes without distinct webbing. (**O.**)

B. Eyes not back of the middle of the head; bill in no case twice as long as the tarsus. (**C.**)

    C. Front toes not webbed[4] (at most with one minute web). (**J.**)

    C. Front toes with at least one distinct web.[6] (**D.**)

D. Tail more than half as long as the wing and with the under feathers at least one inch shorter than the middle ones; wing over 6; bill, 1–1¼..............................24. **Bartramian Sandpiper**.

D. Tail about half as long as the wing; wing, 3¾–4¾; bill, tarsus, and middle toe and nail each about 1 long; bill narrow and somewhat decurved near tip;[7] breast much spotted......26. **Spotted Sandpiper**.

D. Tail less than half as long as the wing. (**E.**)

    E. Wing, 3–4¼ long; bill grooved at tip[8]...........................
......14. **Semipalmated Sandpiper** and 15. **Western Sandpiper**.

    E. Wing, 6¼–7½ long; bill straight and 1–1½ long, just about one fifth the length of the wing.............................23. **Ruff**.

    E. Bill over one fifth the length of the wing. (**F.**)

F. Bill slightly broadened near tip;[9] bill and tarsus about equal and 1½–1¾ long; wing, 5–5½......................5. **Stilt Sandpiper**.

F. Bill slightly broadened near tip; bill over 2, and tarsus under 2 long; wing, 5¼–6 long..........3. **Dowitcher**, 4. **Long-billed Dowitcher**.

F. Bill not broadened near tip. (**G.**)

    G. Wing, 8¼–9½; bill, 3¼–5½ long, and bent upward toward tip;[1] rump and upper tail coverts regularly barred with brown.........
....17. **Marbled Godwit**.

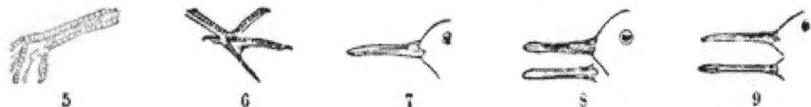

   5   6   7   8   9

G. Wing, 7½–8½; bill, 2¾–3½ long, and bent upward toward tip;[1] rump and tail black, upper tail coverts white between. ....18. **Hudsonian Godwit.**

G. Bill not over 2¾ long and not conspicuously bent upward, usually straight. **(H.)**

H. Wing with a large white patch at the base of the otherwise black primaries; axillars[2] black.............................22. **Willet.**

H. Wing without white patch; axillars barred. **(I.)**

I. Wing, 7–8; bill, 2–2½.
— Upper tail coverts slightly, and tail heavily, barred.......................................19. **Greater Yellow-legs.**
— Lower back and rump white; tail slightly barred. **Green-shank** (253. *Tótanus nebulàrius*) of the Old World has been found in Florida.

I. Wing, 6–7; bill, 1¼–2; legs yellow in life........20. **Yellow-legs.**

I. Wing, 4½–5¼; bill, 1–1¼; legs dusky in life. 21. **Solitary Sandpiper.**

J. Inner web of the outer primary beautifully speckled with blackish.................................25. **Buff-breasted Sandpiper.**

J. Inner web not mottled. **(K.)**

K. Wing over 6; middle pair of tail feathers not lengthened..6. **Knot.**

K. Wing under 6; middle pair of tail feathers acute and abruptly lengthened.[3] **(L.)**

L. Bill, ¼ inch longer than tarsus. **(N.)**

L. Bill not over ⅛ inch longer than tarsus, in most species no longer. **(M.)**

M. Wing, 5–5¾; bill, 1–1¼; middle upper tail coverts black slightly margined with reddish......................8. **Pectoral Sandpiper.**

M. Wing, 4¾–5⅜; middle upper tail coverts white.................................9. **White-rumped Sandpiper.**

M. Wing, 4¼–4⅞; middle upper tail coverts fuscous.................................10. **Baird's Sandpiper.**

M. Wing, 3–4; tarsus about ¾..................11. **Least Sandpiper.**

N. Middle upper tail coverts unbarred, black or blackish; bill about straight.............................7. **Purple Sandpiper.**

N. Middle upper tail coverts unbarred, dusky, or grayish; bill straight to about the middle and then curved downward[4].............................12. **Red-backed Sandpiper.**

N. Middle upper tail coverts mainly white; bill somewhat curved downward throughout its length[5]..................13. **Curlew Sandpiper.**

O. Tibia naked at the joint; crown striped lengthwise; outer web of the primaries without bars....................2. **Wilson's Snipe.**

O. Tibia entirely feathered; crown banded crosswise...1. **Woodcock.**

1

2

3

4

5

P. Wing, 10–12; bill, 5–8½ (very young, 2½–5); axillars[2] rich, dark buff, usually without any bars.............27. **Long-billed Curlew.**

P. Wing, 8½–10¼; bill, 2¾–4¼; axillars barred..28. **Hudsonian Curlew.**

P. Wing, 7¾–8¼; bill, 2–2⅞; axillars barred.......29. **Eskimo Curlew.**

1. **American Woodcock** (228. *Philòhela mìnor*). — A common, muddy-wood-living, long-billed, short-legged, much-mottled, brown snipe, with eyes far back on the head, and the back of the crown with two dark cross stripes; the three outer primaries are abruptly shorter than the fourth and are peculiarly narrow and stiff. The soft muddy places where these game birds get their food by the use of their long, pliable, sensitive bills are

American Woodcock

usually in or near woody tracts. These birds are particularly noted for their nocturnal, spiral flights in the air, which have been called "sky dances." They migrate by night to places where soft ground is to be found.

Length, 11; wing, 5¼ (4½–5¾); tail, 2¼; tarsus, 1¼; culmen, 2½–3. Eastern North America, north to the British Provinces and west to Kansas; breeding throughout. The **European Woodcock** (227. *Scólopax rusticola*) is a larger bird, but similar in appearance. It does not have the three narrow outer primaries. Wing, 7–8; culmen, 3–3¼. Accidental in eastern North America.

2. **Wilson's Snipe** (230. *Gallinàgo delicàta*). — A common

Wilson's Snipe

grassy-meadow-living. long-billed, very much mottled, brownish snipe, with a buff breast and white belly. The eyes are above the ears, and the mottling of the head is so arranged as to give a central, lengthened. light band between two darker ones. This is one of the most noted game birds, because only expert gunners can shoot it as it starts from the ground in its crooked but swift flight. It is found only where the ground is so water soaked as to give it a chance to probe with its soft, sensitive bill, and where clumps of vegetation will enable it to hide from view. ("English Snipe.")

Length, 11 ; wing, 5 (4½-5½); tail. 2½ ; tarsus, 1½ ; culmen, 2½-2¾. North America ; breeding from the northern United States northward, and wintering from Illinois and South Carolina to northern South America.

3. **Dowitcher** (231. *Macrorhámphus gríseus*). — A common, large, shore-living. long-billed, long-legged. generally variegated. brownish-bodied snipe, with darker wings, lighter.

Dowitcher

much-barred tail, and nearly white belly. In winter, the upper parts and breast are plain gray with almost no traces of black or bay, while in summer the upper parts are much mottled with these colors. During migrations this bird is found in large flocks on the mud flats. exposed by the falling tide. (Red-breasted Snipe, summer; Gray Snipe, winter.)

Length, 11 ; wing, 5¾ (5½-5½); tarsus, 1⅜ ; culmen, 2-2½. Eastern North America ; breeding in the Arctic regions, and wintering from Florida to Brazil.

4. **Long-billed Dowitcher** (232. *Macrorhámphus scolopáceus*). — In winter this bird and the last are practically alike except in

size, but in summer the long-billed has the breast and belly more uniformly rufous, and the sides more heavily barred with black. This is the dowitcher of the interior of the United States and is rare on the Atlantic coast, though it can be found there quite regularly in the late autumn. (Western Dowitcher; Red-bellied Snipe.)

Length, 12; wing, 5½-6; tarsus, 1⅜; culmen, 2⅛-3¼. Western North America; breeding in the Arctic regions, migrating south through the western United States (including the Mississippi Valley), and wintering in Mexico and possibly South America.

5. **Stilt Sandpiper** (233. *Micropálama himántopus*). — A rare, very long-legged, long-billed, very much mottled sand-piper, with the center of each of the feathers blackish (in general) and the edges brown-ish-gray. The tail, throat, and line over the eye are much lighter. The colors are much grayer in winter, the under parts being white. It is slow moving as compared with other sand-pipers, and is more apt to squat than fly when disturbed.

Stilt Sandpiper

Length, 7½-9¼; wing, 5⅜; tail, 2¼; tarsus. 1⅜; culmen, 1½-1¾. Eastern North America; breeding far north, and wintering from the West Indies to South America.

6. **Knot** (234. *Trínga canútus*). — A very large and, as usually seen in the United States, mottled, gray-backed, white-bellied, plover-like sandpiper, with more or less of a red, robin-like breast. The back and wings are more beautifully marked in

the summer than in the winter with black, brown, and buff. The *young* has the breast finely spotted or streaked with blackish, and the flanks barred or streaked with the same. The knot is found on muddy flats and sandy beaches, probing the ground, like the true snipe, for its food, which consists of crustaceans and mollusks. The knots bunch very closely when decoyed, and so numbers can be killed by a single discharge. (Robin Snipe.)

Knot

Length, $10\frac{1}{2}$; wing, $6\frac{1}{2}$; tail, $2\frac{3}{8}$; tarsus, $1\frac{1}{4}$; culmen, $1\frac{3}{8}$. Nearly all coasts; breeding in the Arctic regions, and wintering from Florida to South America.

7. **Purple Sandpiper** (235. *Tringa maritima*).—A northern sandpiper, with grayish-purple to ashy head, breast, and back; white throat, and whitish, somewhat streaked belly. The ashy breast is one of the most constant of its peculiarities. The bill is $\frac{1}{4}$ inch longer than the tarsus, and the tibia is feathered to the joint. It has a fondness for rocky shores, where it secures its food from among the seaweeds attached to the stones.

Length, 9; wing, 5 ($4\frac{7}{8}$–$5\frac{3}{8}$); tail, $2\frac{1}{2}$; tarsus, $\frac{7}{8}$; culmen, $1\frac{1}{4}$. Northern hemisphere; breeding in the Arctic regions, and wintering southward to the Middle States and rarely to Florida.

8. **Pectoral Sandpiper** (239. *Tringa maculata*).—A short-necked, mottled, dark-brown-backed, white-bellied, streaky buff-breasted sandpiper, with black upper tail coverts slightly tipped with buff. The back has much black mixed with the brown and buff, the centers of the feathers being black. This is an inhabitant of grassy meadows rather than beaches, and

the name *krieker* is derived from its notes. (Krieker; Grass Snipe.)

Length, 9; wing, $5\frac{1}{4}$ (5–$5\frac{5}{8}$); tarsus, $1\frac{1}{8}$; culmen, $1\frac{1}{8}$. America; breeding in the Arctic regions, and wintering in South America.

Pectoral Sandpiper

9. **White-rumped Sandpiper** (240. *Tringa fuscicóllis*). — A short-billed, dark-brownish-colored, much mottled sandpiper, with white upper tail coverts, throat, and middle of belly. The above is the summer plumage; in winter, the upper parts are slightly streaked, brownish-gray. These birds are social, and frequent the sandy beaches and marshy shores of the coast, as well as the upland lakes of the interior.

Length, $6\frac{3}{4}$–8; wing, $4\frac{7}{8}$; tarsus, $\frac{7}{8}$; culmen, nearly 1. Eastern America; breeding in the Arctic regions, and wintering in the West Indies, Central and South America.

10. **Baird's Sandpiper** (241. *Tringa bairdii*). — This bird is similar to the last, but has the upper tail coverts blackish instead of white. In winter it has a more buffy breast and lighter upper parts.

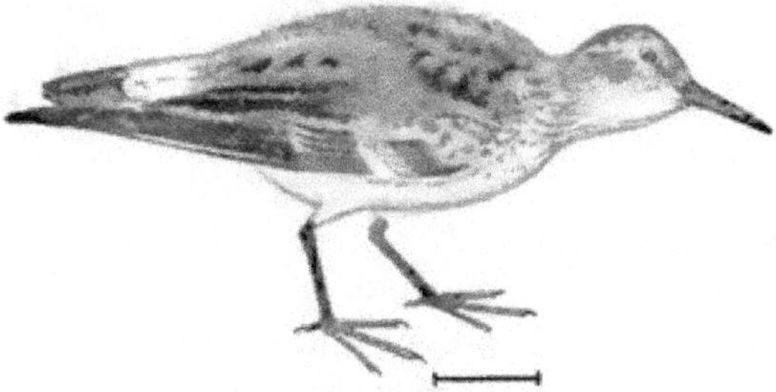

White-rumped Sandpiper

Length, $7\frac{1}{4}$; wing, $4\frac{3}{4}$ ($4\frac{1}{2}$–$4\frac{7}{8}$); tail, $2\frac{1}{4}$; tarsus, $\frac{7}{8}$; culmen, $\frac{7}{8}$. America; mainly in the interior of North and the western portion of South America; breeding in the Arctic regions, and wintering in South America; rare on the Atlantic coast.

11. **Least Sandpiper** (242. *Tringa minutilla.*) — A common, very small, mottled, brownish-backed sandpiper, with the under parts from bill to tail white, streaked with black on the breast. The toes without webbing distinguish this species from No. 14, with which it often associates along our shores and beaches. This species can be seen also on grassy meadows. (Meadow Oxeye; Peep.)

Length, 6; wing, 3½ (3¼–3¾); tail, 1⅞; tarsus, ¾; culmen, ¾. America; breeding north of the United States, and wintering from the Gulf States to South America.

12. **Red-backed Sandpiper** (243ª. *Tringa alpina pacifica*). — A brownish-red-backed, black-bellied sandpiper, with a spotted buff breast and a long bill which is decurved near the tip. In winter it lacks the black patch on the belly and has usually an ash-gray back, a pale gray, somewhat streaked breast, and a white belly. This fearless little snipe is found mainly on beaches and mud flats, though it occasionally visits grassy meadows. (Black Breast, spring; Winter Snipe, autumn.)

Length, 8; wing, 4¾ (4½–5); tail, 2⅛; tarsus, 1⅛; culmen, 1½. North America and eastern Asia; breeding in the Arctic regions, and wintering in the South Atlantic and Gulf States. The **Dunlin** (243. *Tringa alpina*) of the Old World has been seen on Long Island. It is smaller and less brightly colored. Wing, 4⅛–4½.

13. **Curlew Sandpiper** (244. *Tringa ferruginea*). — A very rare, European, rather long-billed, brick-red-colored sandpiper, with black primaries and spotted white tail coverts. In winter it is much grayer. The bill is decurved, curlew-like, from end to end.

Length, 7–9; wing, 4⅞ (4¾–5¼); tarsus, 1¼; culmen, 1½. Old World in general; occasional in eastern North America and Alaska.

14. **Semipalmated Sandpiper** (246. *Ereunètes pusillus*). — A common, very small, short-billed, mottled, brownish-backed sandpiper, with the under parts from chin to tail white, streaked or spotted with black on the breast. The toes have plain webbing at the base. In winter, this bird is more ashy. This species, in habits, form, size, and color, appears like

No. 11, with which it often associates, but the former is more common on sandy beaches, the latter on grassy plains. Large numbers congregate together in flocks, and when disturbed fly in a compact mass. (Sand Oxeye; Peep.)

Semipalmated Sandpiper

Length, 6¼; wing, 3¾ (3¼–4); tail, 2; tarsus, ⅞; culmen, ¾. Eastern North America: breeding north of the United States, and wintering from the Gulf States to Brazil.

15. **Western Sandpiper** (247. *Ereunètes occidentàlis*). — This bird is much like No. 14 in every way. In summer it can be distinguished from it by the brownish-red edgings to the feathers of the back, and usually also by the heavier spots on the breast; in fall and winter the best method is by comparison of the length of the bill. No. 14 rarely has a bill ⅞ long, while this species has one varying from ⅞–1¼. Its range is mainly through the west, though occasionally it is to be found with the eastern species (No. 14) on the Atlantic coast.

Sanderling

Length, 6¼; wing, 3¾ (3⅝–3⅞); tail, 2; tarsus, ⅞; culmen, 1. Western North America; breeding far north, and wintering in Central and South America. Occasional in the eastern United States.

16. **Sanderling** (248. *Cálidris arenària*). —

A common, three-toed, beach-living, mottled-brownish sand-piper, with short bill and unspotted white belly. In winter the reddish of the back is replaced by grayish. This bird often associates with No. 14 on the beaches, but it is larger, lighter in color, and usually less spotted on the breast, which in summer is brownish in tint. (Ruddy "Plover"; Surf Snipe.)

Length, 8; wing, 4? (4½-5); tail, 2½; tarsus, 1; culmen, 1. Nearly all beaches of all countries; breeding in the Arctic regions, and wintering in America south to southern South America.

17. **Marbled Godwit** (249. *Limòsa fèdoa*). — A very large, shy, long-billed, long-legged, generally brownish-red-colored, mottled snipe, with the upper parts much darker, usually blackish marbled with buffy. The inner web of the outer primaries and both webs of the others are buffy, speckled with black. The mottlings, barrings, and streaks are found everywhere except on the throat, which is whitish. The bill is curved up-ward to a slight extent. The *young* has the lower parts less barred. This is a western bird rarely found on the Atlantic coast. (Brown Marlin.)

Length, 16-22; wing, 8? (8½-9½); tail, 3-4; tarsus, 3; culmen, 3½-5½. North America; breeding in the interior from Iowa and Nebraska north-ward, and wintering in Mexico, Central America, and Cuba.

18. **Hudsonian Godwit** (251. *Limòsa hæmástica*). — A bird similar to the last, but smaller, and with the upper tail coverts white, and the tail black, with a narrow tip of white. It is rare on the Atlantic coast, migrating chiefly through the in-terior. (Ring-tailed Marlin.)

Length, 14-17; wing, 8½ (7¾-8¾); tail, 3½; tarsus, 2½; culmen, 2½-3½. Eastern North America; breeding in the Arctic regions, and wintering in South America.

19. **Greater Yellow-legs** (254. *Tòtanus melanoleùcus*). — A rather common, large, long, yellow-legged, long-billed, mottled, brownish-gray-backed, white-rumped snipe, with the white un-der parts, spotted on the breast and sides. In winter the back

is grayer, and the under parts are less spotted. (Greater Tell-tale; Long-legged Tattler.)

Length, 12–15; wing, 8 (7¼–8½); tail, 3¼; tarsus, 2½; culmen, 2–2¼. America; breeding from Iowa northward, and wintering from the Gulf States to Patagonia.

20. **Yellow-legs** (255. *Tótanus flávipes*). — A bird similar in habits and appearance to the last, but appreciably smaller, though the legs are proportionately longer. Both of these species occur wherever there is water, and during migrations are abundant. though more common on coasts than along rivers. This is usually more abundant than No. 19, and more easily decoyed by the hunter; it is especially plentiful in the late summer and autumn during its southward migration. The notes of both this species and the last are a clear, whistling, *wheu-wheu-wheu.* (Summer Yellow-legs; Lesser Telltale.)

Greater Yellow-legs

Length, 10–12; wing, 6½ (6–7); tail, 2½; tarsus, 2; culmen, 1¼–1¾. America; breeding north of the United States, and wintering from the Gulf States to Patagonia. In the United States more common east than west.

Solitary Sandpiper

21. **Solitary Sandpiper** (256. *Tótanus solitárius*). — A common, small, dark, olive-brown-backed, white-bellied sandpiper, with the neck and back spotted with white. The throat

and belly are pure white, and the sides of head, neck, and breast white or slightly buffy, streaked with black. In winter the back is less distinctly speckled. This is an inhabitant of the woody borders of ponds, lakes, and streams of inland places, and is rarely found near salt water. As its name indicates, it is generally solitary in its habits, though sometimes a few (3–6) are to be found together. (Solitary Tattler.)

Length, 8½; wing, 5⅛ (4⅞–5¾); tail, 2⅛; tarsus, 1⅛; culmen, 1⅛. North America; breeding along the northern border of the United States and northward, and wintering in the Southern States and southward to central South America.

22. **Willet** (258. *Symphēmia semipalmata*). — A large, shy, brownish-gray-backed, white-bellied snipe, with a distinct, large white patch on the wings, and white upper tail coverts. In summer the head, neck, and sides are much streaked with white, and the ashy tail is barred with blackish. It is found on both fresh and salt water marshes and shores. The name comes from the shrill-whistled call notes, *pilly - will-willet.*

Length, 16; wing, 8 (7½–9); tail, 3; tarsus, 2–3; culmen, 2–2¾. Eastern North America; breeding from Florida to New Jersey and locally to Maine, and wintering in the West Indies to South America. The **Western Willet** (258ᵃ. *S. s. inornata*) of western North America can hardly be distinguished from the eastern form. It averages larger and with a longer bill, and is in general a grayer bird. It breeds from Texas to Manitoba; mainly found in the Mississippi Valley and westward, but occasionally along the coasts of the South Atlantic and Gulf States.

Willet

23. **Ruff** (260. *Pavoncélla púgnax*). —This Old World species has occasionally been found in eastern United States. The
female is a large, grayish, brown-backed, shore bird, with an
ashy breast and white belly. The back in summer is distinctly barred or streaked with black. The *male* in summer
has a very peculiar and large ruff around the neck and on the
breast, which may be of many colors — chestnut, black, black
and white, white and brown, etc.

Length, 9½–13; wing, 5¾–8; tail, 2½–3; tarsus, 1½–2; culmen, 1–1¼.
The smaller numbers refer to the female. It has been found in different
states from Maine to New Jersey and west to Ohio.

24. **Bartramian Sandpiper** (261. *Bartrámia longicaúda*). — A
large, shy, comparatively long-tailed, plain-and-upland-living,
beautifully mottled,
buff and dark brown,
plover-like sandpiper.
The throat, neck, and under
parts are creamy-buff, the
sides of head and neck brighter and streaked with dark
brown, and the breast with
some arrow-headed spots of
the same. The tail extends
beyond the wings when closed,
and the outer primaries are
barred with black and white.
Its notes have been described
as most weird and mournful.
It is seldom found near the

Bartramian Sandpiper

water and, if near it, probably never wades. In habits, it is
much more of a plover than a sandpiper, and has received
many names to indicate this fact. (Upland "Plover"; Field
"Plover"; Grass "Plover.")

Length, 11–13; wing, 6½ (6¼–7); tail, 3½; tarsus, 2; culmen, 1¼.
North America, mainly east of the Rocky Mountains, north to Nova
Scotia and Alaska; breeding throughout, and wintering south of the
United States to Brazil.

25. **Buff-breasted Sandpiper** (262. *Tryngìtes snbrnficóllis*). —
A small. short-billed, buff-colored. field-and-grassy-plain-living,
plover - like sand-
piper, with the back
and wings a mottled
brownish-buff, dark-
est on the wings.
A peculiar black
speckling on the in-
ner webs of all the
primaries is the dis-
tinguishing mark of
this small species.
It is rare east, com-
mon west.

Buff-breasted Sandpiper

Length, 7–9 ; wing, 5¼ (5–5½) ; tail, 2½ ; tarsus, 1¼ ; culmen, ⅜. North
America, especially in the interior ; breeding in the Arctic regions, and
wintering in South America.

26. **Spotted Sandpiper** (263. *Actìtis maculària*). — A common,
small, brownish-gray-backed sandpiper, with the white under
parts everywhere spotted with black. This is an inhabitant of
the margins of all rivers, ponds. and lakes, as well as of the
ocean. In summer,
it is about our only
fresh-water sand-
piper. It is a rapid
runner and a good
"teeterer." Its sharp
notes *peet-weet* are
given when flushed.
It usually returns to
its starting point, at

Spotted Sandpiper

least after several flushings. (Tilt-up; Teeter Snipe; Peet-weet.)

Length, 7–8 ; wing, 4¼ (4–4½) ; tarsus, 1 ; culmen, 1. America, from
Alaska to southern Brazil ; breeding throughout temperate North America,
and wintering in the West Indies to South America.

27. **Long-billed Curlew** (264. *Numénius longiróstris*). — A very large, long-legged, much-mottled, dark-brown-backed, shore bird, with buffy under parts, and a sickle-like, downwardly curved, exceedingly long bill. The head and neck are peculiarly streaky. These birds, though mainly found along muddy shores and on grassy meadows, are known to live and breed in upland regions at a distance from water. (Sickle-bill.)

Length, 20–26; wing, 10½ (10–11½); tail, 4; tarsus, 3; culmen, 2¼ (young), 5½–8½ (adult). United States; breeding north to the South Atlantic States (casually to New England), and in the interior to Manitoba, and wintering from the Gulf States to the West Indies.

28. **Hudsonian Curlew** (265. *Numénius hudsónicus*). — A large, common, much-mottled, generally brownish, shore bird, with whitish belly, and a long, sickle-like, downwardly curved, slender bill. This is a smaller, but more common bird than the last, and has similar habits and frequents like places. (Jack Curlew.)

Hudsonian Curlew

Length, 16–18; wing, 9½ (9–10¼); tail, 3½; tarsus, 2¼; culmen, 3–4. America; breeding in the Arctic regions, and wintering mainly south of the United States.

29. **Eskimo Curlew** (266. *Numénius boreális*). — A large, slender-billed, long-legged, much-mottled, brownish curlew, with a decidedly curved bill like the last two species. The under parts are buffy, with a darker and very streaky breast. This, the smallest of the curlews, is more abundant in the interior, and frequents dry uplands and fields in preference to muddy shores. It often appears in great flocks on the western prairies. (Small Curlew; Dough-bird; Fute.)

Length, 12–15; wing. 8¼–8¾; tail, 3; tarsus, 1¾; culmen, 2¼–2¾. Eastern North America; breeding in the Arctic regions, and wintering in South America.

FAMILY XL. AVOCETS AND STILTS (RECURVIRÓSTRIDÆ)

This is a small family (11 species) of extremely long-legged, long-necked, slender-billed, wading birds. Their bills are long as well as slender, and have more or less of an upward bend. The **Avocets** swim with great ease, and are tame birds, generally found in flocks. Their food consists of water-insects, and crustaceans, which they obtain mainly in shallow water, swinging the bill from side to side like a man mowing. The **Stilts** are much like avocets, but have even longer legs, and are not so well fitted for swimming: so, though their food consists of the same kind of insects and crustaceans, they obtain nearly all of it by wading.

Key to the Species

* Bill over 3 long, and decidedly curved upward; the three front toes webbed .... 1. **American Avocet.**

* Bill less than 3 long and but slightly curved upward; only two of the toes connected by webbing .... 2. **Black-necked Stilt.**

American Avocet

1. **American Avocet** (225. *Recurviróstra americàna*). — A very long-legged, slender - billed, almost white-bodied, wading bird, with dark wings, having large, white bands formed of the coverts and secondaries. The long bill is decidedly curved upward. In summer the head and neck are cinnamon-

red. Common in the interior along the shores of shallow ponds, rare eastward. (Blue Stocking.)

Length, 16-20 ; wing, 8½ (7½-9½) ; tail, 3½ ; tarsus, 3¾ ; culmen, 3¼. North America ; breeding from Illinois (rarely from Texas) north to the Great Slave Lake and wintering along the Gulf coast to Central America.

2. **Black-necked Stilt** (226. *Himántopus mexicànus*). — An exceedingly long-legged, long-billed, black and white wading bird of the shallow ponds. The black begins back of the forehead and extends along the neck and back to the tail ; the wings are also black. The white in- cludes the fore- head, all lower parts, and most of the tail. The black is glossy and somewhat greenish. This graceful bird is especially fond of wading in shallow, salt- marsh ponds. During the breed- ing season it is very noisy, with

Black-necked Stilt

a sharp *click*-like note, which is often given out while on the wing. Its long wings enable this bird to fly well, and it has the habit common in the whole order (*Limícolæ*) of exhibiting alternately the upper and lower side of the body. With this bird, which is so black above and white below, this change of position adds much to the beauty of its movement. (Lawyer ; Long-shanks.)

Length, 13½-15' ; wing, 9 (8½-9½) ; tail, 3 ; tarsus, 4¼ ; culmen, 2½. United States ; breeding from Texas, irregularly, in the interior to the northern border ; rare on the Atlantic coast north of Florida. South in winter to Brazil.

FAMILY XLI. PHALAROPES (PHALAROPÓDIDÆ)

This is a small family (3 species) of small, brightly colored, long, slender-billed, long-legged, swimming and wading birds, which, on shore, appear like sandpipers. They have lobed toes[1] like the grebes and coots. In this group the sexual characteristics are almost completely reversed. The female is the larger and the brighter colored and does the courting of her mate. When the eggs are laid her duties are about over; the male performs most, if not all, of the duties of incubation. Soon after the young are hatched they are able to swim and find their own food.

Key to the Species

* Bill over 1½ long........................... 3. **Wilson's Phalarope.**
* Bill under 1½ long. (A.)
   A. Bill stout and with a flattened tip;[2] wing over 4⅜ long
   ....1. **Red Phalarope.**
   A. Bill very slender and not flattened; wing under 4⅜ long
   ....2. **Northern Phalarope.**

1. **Red Phalarope** (222. *Crymóphilus fulicárius*). — In summer a red-bodied, gray-winged, black-backed, small, ocean-swimming bird with much black and white on the head. In winter a grayish-backed, white-bellied bird with washings of red on head, wings, rump, and tail. This is mainly an inhabitant of the ocean at some distance from land, and rarely comes to shore except after storms. In the autumn it is occasionally seen on the western lakes and rivers. It keeps in flocks, swimming like a duck or walking on floating seaweed

Red Phalarope

as though it were land. The toes have broad, rounded lobes. (Gray Phalarope.)

Length, 7¼-8¼; wing, 5¼ (5-5½); tail, 2½; tarsus, ?; culmen, ?. Northern parts of the northern hemisphere; breeding far north, and wintering irregularly south to the Middle States, Ohio Valley, and Cape St. Lucas.

2. **Northern Phalarope** (223. *Phaláropus lobátus*). — In summer a common, small, ocean-swimming, slender-billed, brightly marked and colored bird, with much of red, black, white, and gray in its plumage. In winter the upper parts are grayish and white. This bird is often seen in great numbers on the ocean, scores of miles from shore, but is rarely seen on land except in its breeding region

Northern Phalarope

of the far north. Like the last species, its toes are furnished with broad, rounded lobes. (Red-necked Phalarope.)

Length, 7-8; wing, 4¼ (4-4½); tail, 2; tarsus, ?; culmen, ?. Northern hemisphere; breeding in the far north, and wintering south to the tropics.

3. **Wilson's Phalarope** (224. *Stegánopus trícolor*). — This inland phalarope has its back ashy colored, with two stripes extending from the bill past the eyes along the sides of the back to the rump, black in front, changing to chestnut near the tail. Its lower parts are

Wilson's Phalarope

white, with a chestnut tint on the sides of the neck.   The *male* is much duller.   The small size, lobate toes, slender bill, and swimming habits will readily distinguish this bird from any other in the inland regions.

Length, 8–10; wing, 5 (4¾–5¼); tail, 2¼; tarsus, 1¼; culmen, 1¼. Temperate America, mainly inland; breeding from Illinois and Utah, north into the British Provinces, and wintering south to Patagonia.

ORDER X. RAILS, CRANES, ETC. (PALUDÍCOLÆ)

An order of wading birds, differing widely in external peculiarities, but associated together because of structural characteristics.

FAMILY XLII.   RAILS, GALLINULES, AND COOTS (RÁLLIDÆ)

This large family (180 species) of mainly marsh-living birds is readily divided into three groups, both by habits and external peculiarities.   (1) The **Rails** form the largest and most characteristic of these subfamilies.   They have short bills, narrow, compressed bodies, long toes, and short, upwardly turned tails.   They inhabit marshes closely covered with reeds and rushes, and their peculiarly narrow bodies fit them to pass without trouble anywhere they wish between the plants, and their long toes enable them to walk with ease and safety over the softest mud, or even over floating weeds.   (2) The **Gallinules** are brightly marked, chicken-like birds of marshes and reed-grown borders of ponds and lakes.   They swim well and appear in their swimming like coots, which they also resemble in having a horny shield or plate on the forehead;[1] but they are like the rails in having long toes without lobes along their edges.   (3) The **Coots** are swimming birds the size of small ducks, with the legs much longer, and the toes lobed[2] instead

      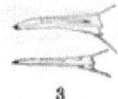

1   2   3

of webbed. There is a horny plate on the forehead.[1] They inhabit creeks and rivers which are surrounded with just such marshes and reed-grown shores as are the dwelling places of rails. The coots are nearly exclusively swimming birds, almost as much so as ducks. The rails swim but little, and the gallinules are intermediate.

Key to the Species

* Forehead with a shield-like, horny extension of the bill ;[1] under tail coverts white ; wing over 6¼ long. (E.)
* No horny extension of the bill on the forehead. (A.)
  A. Bill slender, decurved,[3] 2 or more long. (D.)
  A. Bill slender, decurved,[3] 1¼–1¾ long..............4. **Virginia Rail.**
  A. Bill stout, not decurved, 1 or less long. (B.)
B. Wing over 5 long ; Old World species, rare in America. ....
.... 8. **Corn Crake.**
B. Wing under 5 long. (C.)
  C. Feathers of the back black with broad, buffy borders ....
.... 6. **Yellow Rail.**
  C. Back olive-brown ; wing over 4 long............... ....5. **Sora.**
  C. Back blackish with round, white spots ; wing, 2¼–3¼...........
.... 7. **Black Rail.**
D. Upper parts rich olive-brown, streaked with black ; flanks black barred with white ; wings generally over 6 long and with rufous coverts....................................... 1. **King Rail.**
D. Upper parts grayish streaked with black ; flanks not black, though barred with whitish ....2 and 3 **Clapper Rails.**
  E. Most of the head and all under parts purplish-blue (mottled with white in the young) ; back olive-green (washed with brownish in the young) .... 9. **Purple Gallinule.**
  E. Generally slate-colored above, with conspicuous white streaks on the flanks ; toes without lobes along their edges ....
.... 10. **Florida Gallinule.**
  E. Slate-colored, with white tips to the secondary quills ; toes with broad, membranous lobes[2]..........................11. **Coot.**

1. **King Rail** (208. *Rállus élegans*). — A large, brightly colored, long-billed, cinnamon-red-breasted, olive-brown- to black-backed, distinctly blotched, fresh-water, marsh rail with sides more or less barred with black and white. The wing coverts are brownish-red. The downy *young* are glossy black. This, like most of the rails, rarely flies when it is possible for it to

run and hide in its sedgy home, and so, though it is brightly marked, it is rarely seen.

Length, 16–19; wing. 6½ (6–7); tarsus, 2½; culmen, 2¼. Eastern United States, in fresh-water marshes; breeding north to Missouri and Connecticut and wintering from Virginia southward. Occasionally north to Wisconsin, Ontario, and Maine.

2. **Clapper Rail** (211. *Rállus crépitans*). — A large, pale-colored, olive-gray, salt-marsh rail with yellowish-brown breast, whitish throat, and more or less white-barred belly and sides. Downy *young* are glossy black. This salt-marsh inhabitant takes the place of the last species of the fresh marshes. In the south it is also found in the mangrove swamps. (Mud Hen.)

Length, 14–16; wing, 5½ (5½–6½); tail, 2½; tarsus, 2; culmen, 2–2½. Salt-water marshes of the Eastern and Southern States; breeding from Connecticut southward and wintering in small numbers over about the same range. Casual north to Massachusetts. The **Louisiana Clapper Rail** (211ᵃ. *R. c. saturátus*) of Louisiana is a darker-colored bird having the back broadly striped with brownish-black and the breast more cinnamon-colored.

3. **Florida Clapper Rail** (211–1. *Rállus scóttii*). — This species differs from No. 2 in having the feathers of the back almost black with olive-gray margins, the neck and breast dark cinnamon-red, and the belly and flanks black. These colors give it much the appearance of the king rail, but it lacks the rufous wing coverts of that bird.

Length, 14; wing, 5½; tarsus, 1⅞; culmen, 2¾. Western coast of Florida.

4. **Virginia Rail** (212. *Rállus virginiánus*). — A small, common, brightly colored, short-tailed, long-billed, cinnamon-breasted, brown-backed, distinctly marked rail of both fresh and salt marshes. The sides are somewhat barred with black and white, the wing coverts brownish-red, belly like the breast, and the throat white. The back proper has very dark centers to the feathers. The common morning and evening note of this bird is a grunting sound much like that of a hungry pig.

Like all the rails, it is a difficult bird to observe because of its ability as a skulker.

Length, 8–10½; wing, 4¼ (4–4⅔); tail, 1½; tarsus, 1¾; culmen, 1¼. North America; breeding from Illinois and Pennsylvania, north to Manitoba and Labrador, and wintering from about the same states southward to Central America.

5. **Sora** (214. *Porzana carolina*). — A common, short-tailed, short-billed, long-legged, olive-brown, marsh bird

Virginia Rail

or rail, with many white lines and dashes on the back and wings. The under parts are slaty-gray, changing to white near

the tail, the flanks being barred with black and white. The feathers of the back have darker centers and lighter edges. The *adult* has black about the base of the bill, on the crown, and down the middle of the neck; the *young* lacks these black marks and

Sora

has the breast washed with cinnamon. Fresh-water marshes inhabited by these birds in summer are vocal during the late afternoon and early night with whistled *ker-wees* and loud rolling *whinnies*. Were it not for these cries, many places inhabited by these birds might remain unexplored. (Carolina Rail; Common Crake.)

Length, 8½; wing, 4⅓ (4–4½); tail, 2; tarsus, 1⅓; culmen, ⅔. North America; breeding from Illinois and New York north to Hudson Bay, and wintering from South Carolina to northern South America.

6. **Yellow Rail** (215. *Porzàna noveboracensis*). — A rare, very shy, small, short-billed, much mottled, brownish-yellow rail with the under parts much lighter and less blotched than the back. The feathers of the back have almost black centers, ochraceous buff borders, and more or less of white bars. The sides are barred with dark and white, and the middle of the belly is almost pure white. Its notes have been compared to those of the frogs. (Yellow Crake.)

Length, 7; wing, 3⅔ (3–3½); tail, 1½; tarsus, 1; culmen, ½. North America; breeding from the Middle States, north to Nova Scotia and Hudson Bay, and wintering south to Cuba. Not so rare east as west of the Rocky Mountains.

7. **Black Rail** (216. *Porzàna jamaicénsis*). — A rare, very small, short-billed, very dark-colored, somewhat speckled rail. Its general color is brownish-black, and the markings are mainly white. There is some reddish-brown on the back neck and slate-color on the head and breast. (Little Black Crake.)

Length, 5¼; wing, 2⅞ (2⅓–3⅛); tail, 1½; tarsus, ⅞; culmen, ½. United States; breeding north to Massachusetts, Illinois, and Oregon, and wintering south to western South America.

8. Corn Crake (217. *Crex crex*). — A large, Old World short-billed, brownish-buff rail with much of brownish-red on the wings. The feathers of the back have almost black centers, the sides are barred with white, and the middle of the belly is wholly white. This bird is of only casual occurrence in eastern North America.

Length, 10½; wing, 5¾; tail, 2; tarsus, 1½; culmen, ⅞. Very rare.

9. **Purple Gallinule** (218. *Ionòrnis martínica*). — A short-tailed, long-legged, brilliantly purplish-blue, chicken-like, marsh bird with very long toes, enabling it to walk on the floating leaves. The back itself is a shining olive-green, the under tail coverts white, bill with much red, and the legs yellow. The *young* has much brownish on the back, white mottlings below and bill without red.

Length, 13; wing, 7¼ (6¾-7½) ; tail, 2¾ ; tarsus, 2¼ ; culmen, 1⅝. South Atlantic and Gulf States ; breeding as far north as southern Illinois and South Carolina, straying rarely to Maine and Wisconsin, and wintering from Florida to Brazil.

10. **Florida Gallinule** (219. *Gallinula galeàta*). — A common, southern, dark slate-colored, chicken-like, marsh bird with long toes like the last species. The front half of the bird is nearly black and the rest brownish, except the white under tail coverts, edge of wing, and streaks on the flanks. The bill is red and the legs are yellow in life. (Common Gallinule; Red-billed Mud-hen.)

Florida Gallinule

Length, 13½ ; wing, 7 (6½-7½) ; tail, 2¾ ; tarsus, 2¼ ; culmen, 1¼. Temperate and tropical America ; breeding north to Canada and wintering from the Gulf States to Brazil.

11. **American Coot** (221. *Fùlica americàna*). — A common, large, noisy, short-tailed, short-billed, long-legged, dark slate-colored, swimming bird, with white under tail coverts, white bill, and blackish head and neck. The long toes have broad, scalloped lobes along their edges. When swimming, this bird bobs its head in a peculiar manner; when disturbed, it partly flies and partly swims just over the surface of the

water, giving out a characteristic pattering noise. (Mud Hen; Crow "Duck"; Blue Peter.)

American Coot

Length, 15; wing, 7½ (7-7¾); tail, 2; tarsus, 2¼; culmen, 1¼. North America, north to New Brunswick and Alaska; breeding locally throughout, and wintering from the Middle States to Central America.

FAMILY XLIII. COURLANS (ARÁMIDÆ)

A small family (2 species) of large, rail-like birds, with habits like those of the herons. We have only the following:

Limpkin

1. Limpkin (207. Áramus gigánteus). — A very large, southern (Florida and Texas), long-billed, chocolate-brown, rail - like bird, with most of the plumage

sharply streaked with white. It is usually to be found along the borders of wooded streams and in the swamps, though occasionally it visits the uplands, and, like the heron, perches in trees. It receives the name "Crying-bird," from the character of its wailing call notes. (Courlan; Crying-bird.)

Length, 26; wing, 12 (11-13½); tail, 6½; tarsus, 3½-5; culmen, 3½-5. Florida and southern Texas, south to West Indies and Central America.

FAMILY XLIV. CRANES (GRÚIDÆ)

A family (18 species) of very large, very long-necked, long-legged, heron-like birds, which is placed in this order (Paludi-colæ) with the rails, because of certain structural similarities, not because of size or general appearance. As in the herons, the head is more or less naked, but the plumage is compact, while that of the herons is peculiarly loose. They are inhabitants of marshes and meadows, and live upon both animal and vegetable food, such as frogs, lizards, snakes, mice, Indian corn, etc. Their voice is peculiarly harsh and resonant, and when a number are together the sounds have been likened to those of a pack of hounds in full cry; they can be heard for the distance of a mile, or even two. In flight the neck is extended as in the case of the ibises or storks.

Key to the Species

* Tarsus, 10¾-13 long; general plumage white in the adult............
....1. **Whooping Crane.**
* Tarsus, 9-10¾ long; bill, 4½-6 long................ ....3. **Sandhill Crane.**
* Tarsus, 6-9 long; bill, 2½-4½ long ....2. **Little Brown Crane.**

1. **Whooping Crane** (204. *Grus americàna*). — A very large, white crane, with dull-red head and black wing quills. The red portion, which consists of the top and side of the head and a little along the side of the throat, is free from feathers but is covered by a growth of black hairs. The *young* is similar, but the head is fully feathered, and the plumage is more or

Whooping Crane

less covered by rusty patches, especially on the back. (White Crane.)

Length, 50; wing, 24 (22-26); tail, 9; tarsus, 11½; culmen, 5½ (5-6). Interior of North America; breeding from Illinois north to the Arctic regions, and wintering in the Gulf States.

2. **Little Brown Crane** (205. *Grus canadensis*). — Almost exactly like the next, but smaller, the general color being slaty or brownish gray.

Length, 35; wing, 18½ (17-20); tail, 7; tarsus, 7½; culmen, 3½ (3-4½). Western North America; breeding in the northern portions and migrating southward, mainly west of the Mississippi into the western United States and Mexico.

3. **Sandhill Crane** (206. *Grus mexicana*). — A very large, slaty to brownish-gray crane, with the whole top of the head bare of feathers, but covered with black hairs on a dull reddish skin. The plumage is more or less washed with rusty. The *young* has the head entirely feathered, and the plumage brown, extensively washed with rusty. During the early spring these birds jump about in the most ludicrous manner, as though dancing an Indian war dance, and they stop only when exhausted. (Brown Crane.)

Length, 44; wing, 21¾ (21-22½); tail, 9; tarsus, 10¼; culmen, 5½ (5-6). Southern half of North America, rare on the Atlantic coast except in Georgia and Florida; breeding locally throughout, even north to Manitoba.

Sandhill Crane

ORDER XI. HERONS, STORKS, IBISES, ETC.
(HERODIÒNES)

An order of large, long-necked, long-legged, long-billed, short-tailed birds, with portions of the heads and sometimes of the necks bare of feathers, but covered more or less with hairs.

FAMILY XLV. HERONS, EGRETS, BITTERNS, ETC. (ARDEIDÆ)

A large family (75 species) of large, long-billed, long-necked, long-legged birds, with the head fully feathered, except a space between the eyes and bill (lores). This is the one family of the order (Herodiònes) well represented in all portions of our region. These birds are abundant along the shores of rivers, lakes, salt-water marshes, and bays. In flying, the head is brought back close to the breast by the folding of the neck. Nearly all other birds belonging to this order fly with the neck stretched forward to its full length, and the legs extended backward. The name "squawk" is applied popularly to a number of the species of herons, because of the peculiar cry which is characteristic of the family. **Bitterns** differ from the true herons in being more solitary in their habits, occupying more exclusively grassy meadows and marshes, and in their vocal performances, which have been appropriately called "booming" and "stake driving."

Key to the Species

* Outer toe shorter than the inner one; hind nail fully half as long as the toe.
    — Wing, 10–13 long.........................1. **American Bittern.**
    — Wing, 4–6 long ....2, and 3. **Least Bitterns.**
* Outer toe as long or longer than the inner one; hind nail less than half as long as the toe. (**A.**)
    A. Bill slender, fully four times as long as it is high at base.[1] (**C.**)
    A. Bill stouter, about three times as long as it is high at base.[2] (**B.**)

             1   2

B. Bill less than a half inch shorter than the tarsus; top and bottom of the bill but slightly convex ....13. **Black-crowned Night Heron.**

B. Bill over a half inch shorter than the tarsus; top and bottom of bill
decidedly convex.[1] ....14. **Yellow-crowned Night Heron.**
  C. Wing, 17-22 long; plumage pure white ....
  ....4. **Great White Heron.**
  C. Wing, 17-21 long; upper part grayish or slaty-blue....
  ....5. **Ward's Heron** or 6. **Great Blue Heron.**
  C. Wing, 14-17 long; plumage pure white.......7. **American Egret.**
  C. Wing, 11-14 long; plumage mostly white or slate colored........
  ....9. **Reddish Egret.**
  C. Wing under 11 long. (**D.**)
D. Wing, 6-8 long; back with much green or greenish ... .
  ....12. **Green Heron.**
D. Wing, 8-11 long. (**E.**)
  E. Plumage pure white .....8. **Snowy Heron** (and *young* of No. 11).
  E. Wing coverts more or less margined with rusty ....
  .... 10. **Louisiana Heron.**
  E. Wing coverts without rusty margins ....11. **Little Blue Heron.**

1. American Bittern (190. *Botaúrus lentiginòsus*). — A very
common, large, solitary, retiring, grassy-marsh-living, stout-
billed, buffy and brown, mottled, heron-like bird, with many

American Bittern

elongated, loose feathers on the
crown and lower neck. There
is a velvety black streak on the
side of the neck. This bird
makes a note which sounds like
driving a stake with a mallet,
or at other times like the working of an old wooden pump.
This "booming" can be heard
a long distance, and during its
progress the bird exhibits most
amazing contortions of its body.
It is noted for its ability to stand
in one position for an indefinite
period, though the other members of the family are also good
at "tableaux vivants." (Stake-driver; Marsh Hen; Bog-bull.)
  Length, 24-34; wing, 10¼ (9¼-12¼); tail, 4; tarsus, 3¾; culmen, 3.
Temperate North America; breeding mainly north of the Carolinas, and
wintering from Virginia southward to Central America.

2. Least Bittern (191. *Ardétta exilis).* — A bird formed similarly to the last and with similar habits, but much smaller. It is a buffy and chestnut-colored bittern with the crested crown, back, and tail glossy black and a patch on the side of the breast blackish. The *female* is similar but the crown and back are a purplish-chestnut rather than black, and the under parts are darker and streaked with brownish.

Least Bittern

Length, 13; wing, 4½ (4½-5½); tail, 2; tarsus, 1¾; culmen, 1¾. Temperate North America; breeding north to Ontario and wintering from Florida south to the West Indies and Brazil.

3. Cory's Least Bittern (191-1. *Ardétta neóxena).* — A bird like the last in everything but color and size. It is a bittern with reddish-chestnut on the breast and under parts as well as on the sides of the head and throat and the wing coverts. The crown, back, and tail are black, with a distinct green gloss. This species has the under tail coverts a dull black, the last has them washed with buffy. Only a few specimens have been observed, and those, curiously, in widely separated localities.

Length, 11; wing, 4¼; tarsus, 1¾; culmen, 1¾. Florida, Ontario, and Michigan.

4. Great White Heron (192. *Árdea occidentális).* — An exceedingly large, southern (Florida), entirely white heron with (in the breeding season) long, narrow, stiffened feathers on the lower neck and back and two narrow plumes on the head. This is a larger bird than the American egret (No. 7) and much larger than the snowy heron (No. 8), but has not the "aigrette" plumes of those species.

Length, 50; wing, 19 (17–21); tarsus, 8¼; culmen, 6¼. Florida, Cuba, and Jamaica.

5. **Ward's Heron** (193. *Árdea wárdi*). — A Florida great blue heron. It is similar to the next but somewhat larger. Generally the lower parts are whiter, the neck darker, and the legs lighter, being olive instead of black.

Length, 48–54; wing, 19¼–20½; tarsus, 8¼; culmen, 6¼–7. Florida; common in the southern half of the state.

6. **Great Blue Heron** (194. *Árdea herodias*). — An exceedingly large, common, generally bluish or slate-colored, crested heron, with many black, white, and yellowish streaks on head, neck, and belly, and chestnut on the bend of the wing. The tibia feathers are brown, the center of crown and throat white, and the sides of the crown black. The *young* has the entire crown black and lacks the plumes of the old bird. As feeders these are solitary birds, though they nest and roost in colonies. Their food is made up of fishes, frogs, snakes, mice, etc. (Blue "Crane.")

Great Blue Heron

Length, 42–50; wing, 19 (18–20); tail, 7½; tarsus, 7; culmen, 4½–6½. North America; breeding north to Hudson Bay, and wintering from Pennsylvania south to the West Indies and northern South America.

7. **American Egret** (196. *Árdea egrétta*). — A very large, pure white heron with about fifty straight "aigrette" plumes on

the back (in the breeding season) reaching beyond the tail. To get these plumes, which are at present fashionable for ladies' hats, this species and the next must be shot in the breeding season; so a few years of this "fashion" have made these most graceful and dainty birds very rare, and a few years more of the slaughter will render them extinct. Some women wearing such plumes try to exonerate themselves from blame on the plea that the birds are killed without their approval, but that being dead no harm can be done by purchasing and using their feathers. They are forgetful of the fact that every use of such a plume continues the fashion, increases the demand, and leads to the further killing of birds in constantly increasing numbers. Hence all who wear the plumes are directly responsible for the slaughter of the birds.

Length, 40; wing, 15 (14–17); tail, 6; tarsus, 6; culmen, 4¼–5. Temperate and tropical America; breeding north to Illinois and New Jersey, straying to New Brunswick and Manitoba, and wintering from Florida to Patagonia.

8. **Snowy Heron** (197. *Ardea candidissima*). — A small, beautiful, crested, pure white heron. with about fifty *recurved*, "aigrette" plumes on the back during the breeding season. The bill and legs are black, and the lores and feet yellow. Becoming exceedingly rare, because killed, like No. 7, in the breeding season. (Snowy Egret.)

Length, 20–27; wing, 9½ (8¼–10½); tail, 4; tarsus, 3¾; culmen, 2–3¼. Temperate and tropical America; breeding north to Long Island, and

Snowy Heron

wintering from Florida south to central South America, casually north to Nova Scotia and British Columbia.

9. **Reddish Egret** (198. *Árdea ruféscens*). — A southern, large, "aigrette" heron, which occurs in two color phases. (1) Pure white throughout, with the exception of the tips of

Reddish Egret

the primaries, which are sometimes speckled with grayish. (2) Slate-colored on the body and chestnut-colored on the neck. The adult, in breeding dress, has about thirty of the "aigrette" plumes. Intermediate forms between the phases are also found.

Length, 27–32; wing, 12½ (12–14½); tail, 4½; tarsus, 5¼; culmen, 3½. Gulf States north to southern Illinois, south to Jamaica and Central America.

10. **Louisiana Heron** (199. *Árdea tricolor rujicállis*). — A small, southern, bluish-slate-colored heron, with white belly and throat line and pur-plish crest and neck. The white rump is concealed by elongated, purplish-white-tipped "aigrette"

Louisiana Heron

plumes, reaching to the tail. As the *young* lack plumes, they show the white lower back and rump, and the back has more or less of brownish washings.

Length, 23–28; wing, 10 (8½–11); tail, 3½; tarsus, 3¾; culmen, 3½–4½. Gulf States, south to Central America and West Indies, casually north to New Jersey and Indiana.

11. **Little Blue Heron** (200. *Árdea cœrùlea*). — A small, common, bluish-slate-colored heron, with the head and neck slightly purplish. The lower neck and back feathers are lengthened and sharply pointed. The legs and feet are black.

The *young* are white, with bluish-slate-colored tips to the primaries and greenish-yellow legs and feet. Of course specimens with all gradations of color, intermediate between that of the young and adult, can be found. The young of this species can at some distance be distinguished from the snowy herons by their greenish instead of black legs.

Length, 20–30 ; wing, 10 (9–11) ; tail, 4¼ ; tarsus, 3½ ; culmen, 2¾–3¼. Eastern United States; breeding north to Illinois and New Jersey, wandering north to Nova Scotia, and wintering from Florida to northern South America.

12. **Green Heron** (201. *Ardea virescens*). — A common, small, dark-chestnut-bodied, greenish backed and crowned heron, with much white in streaks down the front from chin to the lower breast. This solitary heron is found more frequently in wooded borders of streams and ponds than in open places, and is most active in the morning and evening. (Poke.)

Length, 15½–22½ ; wing, 7¼ (6½–8) ; tarsus, 2 ; culmen, 2–2¼. Temperate North America ; breeding north to Ontario and Oregon, and wintering from Florida to northern South America.

Green Heron

13. **Black-crowned Night Heron** (202. *Nycticorax nycticorax nævius*). — A common, stout-billed, night-flying, bluish-gray heron. with the crown and back greenish-black. The crown is furnished with two or three slender plumes. *Young* with much of mottled browns on the back and no plumes on the head. After sunset, these birds leave their roosts to feed, giving out occasionally their harsh *quawk*. They are very social, roosting together in hundreds.

Black-crowned Night Her n

When feeding the young,
they may be seen gather-
ing food in the daytime.
(Quawk.)

Length, 24 ; wing, 12 (11–13);
tail, 5 ; tarsus, 3¼ ; culmen, 3.
America ; breeding north to
Ontario and Manitoba, and wintering from the Gulf States to
southern South America.

**14. Yellow-crowned Night
Heron** (203. *Nycticorax vio-
laceus*). — A common,
crested and plumed, stout-
billed, night-flying, grayish-
blue heron with a buffy
crown, white cheeks and
mainly white plumes on an
otherwise black head. The
neck and lower parts are lighter than the back. The long,
loose feathers of the back
extend beyond the tail.
The colors of the *young* are
mottled browns and there
are no head plumes. These
birds are solitary in their
habits and are never seen
in colonies like the last
species. They are found
singly or in pairs along the

Yellow-crowned Night Heron

borders of wooded streams, and are less strictly nocturnal
birds than the black-crowned night heron.

Length, 22–28; wing, 12 (10¼–12¼); tail, 5; tarsus, 4; culmen, 2¼.
Tropical and warm temperate North America, north to the Carolinas,
lower Ohio Valley and Lower California, casually to Massachusetts and
Colorado ; breeding throughout its United States range.

FAMILY XLVI. STORKS AND WOOD IBISES (CICONÍIDÆ)

A family (25 species) of mainly Old World, stout-billed, heron-like birds with a large portion of the head naked or free from feathers and with the bill neither curved for its whole length nor decidedly widened at tip. Our species have the bill extremely stout at base, it being practically as high as the head.

Key to the Species

* End of bill downwardly curved [1]........... ...
...............................1. **Wood Ibis.**
* End of bill upwardly curved [2]...............
..............................2. **Jabiru.**

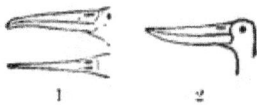

1. **Wood Ibis** (188. *Tántalus loculàtor*). — An exceedingly large, white, ibis-like bird with the head and neck bare of feathers and the very long, stout bill straight for half its length, and curved downwards. The wing quills and the tail are glossy greenish-black. The *young* are more grayish in color, have the breast more or less feathered, and the head and neck a decided grayish-brown.

Length, 35–45; wing, 18½ (17½–19½) ; tarsus, 8; culmen, 7–9. Southern United States ; breeding in the Gulf States ; after the breeding season it sometimes wanders northward to Kansas and New York. South to central South America.

Wood Ibis

2. **Jabiru** (189. *Myctèria americàna*).—A tropical, extremely large, white stork, with immensely large recurved bill; head and neck bare, excepting a hairy patch on the back head. The head and neck are black, with a broad red collar round the lower part. The *young* has some brownish-gray on the back and lower portion of the neck.

Length, 54; wing, 26 (24½–27); tail, 9½; tarsus, 12; culmen, 9½–13. Tropical America; north casually to southern Texas.

Jabiru

FAMILY XLVII. IBISES (IBÍDIDÆ)

A family (30 species) of large, short-legged (for the heron order), shore-living birds, with peculiarly long, downwardly curved bills. They are found only in warm countries, and live in flocks throughout the year. Their food is mainly crustaceans, reptiles, and fish which they find on mud flats at low tide, and on the shores of lakes, bays, and salt-water marshes. The four species here given are all that occur in North America.

Key to the Species

* General color white in the adult (grayish-brown in the young without bright reflections on the back)....................1. **White Ibis.**

* Bright red or scarlet...................................2. **Scarlet Ibis.**

* Chestnut with purplish and greenish reflections in the adult (dark-brown with greenish reflections on the back in the young). **(A.)**

 A. Lores greenish in life; feathers around the bill like the back in color ....3. **Glossy Ibis.**

 A. Lores red; feathers around the bill white ....
 ....4. **White-faced Glossy Ibis.**

1. White Ibis (184. *Guára álba*). — A large, shore-living, white bird, with a long sickle-like, downwardly curved bill. The tips of the outer primaries are black. The bill is evenly curved from end to end. The *young* is grayish-brown on the back, and white on the belly and rump. The flocks of these birds when on the wing are rendered conspicuous by the contrast be-

White Ibis

tween the white of the general plumage and the black tips of the primaries. On account of the peculiar bill these silent birds have a curlew-like appearance, but the bare spot around the eyes distinguishes them. They live in flocks of from five to hundreds throughout the year. (Spanish "Curlew.")

Length, 21–28 ; wing, 11½ (10½–12½); tail, 5 ; tarsus, 3½ ; culmen, 5–7. South Atlantic and Gulf States, south to northern South America ; north to North Carolina, Illinois, Utah, and Lower California, casually to Connecticut.

2. Scarlet Ibis (185. *Guara rúbra*). — A scarlet-colored ibis, with black tips to the secondaries. This is a South American bird, but has been seen a few times in Florida, Louisiana, Texas, and the West Indies.

Length, 28 ; wing, 11 ; tail, 5 ; tarsus, 3¾ ; culmen, 6.

Glossy Ibis

3. Glossy Ibis (186. *Plégadis*

autumnalis). — A bright, chestnut-colored ibis, with brilliant, purplish and greenish reflections on the back, wings, under tail coverts, and the front of the head. The *young* is a blackish-brown bird, with greenish reflections on the back. This is a rare species in the United States.

Length, 24 ; wing, 11 (10–12) ; tail, 4 ; tarsus, 3½ ; culmen, 4¼–5¼. Warmer parts of the Old World, the West Indies, and southeastern United States, wandering north to New England and Illinois.

4. **White-faced Glossy Ibis** (187. *Plégadis guarduna*). — A bird similar to the last. The *young* is so nearly like the

White-faced Glossy Ibis

young of the glossy ibis that the determination must be more or less uncertain, but the *adult* has white feathers around the base of the bill. The lores are red in life, while those of No. 3 are greenish.

Length, 24 ; wing, 10½ ; tail, 4 ; tarsus, 3¼ ; culmen, 3¾–6. Western United States from Texas to California and Oregon ; casually to Kansas and Florida; southward to West Indies, Mexico, and South America.

FAMILY XLVIII. SPOON-BILLS (PLATALEIDÆ)

A small family (6 species) of long-legged, long-necked, heron-like shore birds, with peculiarly broadened, spoon-shaped bills. They all live in warm countries, and are usually found in flocks. Their method of obtaining food is peculiar. The bill is placed in the soft mud

Roseate Spoonbill

and swung from side to side, the food, which consists mainly of mollusks, being thus scraped up.

1. **Roseate Spoonbill** (183. *Ajaja* (*i-à-n-i*) *ajaja*). — A very large, rare, southern, pink or rosy-colored ibis-like bird, with a head bare of feathers, and a bill much broadened at the tip, like a spoon. The sides of neck and end of the tail are buff, and the neck and upper back nearly white. The *young* has the head feathered. These birds are generally in flocks, and the nesting is in colonies. (Pink "Curlew.")

Length, 28-35; wing, 15; tail, 4½; tarsus, 4¼; culmen, 6¾. South Atlantic and Gulf States, south to Patagonia.

ORDER XII. TOOTH-BILLED WADERS (ODONTO-GLÓSSÆ)

An order consisting of the following:

FAMILY XLIX. FLAMINGOES (PHŒNICOPTÉRIDÆ)

A small family (7 species) of large, exceedingly long-legged, long-necked, web-footed, semi-tropical birds, with peculiarly bent bill, the edges of which are furnished with ridges or lamellæ, like those of the ducks.

1. **American Flamingo** (182. *Phœni-cópterus rúber*). — A southern, exceedingly tall, rosy to vermilion-colored wading bird, with black wing quills, and a peculiar, heavy, abruptly bent bill. The toes of the flamingo are fully webbed, and the lamellæ of the bill are used as strainers (as in the case of the ducks) through which the sand and mud are separated from the food. These birds gather in flocks in shal-

American Flamingo

low bays or mud flats, usually near the sea, and with the bill in the soil procure their food, which consists in great part of mollusks and crustaceans. In flying, the neck and legs are stretched out at full length.

Length, 45; wing, 16; tail, 6; tarsus, 13; culmen, 5½. Atlantic coast of the warmer parts of America; southern Florida.

ORDER XIII. LAMELLIROSTRAL OR TOOTH-BILLED SWIMMERS (ANSERES)

An order consisting of the following:

FAMILY L. DUCKS, GEESE, AND SWANS (ANÁTIDÆ)

This, the largest family (200 species) of swimming birds, comprises all our domestic water fowl as well as an important portion of the gunner's prey. Their feathers form the softest material for our pillows and couches. and their flesh the most palatable of foods. These birds are readily separated into five easily recognized groups or subfamilies, viz: swans, geese, sea ducks, river ducks. and fish ducks.

[1] They are all furnished with ridges or teeth along the edges of the bill,[1] which in most cases serve as strainers for removing the mud, sand, etc., from the food; in a few cases they serve as teeth.

(1) The **Swans**, which are the largest of these birds, form the smallest group (10 species). They are large, very long-necked. white (adult), gracefully swimming birds. with a stripe of bare skin extending from the eye to the bill. Because of the position of the legs, far back along the body, their movements on land are very awkward. In feeding they do not dive, but merely tip up the body, or usually simply thrust the head and neck under water. Their food is in good part vegetable, but they eat snails also. Nos. 1 and 2.

(2) The **Geese** form a group intermediate between the swans and the ducks. They are large, long-necked. comparatively long-legged birds, with the space in front of the eye feathered.

They spend much less of their time in the water than ducks do, and the food of most species is almost entirely vegetable. The legs being longer, they are better walkers than ducks. In water they obtain their food by tipping up the tail and thrusting the head and neck as far into the water as possible. In this habit they are like the swans and the river ducks, but unlike the sea ducks and the fish ducks. The hissing, when they are interfered with, is a trait common to both geese and swans. Nos. 3-10.

(3) The **Sea Ducks** are the largest of the subfamilies (nearly 100 species), and they are found in the largest flocks. These are the ducks of the open and deeper waters of large lakes, bays, and coasts; many of the species are found only in salt water. They do not, as a rule, "tip up" like the river ducks, but dive, often to great depths, for their food. This consists mainly of animal matter, such as snails, crustaceans, etc., but not including fish. These ducks have the hind toe bordered with a rounded membrane or lobe-like web.[2] They are generally day feeders, while most of the river ducks feed at night. With the exception of the canvas-backs, considered the best of all ducks for food, the sea ducks are not so palatable as the river ducks. Some species have very rank, coarse flesh, while the river ducks are all good table food if well cooked. Nos. 11-28.

(4) The **River Ducks** (50 species) include most of the ducks of rivers and ponds, and differ from the last subfamily in not diving for their food. They are mere "tip ups," spending a good portion of their time with their tails in air and heads and necks immersed, probing the bottom of shallow places for their food, which consists of both vegetable and animal matter, such as roots, seeds, snails, insects, etc. The hind toe is simple;[3] that is, it has no such lobed membrane as is found on the sea ducks. These ducks are found in the United States, chiefly as migrants, and visit mainly quiet and shallow and usually fresh waters.

When disturbed, they leave the water at a bound, and in a few seconds are beyond the gunner's range.

(5) The **Fish Ducks** are narrow-billed ducks, with the heads generally crested. They have the lobed hind toe,[4] and like the sea ducks, dive for their food. Their prey consists of fish, which they pursue under water. There are but three species in North America, of which two are "fishy" food. Only the hooded merganser is good for table use. The saw-like teeth along the nearly cylindrical bill[3] enable these ducks to capture their prey and give the name sawbills.

Key to the Subfamilies

* Neck as long as the body; tarsus, 4 or more long; wing, 20 or more; adult entirely white.............................**Swans**, below.
* Neck shorter than the body; tarsus under 4 long. (**A.**)
  A. Tarsus, 2–4 long and longer than the middle toe without claw (except in No. 10, a southern species with a hind toe about 1 long); front of tarsus with rounded scales[1] instead of square scutellæ....
  ....Geese, p. 280.
  A. Tarsus not over 2 long and shorter than the middle toe without claw; front of tarsus with distinct scutellæ.[2] (**B.**)
  B. Bill nearly cylindrical, only about as wide as high throughout;[3] head in most cases distinctly crested..................**Fish Ducks**, p. 304.
  B. Bill always wider than high near tip; head rarely crested. (**C.**)
    C. Hind toe with a rounded membranous lobe[4]....**Sea Ducks**, p. 284.
    C. Hind toe without a lobe-like border[2]........**River Ducks**, p. 297.

SWANS (SUBFAMILY CYGNINÆ)

Characteristics given on p. 276

Key to the Species

* Bare skin in front of eye with yellow; back end of nostril much nearer to the tip of bill than it is to the front corner of the eye..........
  ....1. **Whistling Swan.**
* Bare skin in front of eye without yellow; back end of nostril about midway between the tip of bill and the front corner of the eye......
  ....2. **Trumpeter Swan.**

1   2   3   4

1. **Whistling Swan** (180. *Òlor columbiànus*). — An exceedingly. large, very long-necked, swimming bird, with the plumage white throughout. Feet and bill black, with a yellow spot on the lores. *Young* grayish with a brownish head. When feeding, this swan is very noisy, especially at night. Its "notes are extremely varied, some closely resembling the deepest base of the common tin horn, while others run through every modulation of false note of the French horn or clarionet." These different notes are supposed to be given by birds of different ages. Rare on the Atlantic coast north of Virginia.

1. Whistling Swan   2. Trumpeter Swan

Length, 56; wing, 22; tail, 7½; tarsus, 4; culmen, 4. North America; breeding in the Arctic regions, and wintering along the South Atlantic States.

2. **Trumpeter Swan** (181. *Òlor buccinàtor*). — Like the last but larger and without the yellow spot on the bare skin in front of the eye. *Young* with the body grayish tinted and the head and neck somewhat brownish. Rare east of the Mississippi. The habits are about the same as in No. 1, but the notes are more musical.

Whistling Swan

Length, 64; wing, 24 (21–28); tail, 8½; tarsus, 4¾; culmen, 4½. Chiefly in the interior of North America; breeding from Iowa northward, and wintering along the Gulf States. Its habitat extends from the Atlantic to the Pacific but it is very rare along the Atlantic.

GEESE (SUBFAMILY ANSERINÆ)

Characteristics given on p. 276

Key to the Species

* Wing, 8-10 long ; Louisiana to Texas. . . . . .10. **Fulvous Tree-duck.**
* Wing over 11 long. (**A.**)
    A. Serrations on the cutting edge of the upper mandible scarcely visible from the side at all ; if visible then only at the base ; bill, feet, and portions of the head black. (**D.**)
    A. Serrations visible from the side for more than half the length of bill ; bill and feet pale. (**B.**)
  B. Depth of bill at base about ⅓ the length of culmen ; forehead white in the adult. . . . . .5. **White-fronted Goose.**
  B. Depth of bill at base much greater than ⅓ the length of culmen. (**C.**)
    C. General plumage of *adult* white ; *young* grayish-brown with the wing coverts widely margined with white. . . . . .3. **Snow Goose.**
    C. General plumage gray, grayish-brown, or brown without conspicuous white margins to the wing coverts . . . . .4. **Blue Goose.**
  D. Head without white but the side of the neck with white streaks ; belly white. . . . . .7. **Brant.**
  D. Head without white but the side and front of the neck with white streaks ; belly brownish-gray. . . . . .8. **Black Brant.**
  D. Head with a whitish triangular patch on the cheek and throat (these parts are mixed with blackish in the young). . . . . .6. **Canada Goose.**
  D. Head mostly white ; lores black. . . . . .9. **Barnacle Goose.**

3. **Lesser Snow Goose** (169. *Chen hyperborea*). — A large, white-plumaged goose, with black-tipped primaries, and red bill and feet. The *young* has much grayish on the head and

back ; rump. tail. and lower parts white, and white margins to the wing coverts. This, the smaller snow goose, is rarely found east of the Mississippi. The eastern form is given below.

Lesser Snow Goose

Length. 23-28 ; wing, 14½-17 ; tail, 5½ ; tarsus, 2¼-3½ ; culmen, 2-2½. Pacific coast to the Mississippi Valley ; breeding in Alaska, and wintering south to Illinois and California. The **Greater**

Snow Goose (169ª. *C. h. nivalis*) is like the last, but much larger. Length, 28-38; wing, 17-19; tail, 6½; tarsus, 3-3½; culmen, 2¼-2¾. North America; breeding far north, and wintering from Maryland to Cuba. Rare on the Atlantic coast north of Virginia.

4. **Blue Goose** (169-1. *Chen cœruléscens*). — A brownish-gray goose, with the head and upper neck white, and the middle and lower neck blackish. The lower belly is a light gray, or sometimes almost white. The wing coverts have almost no whitish margins. The *young* has the head and neck grayish-brown, with only the chin white.

Length, 26-30; wing, 15-17; tail, 5½; tarsus, 3¼; culmen, 2¼. Interior of North America; breeding on eastern shores, Hudson Bay, and wintering on the Gulf coast. Rare on the Atlantic coast.

5. **American White-fronted Goose** (171ª. *Anser albifrons gámbeli*). — A large, brown-necked, gray-backed, white-bellied goose, with a white forehead on an otherwise brown head. The nearly white breast is peculiarly blotched with black. The *young* lacks the white forehead and the black breast blotches. Although rare on the Atlantic coast, these geese are

American White-fronted Goose

common from the Mississippi Valley to the Pacific, mainly in low, bushy, or wooded regions.

Length, 27-30; wing, 14¼ 17½; tail, 5½; tarsus, 2⅜-3¼; culmen, 1¾-2¼. North America; breeding in the Arctic regions, and wintering south to Mexico and Cuba.

6. **Canada Goose** (172. *Bránta canadénsis*). — A common, very large, grayish-brown-bodied, black-necked, black-tailed goose,

with a broad white patch under the head, extending on the sides back of the eyes. The chin and the rest of the head are

Canada Goose

black. The under parts are much lighter, fading to white around the tail. The *young* has the white cheek and throat patch mixed with blackish. This is the common wild goose of the eastern United States, and the wedge form of the flocks in their migrations through the air has been seen and the noise of their *honking* heard by most persons.

Length, 35-43; wing, 15½-21; tail, 7; tarsus, 2½-3½; culmen, 1¼-2¼. Temperate North America; breeding in the Northern States and British Provinces, and wintering from the Middle States to Mexico. The **Hutchins's Goose** (172ᵃ. *B. c. hutchinsii*) is like the last but smaller. Length, 25-34; wing, 15-18; tail, 5½; tarsus, 2¾; culmen, 1¼-1⅞. North America; breeding in the Arctic regions, and migrating south, mainly through the Mississippi Valley and westward. The **Cackling Goose** (172ᶜ. *B. c. minima*) is still smaller and has a darker and more brownish breast and upper belly. Length, 23-25; wing, 13½-15; tail, 5; tarsus, 2⅜; culmen, 1¼. Western North America; breeding in Alaska, and migrating southward through the Western States, west to Wisconsin.

7. **Brant** (173. *Branta bernicla*). — A large,

Brant

brownish-gray goose, with black head, neck, and breast, except some white scratchings on the sides of the neck just below the head. The lower breast is ashy, fading to white on the belly and longer tail coverts; the wing quills and tail feathers are almost black. The *young* has less white on the neck, but the secondary wing quills are tipped with white. These geese fly in a rather compact mass without the leader so characteristic with the Canada goose.

Length, 23–30½; wing, 12½–13½; tail, 4½; tarsus, 2¼; culmen, 1¼–1½. Northern portions of the northern hemisphere, in North America chiefly on the Atlantic coast. Rare away from salt water.

8. **Black Brant** (174. *Brinta nigricans*). — Like the last, but the lower breast and upper belly are much darker, almost blackish, and the white scratchings are found both on the sides and front of the neck.

Length, 22–29; wing, 13; tail, 4¼; tarsus, 2¼; culmen, 1¼. Arctic and western North America; migrating south to lower California; casual in the Atlantic States.

9. **Barnacle Goose** (175. *Brinta leucópsis*). — An Old World goose with nearly the whole head white to the neck, except a black loral stripe. It is rarely found on our shores.

Length, 24–28; wing, 15–17; tail, 5½; tarsus, 2¾; culmen, 1¾.

Barnacle Goose

10. **Fulvous Tree-duck** (178. *Dendrocygna fúlva*). — An extreme southern, small, duck-like, yellowish-brown goose with

white tail coverts both above and below.   This white is
rendered conspicuous by the black rump and tail; there is
also a black line extend-
ing down the nape and
back neck. The wing is
without a white specu-
lum. This bird nests in
trees, and to give it
power to grasp the limbs
of trees its hind toe is
much lengthened, being
about an inch long.

Length, 20 ; wing, 9 ; tail,
3¼ ; tarsus, 2¼ ; culmen, 1¾.
Southern United States, Louisiana, and Texas to California, and southward to
Mexico.   Also in South
America. Casual in North
Carolina and Missouri. The
Black-bellied Tree-duck
(177. *Dendrocygna autumnalis*) is a similar bird, but

Fulvous Tree-duck

has a large white patch on the wings and a black belly. It is found from
southern Texas westward, and southward into South America.

SEA DUCKS (SUBFAMILY FULIGULINÆ)

Characteristics given on p. 277

Key to the Species

* Wing, 6 or less long ; tail feathers with narrow webs and stiff shafts
  extending beyond the webs ;[1] upper tail cov-
  erts very short. (**N.**)
* Wing, 6–7 long ; upper tail coverts about ½ as
  long as the tail.............18. **Buffle-head.**
* Wing over 7 long. (**A.**)
  A. Feathers at the side and at the top of the bill so extended as to
     leave a bare portion between, which is ⅓ as long as the bill.[2] (**L.**)
  A. No such great extension of both the loral and the frontal feathers
     on the upper mandible. (**B.**)

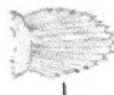

B. Bill peculiarly bulging at base ;[3] nail large and so united with the bill as to give the nail a very indistinct outline. (**J.**)

B. Bill appendaged with a lobe at base formed of the skin of the cheeks ;[4] culmen about 1 long ; speculum[5] violet.........20. **Harlequin Duck.**

B. Bill appendaged with a leathery expansion at the sides near tip ; culmen over 1½ long ; speculum white. The **Labrador Duck** (156. *Camptolaimus labradórius*) might possibly be found, though it is thought to be extinct.

B. Bill of the usual duck form. (**C.**)

  C. Tail pointed (over 6 long, *male* ; about 3 long, *female*); bill black and orange ; nostril within less than ¼ inch of frontal feathers ... .19. **Old Squaw.**

  C. Nostril about ⅓ inch from frontal feathers ; nail of bill narrow and distinct. (**D.**)

D. Bill high at base (over ½ as high as long); under tail coverts white. (**I.**)

D. Bill not so high at base ; under tail coverts dark. (**E.**)

  E. Bill decidedly wider near tip than at base.[6] (**G.**)

  E. Bill with the width near tip about the same as at base (in any case less than ⅛ inch wider). (**F.**)

F. Bill about ⅓ as wide as the length of culmen ....11. **Redhead.**

F. Bill about ⅓ as wide as long......................12. **Canvas-back.**

  G. *Male* with an orange ring around neck ; speculum[5] bluish-gray ; *female* chiefly brown................... 15. **Ring-necked Duck.**

  G. *Male* with white speculum ; *female* with white face. (**H.**)

H. Wing over 8¼ long............. ....13. **American Scaup Duck.**

H. Wing under 8¼ long ....14. **Lesser Scaup Duck.**

  I. *Male* with gloss of the almost black head and throat green ; *female* with a brown head ....16. **American Golden Eye.**

  I. *Male* with the gloss of the dark head and throat purple ; *female* with a brown head...................17. **Barrow's Golden Eye.**

J. Wing, 10½ or more long ; a white wing patch in both sexes......... ....25. **White-winged Scoter.**

J. Wing less than 10½ long. (**K.**)

  K. Culmen, 1⅝ or more long ; the feathers on the culmen reaching about as far forward as those on the sides of the upper mandible ....24. **American Scoter.**

  K. Culmen less than 1⅝ long ; feathers on the culmen reaching about an inch farther forward than those on the side of the upper mandible ....26. **Surf Scoter.**

L. Feathers on the culmen extending forward much farther than those on the side of the upper mandible.................23. **King Eider.**

3

4

5

6

L. Feathers on the side of the bill extending forward farther than those on the culmen. (**M.**)

M. The two bare stripes of bill between the culmen feathers and the side feathers end in sharp points............21. **Northern Eider.**

M. The two bare stripes with the back ends broad and rounded...... ....22. **American Eider.**

N. Outer toe longer than the middle toe; lining of the wings whitish... .... 27. **Ruddy Duck.**

N. Outer toe shorter than the middle toe; lining of the wings blackish.. ....28. **Masked Duck.**

    11. Redhead (146. *Aythya americana*). — A duck similar to the next, and often confounded with it. The head is

Redhead

a lighter color, and has not the blackish blotches, found on crown and chin of that species; the wavy lines of black and white on the back are about equal in width, while in the canvas-back the white ones are wider; the comparative width of bill is greater, being nearly one half the length. The *female* lacks the wavy cross lines of the female canvas-back, so is readily distinguished from that species. It is more like the female ring-neck (No. 15) in coloring, but has a wing over 8 long, while the ring-neck has one less than 8 long.

    Length, 17-22; wing, 9 (8½-9½); tail, 3; tarsus, 1½; culmen, 2¼. North America; breeding from Maine, Michigan, and California northward, and wintering from the Middle States south to Mexico. Found on bays and rivers rather than on coasts.

12. Canvas-back (147. *Aythya vallisnèria*). — A large, chestnut-headed, black-breasted duck, with the back, wings, and lower belly appearing like canvas, with fine wavy cross lines of black and white, the white lines wider. *Female*, with the whole head and neck somewhat of a chocolate or cinnamon color, and the back grayish-brown barred with white, wavy cross lines. Belly whitish. This species has the name among epicures of being the best of all game ducks. Be-

Canvas-back

cause of its destruction for food purposes, it is becoming scarcer each year. The species, 11–15, are somewhat intermediate between river and sea ducks, and are more frequently found on rivers and bays than on open seas; when on shallow waters they merely "tip up" in feeding.

Length, 20–24; wing, 9 (8¾–9¼); tail, 3; tarsus, 1⅔; culmen, 2¼. North America; breeding from the northwestern states northward, and wintering from the Middle States to Cuba and Mexico.

13. American Scaup Duck (148. *Aythya marila néarctica*). — A large, common, black-headed, "canvas"-backed black-breasted, black-tailed duck, with white speculum and belly; the head shows, in proper light, greenish reflections. The back, sides, and lower belly are covered with many black and white wavy cross-bars. The *female* is mainly umber-brown colored, with a white speculum, belly, and band around base of bill; the back and sides are generally waved with white bars. The name *scaup* is derived from the sound of its notes. A very common bay duck. (Greater Scaup Duck; Black-head; Blue-bill.)

American Scaup Duck

Length, 19; wing, $8\frac{5}{8}$ ($8\frac{1}{4}$–9); tail, 3; tarsus, $1\frac{1}{2}$; culmen, 2. North America; breeding from Manitoba (rarely Minnesota) northward, and wintering from Long Island to northern South America.

14. **Lesser Scaup Duck** (149. *Áythya affinis*). — A duck smaller than the last, but with nearly the same coloring, excepting that the reflections from the head of the *male* are purplish. The *female* can be separated from the last only by the difference in size. The habits of the two species are much the same, but this one is more frequently found in the fresh waters of bays and rivers. (Little Black-head.)

Length, 16; wing, $7\frac{3}{4}$ ($7\frac{1}{2}$–$8\frac{1}{4}$); tail, $2\frac{1}{4}$; tarsus, $1\frac{3}{8}$; culmen, $1\frac{3}{4}$. North America; breeding mainly north of the United States, and wintering from Virginia to Cuba.

15. **Ring-necked Duck** (150. *Áythya collàris*). — A small, white-bellied, black duck, with an indistinctly outlined chestnut collar around the neck. The speculum is gray, and the lower belly and sides have wavy cross lines of black. The *female* is rusty-brown, with white belly and gray speculum. The wing is less than 8 long. This is especially a fresh-water duck, probably more so than any other one of the genus.

Length, 17; wing, $7\frac{1}{2}$; tail, $2\frac{1}{4}$; tarsus, $1\frac{3}{8}$; culmen, $1\frac{7}{8}$. North America; breeding in the interior from Iowa northward, and wintering from the Middle States to Central America. Not common on the Atlantic coast north of Virginia.

16. **American Golden-eye** (151. *Clangùla clangùla americàna*). — A brightly marked, dark-green (almost black) headed,

black and white duck. The back, tail, and primaries are black ; a spot at base of bill, neck, under parts, and much of the wings white. The white spot on the head is rounded and about a half inch high. The *female* is a brown-headed, grayish-backed, white-bellied duck with white speculum on wings. This duck

Ring-necked Duck

receives its name "Whistler" from the unusually loud sound produced by its wings when flying. (Whistler; Garrot.)

Length, 19; wing, 8¾ (8-9½); tail, 3½; tarsus, 1½; culmen, 1⅜-2. North America; breeding from Maine northward, and wintering throughout most of the United States to Cuba and Mexico.

17. Barrow's Golden-eye (152. *Clangula islándica*). — A duck similar to the last, but the head and throat are a dark, glossy, purplish-blue instead of green, and the white spot at

base of bill is elongated and more or less pointed at the ends.
measuring along the bill an inch. The *female* is so like the
last that it cannot always be distinguished from it. This is
the more northern species. (Rocky Mountain Garrot.)

American Golden-eye   Barrow's Golden-eye

Length, 21; wing, 9 (8½-9½); tail, 4; tarsus, 1⅓; culmen, 1⅛-1¾.
Northern North America; breeding from the Gulf of St. Lawrence and
Colorado northward, and wintering south to New York, Illinois, and
Utah.

Buffle-head

18. **Buffle-head** (153. *Chari-
tonétta albéola*).—A very small,
common, black-backed, gray-
tailed, white duck, with a fluffy
head peculiarly marked with
black, purple, green, and white.
The white forms a broad patch
across the top of the head, and
ends back of the eyes. The
wing is mainly white excepting the black primaries. The
female does not have the full
fluffy head of the male; the
head and back are a rich
brown, fading through grays
to a white breast and fore-

belly. There is a distinct patch of white on the cheeks and a
white speculum on the wings. This bird is noted as a diver,
being compared to the grebes. (Dipper; Spirit-duck; Butter-
ball.)

Length, 14½; wing, 6½ (6–7); tarsus, 1½; culmen, 1. North America;
breeding from Maine and Montana northward, and wintering from the
Middle States to West Indies and Mexico.

19. **Old-squaw** (154. *Harélda hyemális*). — In winter it is
a long-tailed, brown duck, with a white belly, head, and neck,
except a brown patch
on the side of the
head, gray around the
eyes, and light gray
shoulder feathers. In
early spring it is
sometimes found in
more or less of breed-
ing dress, when the
whole upper parts,
including neck and
breast, are rich browns, excepting a large patch of light gray
around the eyes. The *female* lacks the two long tail feathers of
the *male*, and is a white-bellied, blackish-brown-backed, white-
headed duck, with blackish spots on cheeks, crown, and chin.
The scolding or talking notes of this bird have given it many of
the common names. (Long-tail; South-southerly; Old-wife.)

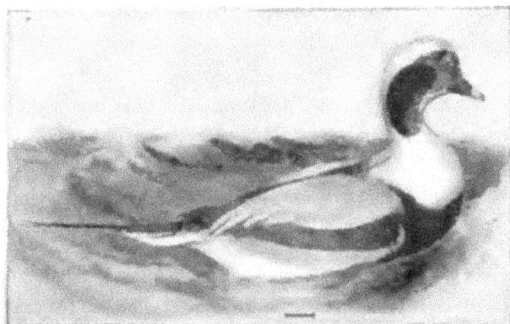

Old-squaw

Length, 15–23; wing, 8¼ (8½–9); tail, *female* 2½, *male* 8; tarsus, 1½;
culmen, 1¼. Northern hemisphere; breeding in the Arctic regions and in
America, wintering south to Virginia and Kentucky, rarely to Florida
and Texas.

20. **Harlequin Duck** (155. *Histriónicus histriónicus*). — A
northern, rich, blue-slate-colored duck, with fantastically
arranged white marks, brown belly, and chestnut sides.
There are two white collars, one above and one below the
breast; three white patches on the side of head and neck,
one at base of bill, one on cheek, and one on side of neck;

a mahogany-colored stripe on side of crown, and several white blotches on wings. *Female*, grayish-brown, with the front of head and a patch on the cheek whitish. Belly lighter than the back. A most expert diver, living on fish and other water animals, and forming but poor food for human beings.

Length, 16½; wing, 7½ (7–8); tail, 3½; tarsus, 1¼; culmen, 1⅛. Northern North America ; breeding from Newfoundland northward, and wintering south to the Middle States and California.

21. Northern Eider (159. *Somatèria mollíssima boreàlis*).— This more northern eider duck, which is

Harlequin Dock

rarely found as far south as Massachusetts, has the bare portions of the bill extending backward by the sides of the culmen in two narrow, rather sharp points; in the next species these points are broad and rounded. The colors are practically the same as those of the American eider, given below.

Length, 24 ; wing, 11 ; tail, 4 ; tarsus, 1¾ ; culmen, 1⅞. Northern North America ; wintering south to coast of Massachusetts.

22. American Eider (160. *Somatèria drésseri*). — In breeding plumage, it is a large, mainly white duck, with the lower parts from breast, the tail, and lower back black. The head is

1.

2.

1. Northern Eider   2. American Eider

greenish tinted, and has a large V-shaped patch of black on the crown, and the breast is creamy tinted. The *female* (also

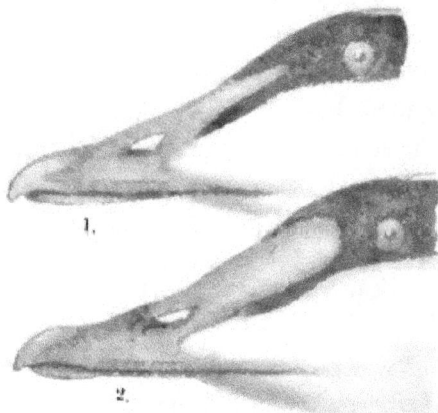

the *male* in certain stages) is rusty-brown to buffy, mottled and barred with black, the mottling including the head and throat. Both of these eiders have practically the same habits; they are true sea ducks, spending most of their time some distance from shore, diving for mussels, which form their principal food.

Length, 24; wing, 11; tail, 4; tarsus. 1¼; culmen, 1⅞. Atlantic coast of North America; breeding from Maine to Labrador, and wintering south to New Jersey and west to the Great Lakes.

American Eider

23. King Eider (162. *Somatèria spectàbilis*). — A large, distinctly blotched, black-bodied duck, with mainly white head, neck, and breast. The crown is bluish-gray, cheeks somewhat green, and breast buff. There is a black band at base of upper mandible and a V-shaped mark under the throat; white wing coverts and side of rump. The *female* is rusty-brown, mottled and barred with darker, but with head and throat almost un-streaked. The king

King Eider

eider can be best distinguished by the feathering at the side of the bill, which does not reach forward to the nostril.

Length, 24 ; wing, 11 ; tail, 4 ; tarsus, 1⅜ ; culmen, 1⅛. Northern hemisphere ; breeding in the Arctic regions, and wintering in America south to Great Lakes, and casually to Virginia or even to Georgia.

24. American Scoter (163. *Oidèmia americàna*). — A large, northern, winter, black duck, with the upper parts slightly iridescent and the lower parts slightly brownish. The bill of the *male* has a peculiar hump back of the nostrils, which is lacking in the *female*. The *female* is dusky-brown in color, lighter below, with some dull white about throat, lower part of head, and belly. This and the next two species, popularly called "coots," are very poor food for man, being extremely "fishy." All these scoters are alike in habits, living mainly at sea, over beds of bivalves, for which they dive. (Black Coot.)

Length, 20 ; wing, 9 (8¼-9½); tail, 4 ; tarsus, 1¾ ; culmen, 1⅞. Northern North America, living mainly along coasts and on large inland waters; breeding from Labrador westward, and wintering south to New Jersey, Great Lakes, Colorado, and California.

25. White-winged Scoter (165. *Oidèmia dèglandi*). — A black duck with white speculum on the wings and a white spot below the eye. The feathers on the side of upper mandible reach almost to the nostril, about as far as do

White-winged Scoter

those on the culmen. This is the best feature by which to distinguish this scoter. The *female* (also the *male* and *young* in winter) is sooty-brown, lighter and grayer below, with white speculum, and more or less of whitish spots on the head. (White-winged Coot.)

Length, 22; wing, 11 (10½–11½); tail, 4½; tarsus, 2; culmen, 1½. Northern North America; breeding in Labrador and westward, and wintering south to Virginia, southern Illinois, and Lower California.

26. Surf Scoter (166. *Oidēmia perspicillàta*). — A black duck, with a square white blotch on the crown and a triangular one on the back neck. The orange and yellow bill has a round black spot on the side back of the nostril. The feathers on the culmen extend forward almost to the nostril, while those on the side of bill do not. The *female* is almost everywhere sooty-brown, paler below, and whitish on the belly; the sides of the head have whitish spots at base of bill and on cheeks. The *female* has not such a bulging base of bill nor such an extension of feathers on the culmen. (Sea Coot.)

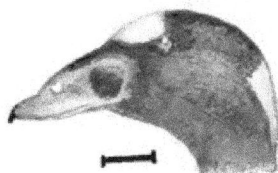

Surf Scoter

Length, 20; wing, 9½ (9–10); tail, 4; tarsus, 1¾; culmen, 1½. Northern North America, on coasts and inland waters; breeding from the Gulf of St. Lawrence northward, and wintering south to Virginia and the Ohio River, and casually to Florida.

27. Ruddy Duck (167. *Erismatùra jamaicénsis*). — A common, and, in full dress, brightly colored, black-crowned, white-cheeked, chestnut-backed duck, with wavy white and gray breast and under parts, and a short, black tail of narrow, stiff, sharp-pointed feathers. The *female* (also the *male* as usually found) has a dull reddish-brown back, grayish-white

Ruddy Duck

cheeks with a dusky bar extending back from the bill, and the lower parts mottled buffs and browns. The species can be readily separated from all others (except the next, which is very rare), by the peculiar tail feathers almost exposed to their bases. This is a good diver and often escapes pursuit by diving backwards and swimming under water to some secure place where it can hide. In flying, its rounded form and rapid wing movements enable one to distinguish it from other ducks. In rising from the water it makes use of its feet, running, as it were, on the surface of the water for some distance, before it is able to sustain itself in the air. If there is not room for this surface running, it will dive and hide rather than attempt flight. In swimming, it frequently holds its tail erect, and this attitude gives it a peculiar appearance.

Length, 15; wing, 5¼ (5½–6); tail, 3½; tarsus, 1¼; culmen, 1¼. North America south to northern South America; breeding mainly north of the United States, but locally even south to Central America.

Masked Duck

28. Masked Duck (168. *Nomonyx dominicus*).— A small, tropical, stiff-tailed duck which has accidentally drifted into the United States a few times. It is a chestnut-red duck, with black on the crown and back, and white on the wings at the coverts. The *female* is a mottled, dusky, yellowish-brown and rusty duck, with two blackish stripes on each side of the head. The inner secondaries are so lengthened as to fold over the primaries in the closed wing.

Length, 13; wing, 5½; tail, 3½; tarsus, 1; culmen, 1⅝. Tropical America north to the Gulf coast of Texas and accidental in Wisconsin, New York, and Massachusetts.

RIVER DUCKS (SUBFAMILY ANATINÆ)

Characteristics given on p. 277

Key to the Species

* Bill decidedly broadened toward tip, being nearly twice as wide as at base [1]..38. **Shoveller.**
* Bill little if at all widened toward tip. (**A.**)
  A. Tail feathers broad and rounded at tip ; head more or less crested ; crown green or greenish with purple reflections ; throat white.....
  ....40. **Wood Duck.**
  A. Head not crested. (**B.**)
  B. Central tail feathers very much lengthened, making tail over 7 long (*male*), or central feathers broad and sharp-pointed (*female*) ; neck unusually long...39. **Pintail.**
  B. Tail and neck not lengthened. (**C.**)
    C. Bill decidedly shorter than the head ; wing, $9\frac{1}{2}$–$11\frac{1}{2}$ long ; belly white. (**I.**)
    C. Bill about as long as the head, or longer. (**D.**)
  D. Wing less than $8\frac{1}{2}$ long. (**H.**)
  D. Wing over $9\frac{1}{2}$ long. (**E.**)
    E. Speculum [2] white or grayish white......32. **Gadwall.**
    E. Speculum a rich purple with a black border. (**F.**)
  F. Speculum bordered at both ends with narrow black and white bands.
  ....29. **Mallard Duck.**
  F. Speculum with only a black border, no white. (**G.**)
    G. Throat blackish or buffy, without streaks.......31. **Florida Duck.**
    G. Throat finely streaked with black. ....30. **Black Duck.**
  H. Wing coverts leaden gray without blue......35. **Green-winged Teal.**
  H. Wing coverts sky blue..36. **Blue-winged Teal.** 37. **Cinnamon Teal.**
    I. Head and throat mainly buffy, finely barred with black..........
  ....34. **Baldpate.**
    I. Head and throat with much brown or reddish brown............
  ....33. **Widgeon.**

29. Mallard (132. *Anas bóschas*).— A large, brilliantly colored, bright-green-headed, chestnut-breasted duck, with a white ring around the lower neck. The belly and sides are nearly white, barred with many fine, wavy lines of black ; the back is brown ; upper tail coverts black and some of them recurved. The speculum is rich purple, bordered by both black and white bands. *Female* very different except the speculum ; the colors peculiarly mottled buffy and brownish blacks. This species is far more common in the interior than on the coast. It is the original form of the common domestic duck, and its voice is the same *quack*. (Green-head.)

Length, 23; wing. 11 (10¼–12); tail, 3½; tarsus, 1⅜; culmen, 2⅜. Northern hemisphere; breeding from the Gulf States northward, and wintering south to Central America.

30. **Black Duck** (133. *Anas obscùra*). — A very dark-colored, almost black duck, with a black-bordered rich purple speculum. The head is lighter, the cheeks being a streaky buff. There is no decided white except under the wings, but there are buffy margins to most of the feathers. This is more common along the coasts

Mallard

than the last, and can always be separated from the female of that species by the lack of white border to the speculum. In habits and voice it is like the mallard. (Dusky Duck.)

Length, 22; wing, 11 (10–11½); tail, 3½; tarsus, 1¾; culmen, 2⅛. Eastern North America west to the Mississippi Valley; breeding from New Jersey and Illinois to Labrador, and wintering from the Middle States to Cuba.

31. **Florida Duck** (134. *Anas fulvigula*).

Black Duck

In habits, voice, and coloring this species is similar to the black duck, but more buffy; sides of head and whole throat buffy without streaks; the speculum is greenish-purple.

Length, 20; wing, 10½; tail, 3½; tarsus, 1⅝; culmen, 2. Florida. The **Mottled Duck** (134ᵃ. *A. f. maculosa*) of Texas to Kansas differs from the Florida duck in having the buffy cheeks streaked with brown and the rest of the plumage more mottled.

32. Gadwall (135. *Anas strépera*).—A buffy-headed, mottled-gray-bodied duck, with middle wing coverts chestnut, greater wing coverts black, speculum white, and belly nearly white. The breast and neck have a scaled appearance, because of the white edges and centers of the feathers. *Female* mottled browns with a nearly white speculum and white belly; there is almost no chestnut on the wing coverts; the axillars and under wing coverts are pure white. This is a common species in the interior, but rare north of Virginia on the coast. (Gray Duck.)

Gadwall

Length, 20; wing, 10½ (10–11); tail, 4½; tarsus, 1⅝; culmen, 1⅝. Northern hemisphere: breeding in America from Kansas and Gulf of St. Lawrence northward, and wintering from Virginia south to Florida and Texas.

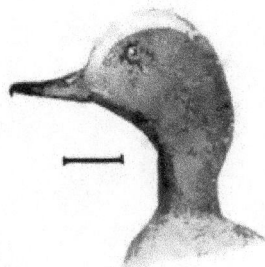

Widgeon

33. Widgeon (136. *Anas penélope*).—A rare duck from the Old World, with the head and throat reddish-brown except a whitish crown and blackish throat; the sides and back covered with many black lines, and the lower breast and belly white. The *female* is like the female of the next species, but the head and throat are a decided brown and the greater wing coverts brownish-gray.

Length, 19; wing, 10½ (10–11); tail, 4; tarsus, 1½; culmen, 1¾. Northern parts of the Old World, occurring occasionally in the eastern United States.

34. **Baldpate** (137. *Ànas americàna*). — A brownish-backed, reddish-breasted, white-bellied duck with a speckled, light-colored, mainly buffy head and neck. From the eye backward on the side of the head there is a glossy green patch and the crown is almost white. The wing coverts are largely white, the speculum green with a black border, and the under tail coverts abruptly black. The *female* has a light, speckled, buffy head and neck similar to the male, but it lacks the white crown and the green eye patch. The great amount of white on the wing coverts and belly distinguishes this from other ducks. This duck ranks high among sportsmen on account of the delicacy of the flesh. It feeds upon the same "wild celery" as the canvasback, but it cannot dive, so it watches the diving ducks and filches their prey the moment their heads appear. (American Widgeon.)

Baldpate

Length, 20; wing, 10½ (10–11); tail, 4¼; tarsus, 1½; culmen, 1¼. North America; breeding mainly north of the United States, and wintering from Virginia to northern South America.

35. **Green - winged Teal** (139. *Ànas carolinénsis*). — A very small, common, chestnut-headed, wavy-lined, gray duck, with shining green patch on the side of the head, a green speculum

Green-winged Teal

on wing, and a white crescent on the side of body in front of the
wing. *Female* is principally buff and dark browns, blotched on
the body and speckled on the head and neck. The wing markings are about the same as those of the male.

Length, 14 ; wing, 7 (6¼-7½) ; tail, 3 ; tarsus, 1¼ ; culmen, 1¼. North
America ; breeding chiefly north of the United States, and wintering from
Virginia to Kansas and south to Central America. The **European Teal**
(138. *Anas crecca*) is so nearly like the last that the *female* cannot be
distinguished, but the *male* lacks the white crescent in front of the wing.
Old World, occasionally found in eastern North America.

36. **Blue-winged Teal** (140. *Anas discors*). — A small, common, black-headed, spotted, brown-bodied duck, with a bright
patch of light blue on the
wing coverts and a white
crescent on the side of the
head in front of the eye.
The speculum is dark green.
The *female* (also the *male*
in summer) has the wings
nearly as above given, but
the head is very different,
being blackish and buffy
spotted or dotted, and the
throat is about white. These
birds fly in small dense
flocks.

Length, 15½ ; wing, 7¼(7-7½) ;
tail, 3½ ; tarsus, 1¼ ; culmen, 1¾-
1½. North America, more abundant eastward ; breeding from

Blue-winged Teal

Kansas and Illinois northward, and wintering from Virginia south to
northern South America.

37. **Cinnamon Teal** (141. *Anas cyanóptera*). — A duck similar to the last, but the *male* has a richer and more glossy
chestnut color below. The *female* (also the *male* in summer)
has the plumage darker and only a small portion of the upper
throat unstreaked. The belly is usually heavily spotted and

the breast deeply tinged with light brown. The bill of this species is larger than that of No. 36, the culmen ranging from $1\frac{5}{8}$ to $1\frac{7}{8}$.

Length, $16\frac{1}{2}$; wing, $7\frac{1}{2}$ ($7\frac{1}{4}$–8); tail, $3\frac{1}{2}$; tarsus, $1\frac{1}{4}$; culmen, $1\frac{5}{8}$–$1\frac{7}{8}$. Western America east to the Rocky Mountains and south to Patagonia. Casual eastward, Illinois, Florida.

38. **Shoveller** (142. *Spátula clypeáta*). — A large, broad-billed, bright-colored, white-breasted, chestnut-brown-bellied, dark-headed duck with blue wing coverts and green speculum. The bill is spoon-shaped, being nearly twice as wide near the tip as at the base. The *female* is mainly dark-brown blotched on a buff ground: the middle of the belly is lightest and the back darkest. The wings are much like those of the male. The large, spoon-shaped bill distinguishes the species. (Broad-bill; Spoonbill Duck.)

Cinnamon Teal

Length, 17–21; wing, $9\frac{1}{4}$ (9–10); tail, 3; tarsus, $1\frac{1}{2}$; culmen, $2\frac{1}{4}$. Northern hemisphere; breeding in America from Texas to Alaska, and wintering from New Jersey and southern Illinois to northern South America. Not abundant on the coast north of the Carolinas.

Shoveller

39. **Pintail** (143. *Dáfila acúta*). — A sharp-tailed, dark-brown-headed, wavy-gray-backed duck, with a long neck, having a white stripe on the side and a black line above. The speculum is greenish-purple, usually bordered by black and white. The breast and belly are white, with the sides strongly marked with wavy black lines; the

central tail feathers are much lengthened and glossed with green. The *female* has a streaky blackish and buffy head, whitish throat, dark buffy breast, spotted with blackish, and very much spotted and barred sides and back. The sides and back have many whitish, crescent-shaped marks. The speculum is grayish-brown bordered with white. The central tail feathers are

Piutail

broad with acute points; the under wing coverts are dusky. The *male* in summer is somewhat like the female in coloring, except the wings. (Sprigtail.)

Length, 21–30; wing. 10¼ (9½–11¼); tail. *male* 9, *female* 4; tarsus, 1¾; culmen, 2. Northern hemisphere. In North America breeding from Iowa to the Arctic Ocean, and wintering from Virginia southward.

Wood Duck

40. **Wood Duck** (144. *Aix spónsa*).— A common, distinctly crested, brilliantly colored, woodland-living duck, with greens, blues, buffs, browns, blacks, and whites in the plumage. Any more elaborate description of this very beautiful duck would be useless; it must be seen to be appreciated. The *female* is a

slightly crested, somewhat iridescent, grayish and slaty-brown duck, with lower breast streaked with buff; the throat is white and there is a white stripe from the eye backward; the head is purplish-brown on the crown, and ashy-brown on the sides. The forest-bordered fresh waters form the home for this bird, and a hole in tree or stump its nesting quarters. (Summer Duck.)

Length, 18½; wing, 9; tail, 4½; tarsus, 1⅜; culmen, 1⅜. Temperate North America; breeding from Florida to Hudson Bay, and wintering from the Middle States to Mexico.

FISH DUCKS (SUBFAMILY MERGINÆ)

Characteristics given on p. 278

Key to the Species

* Wing, 9¼-11¼ long; frontal feathers extending beyond those on the side of the bill .... 41. **American Merganser.**
* Wing, 8¼-9½ long; frontal feathers not extending beyond those on the side of the bill .... 42. **Red-breasted Merganser.**
* Wing, 7-8¼ long; crest on head high and flattened sideways.........
.......43. **Hooded Merganser.**

American Merganser

41. **American Merganser** (129. *Mergánser americánus*). — A slightly crested, slender-billed, dark-green-headed, fish duck, with the back and wings black and white, the tail gray, and the under parts, including breast, buffy white. The *female* is reddish-brown-headed, gray-backed and whitish-bellied, with a white patch on throat, and white speculum. The head color ex-

tends farther down the neck in this species than in the
next, and the distance from the nostril to end of bill is less,
being in this species 1½ inches, in the next 1¾. These fish-
eating ducks inhabit both fresh and salt waters, are great
divers, and can pursue and catch their food while under the
surface. (Goosan-
der; Shelldrake;
Sawbill.)

Length, 25 ; wing,
10½ (9½-11½) ; tail, 5 ;
tarsus, 1⅞ ; culmen, 2.
North America ; breed-
ing from Pennsylvania
and Colorado north-
ward, and wintering
from Maine, Illinois,
and Kansas southward.

Red-breasted Merganser

**42. Red - breasted
Merganser** (130. *Mer-
gánser serrátor*).— A common, crested, dark-green-headed, reddish-breasted, fish duck, with the back made up of white, black,
and gray. The reddish breast is streaked with blackish, the
head and neck are green-glossed, and the rump and sides barred
with black and white. The *female* has the head and upper
neck cinnamon-brown, the back gray, and the breast and belly
white. The speculum is white, and the throat whitish.

Length, 23 ; wing, 9 (8½-9½); tail, 4 ; tarsus, 1⅞ ; culmen, 2⅜. Northern portion of northern hemisphere ; breeding in America from the
northern border of the United States northward, and wintering throughout most of the United States.

43. Hooded Merganser (131. *Lophódytes cucullátus*).— A small,
strongly crested, fish duck, with black and white head, black
back, white belly, and cinnamon-red sides. The head and
neck are black except a large, central, fan-shaped part of the
very flat, high chest, which is white. The black and white of
the lower neck and breast are so arranged as to give the appearance of two white collars, wide and touching in front, narrow

and widely separated behind. The *female* is smaller, has a smaller, rusty-brown crest and a grayish-brown back, with nearly white belly and grayish breast. The other fish ducks prefer running, dashing waters, this one the quiet pools and lakes; the others are "fishy," but this is palatable.

Length, 18; wing, $7\frac{1}{2}$ (7-8); tail, 4; tarsus, $1\frac{1}{4}$; culmen, $1\frac{1}{4}$. North America, south to Mexico and Cuba; breeding mainly throughout, and wintering in most sections of the United States.

Hooded Merganser

ORDER XIV. TOTIPALMATE SWIMMERS (STEGANÓPODES)

An order of swimming birds with the four toes connected by webbing; nostrils small or none; bill without lamellæ; throat usually furnished with a pouch.

FAMILY LI. MAN-O'-WAR BIRDS (FREGÁTIDÆ)

A very small family (2 species) of very large, marine birds of tropical seas, with long, forked tails, and unexcelled length of wing. They surpass all other birds in their power of flight, and are found hundreds of miles from shore, apparently independent of solid earth. They poise for hours on motionless wings, facing the wind, sometimes at great heights, above the storms. Their legs are so small and weak that they can scarcely swim or walk, and they cannot dive. They obtain all their food while on the wing, gracefully darting beneath the surface of the water for fish, or often capturing those which, chased by enemies below, leap for a moment into the air. They often pursue and steal the captured food of gulls, terns, and other birds.

1. **Man-o'-War Bird** (128. *Fregáta áquila*). — A tropical, large, long-winged, black, ocean bird, with long, deeply forked tail. The *female* is a dark brown bird with the breast and upper belly white. The *young* is like the female, but also has the head and neck white. This bird spends most of its time on the wing, and usually over the water. It is a kind of sea buzzard.

Man-o'-War Bird

The man-o'-war birds nest together in thousands in low bushes near the coast.

Length, 40; wing, 25 (22–27); tail, 18; tarsus, 1; culmen, 4¼. Tropical and subtropical coasts. In America, north to Florida and Texas, and casually to Ohio, Kansas, Nova Scotia, etc.

FAMILY LII. PELICANS (PELECÁNIDÆ)

A small family (12 species) of very large, short-tailed birds, with very long, peculiarly pouched bills, the pouch being used like a dip net for catching its fishy food. Under the skin there are great air sacs like those of the gannets. This makes them peculiarly buoyant on the water, and gives them great grace of movement. In the air, also, their movements are easy and strong, but not very rapid. They give a few flaps of the wings, then sail a short distance, then again give a few flaps of the wings. They are usually in flocks, and it is interesting to see the alternate flapping and sailing of the whole as though directed by a leader. These birds nest in large colonies, and are found in all the warmer parts of the world. Some are exclusively marine, and some are found far from the coast.

Key to the Species

* Tarsus over 3½ ; plumage mainly white...1. **American White Pelican.**
* Tarsus under 3½ ; plumage with much brown..... 2. **Brown Pelican.**

American White Pelican

1. American White Pelican (125. *Pelecánus erythrorhýnchos*). — An exceedingly large, white, swimming bird, with a very long, pouched bill and black primaries. The *young* is similar, but with some brownish-gray on the top of the head. In the breeding season, there is a peculiar crest on the bill. This species procures its food mainly by swimming and dipping; the next by darting from the air into the water. This species is found both along the coast and in the center of the continent a thousand miles from salt water; the next is almost exclusively marine.

Length, 60; wing, 22 (20–25); tail, 6; tarsus, 4½; culmen, 11–15. North America, rare or accidental on the Atlantic coast, common on the Pacific; breeding from Minnesota northward far into the British possessions, and wintering from the Gulf coast to Central America.

2. Brown Pelican (126. *Pelecánus fúscus*). — This is a bird similar to the last, but smaller.

Brown Pelican

It is a yellow-headed, gray-backed pelican, with blackish-brown lower parts. In breeding plumage, there is a seal brown stripe along the whole of the back neck. During the rest of the year the whole neck is whitish. These birds fly low over the water, just beyond the breakers, usually in small flocks. They only casually stray into the interior.

Length, 50; wing, 19 (18–21); tail, 6½; tarsus, 2⅗; culmen, 9½–12¼. Atlantic coasts of tropical and subtropical America, North Carolina; accidental in Illinois.

FAMILY LIII. CORMORANTS (PHALACROCORÁCIDÆ)

A family (30 species) of large, generally distributed, mainly salt-water birds, though occasionally found along the shores of fresh-water lakes. They are long-necked, large-tailed, short-legged, hooked-billed birds, which when standing are forced to take nearly an erect position and make use of the tail as a partial support. They pursue their prey of fish by swimming under the water, and in doing this make use of their wings as well as feet, and are thus like the darters and auks. They dive from the surface of the water, instead of from the air like the gannets.

Key to the Species

* Wing, 13 or more long ; tail of 14 feathers ; pouch notched behind. . . .
. . . . .1. **Cormorant.**
* Wing, 11–13 long ; tail of 12 feathers. . . .2. **Double-crested Cormorant.**
* Wing, 9–11 long ; tail of 12 feathers. . . . . .3. **Mexican Cormorant.**

1. **Cormorant** (119. *Phalacrocorax cárbo*). — A diving, marine, narrow-billed, rounded-tailed, very dark-colored cormorant, with a white patch on the flanks. In the breeding season there is a large, white patch on the head back of the eye. The plumage of the back and wings is bronzy, with more or less of iridescent colors. The *young* has much brown on the back, neck, and head, and the throat and breast are grayish-brown, changing to white on the belly. (Shag.)

Length, 36; wing, 13½ (12½–14); tail. 7½; tarsus, 2; culmen, 2½–3. Coasts of the North Atlantic of both Old and New Worlds; breeding from the Bay of Fundy northward, and wintering casually south to the Carolinas.

2. **Double-crested Cormorant** (120. *Phalacrò-corax dilòphus*). — A common, double-crested, black cormorant, with a greenish iridescence to the feathers of the head, neck, and body, and coppery-gray to those of the back and wings. Bare skin on sides of the head, around the eyes orange (in life). There is a tuft of curling feathers on each side of the head, above the eyes, forming the "double crest." This is the "cormorant" of the Middle States. The *young* has a white breast changing to gray on

Cormorant

the throat, and black on the lower belly. It is like the last species in being much browner on the head, back neck, and upper back than is the adult.

Length, 32; wing. 12½ (12–13); tail, 6½; tarsus, 2; culmen. 2–2¼. Eastern North America; breeding from the Bay of Fundy and Dakota northward, and wintering from Maryland and southern Illinois southward. The **Florida Cormorant** (120ᵃ. *P. d. floridànus*) is much like the last in color, but smaller. Wing, 11¾ (11¼–12½). Common on the Gulf coast, South Atlantic and Gulf States, north to southern Illinois.

Double-crested Cormorant

3. **Mexican Cormorant** (121. *Phalacròcorax mexicànus*). — A small, southwestern cormorant with intense violet-purplish luster on the black of the body. The pouch on the neck is

orange, with white edges. This species is found along the western Gulf coast and has been seen as far north as Kansas and southern Illinois.

Length, 25; wing, 10½ (10–10½); tail, 6½; tarsus, 1½; culmen, 1¾. West Indies and Central America, to the southern United States.

FAMILY LIV. DARTERS (ANHÍNGIDÆ)

A small, tropical family (4 species) of very long-necked, short-legged swimming birds of fresh-water swamps. When alarmed, they have the habit, like the grebes, of sinking quietly backward into the water and swimming to a safe place, keeping only the head and neck above the surface. When in this position, they present the appearance of water-snakes, whence they derive one of their common names. Even when perching on limbs of trees above the water, they can, when disturbed, drop into and sink noiselessly under the water, making hardly a ripple on the surface. They resemble the cormorants in appearance, and like them and the auks, they use their wings in swimming under water.

1. **Anhinga** (118. *Anhinga anhinga*). — A southern, very long-necked, slender-billed, short-legged, swimming and diving bird; glossy, greenish-black, with grayish wings and tail. The wing coverts and shoulders are much dotted and blotched

Anhinga

with white, and the rounded tail is tipped with whitish. The *female* is similar but has the head, neck, and breast brownish. A common bird in the swamps of the Gulf States. (Snake-bird; Water Turkey.)

Length, 34; wing, 14; tail, 11; tarsus, 1¼; culmen, 3¼. Tropical and subtropical America, north to North Carolina and Kansas.

FAMILY LV. GANNETS (SULIDÆ)

A small family (8 species) of large, heavy, sea-birds, which, except when migrating, are never found far from land. In their movements through the air they alternate their flapping with short periods of sailing. They are large-bodied birds,

Booby

but have such extensive air cavities under the skin as to render them very light on the water; thus they swim with great ease. Associated in small flocks, these birds fly with outstretched neck, usually at some height above the waves, and, when a fish is seen, close the wings and shoot downward like an arrow to secure the prey.

1. **Booby** (115. *Sula sula*). — A dark brown gannet, with white breast and belly. The head and neck are sometimes streaked with lighter brown and the breast is tinted with darker brown. The *young* has even the lower parts brownish, though not so dark as the back. An inhabitant of barren shores.

Length, 30; wing, 15¾ (14–16½); tail, 8¼; tarsus, 1½; culmen, 3¾. Atlantic coast of tropical and subtropical America, north to Georgia in summer.

2. **Gannet** (117. *Sula bassàna*). — A white gannet, with yellowish head and neck and nearly black primaries. *Young*, mottled grayish-brown above and white on the breast and belly, with grayish-brown edges to the feathers. The mottlings of the back consist of wedge-shaped white spots on the feathers. (Solon Goose.)

Length, 36; wing, 19 (17–21); tail, 10; tarsus, 2; culmen, 4. Coasts of the North Atlantic;

Gannet

breeding in America from Nova Scotia northward, and wintering from Virginia to the Gulf of Mexico.

FAMILY LVI. TROPIC BIRDS (PHAËTHÓNTIDÆ)

A small family (3 species) of tropical, tern-like, marine birds, with peculiarly elongated central tail feathers. They are graceful birds, capable of strong, rapid flight; sometimes they are seen far from the coast, though usually found near the shore. They live almost entirely on the wing, and catch their prey, which consists almost exclusively of fish, by dropping suddenly down upon it from the air.

1. **Yellow-billed Tropic Bird** (112. *Phaëthon flaviróstris*). — An exceedingly long-tailed, long-winged, white sea-bird, with black on the outer quills and shoulder feathers of the wings. The shafts of the tail feathers are also black. Bill yellow and tail feathers tinged with salmon. The *young* lacks the elon-

gated central tail feathers, and has the upper parts somewhat irregularly barred with black. The tail feathers are marked with a black spot near the tip.

Yellow-billed Tropic Bird

Length, 30; wing, 11; tail, 20 or less; culmen. 2½. West Indies to Central America, north to Florida and Bermuda, accidental in New York and Nova Scotia.

ORDER XV. TUBE-NOSED SWIMMERS (TUBI-NARES)

An order of marine birds with tubular nostrils; practically, as far as our own birds are concerned, consisting of but the following:

FAMILY LVII. FULMARS, SHEARWATERS, AND PETRELS (PROCELLARIIDÆ)

This is a large family (70 species) of strong, swiftly flying birds, belonging strictly to the open ocean, and rarely seen near the shore except for breeding purposes. The fulmars and shearwaters are large birds, but some of the petrels are very small. The fulmars are much like gulls in appearance, but their method of flying is very different. They flap their wings more like owls, and in scudding they hold them very straight, at right angles with the body; they sail close to the waves for great distances, apparently without moving their wings. The flight of the petrels is peculiarly light and airy, more like that of butterflies than like the flight of birds. They often gather in flocks around vessels at sea and follow them for miles. Though they spend most of the time near the surface of the water, they do not appear to swim, but are constantly on the wing, beating to and fro about the ship. The shearwaters derive their name from their habit of strongly and swiftly "shearing the crests of the waves and skimming the billows with marvelous ease and without visible motion of the pinions." (Dr. Coues.)

Key to the Species

* Under mandible not hooked at tip;[1] wing, 11-14 long......1. **Fulmar.**
* Under mandible hooked at tip much like the upper,[2] or else with wings under 7 long. (**A.**)
  A. Wings, 4-7 long. (**E.**)
  A. Wings, 7-15 long. (**B.**)
B. Wings, 13½-15 long; culmen over 2 long.....2. **Cory's Shearwater.**
B. Wings, 11-13½ long; culmen, 1½-2 long. (**D.**)
B. Culmen under 1½ long. (**C.**)
  C. Wing, 11-12 long......................6. **Black-capped Petrel.**
  C. Wing, 7-10 long ....4. **Audubon's Shearwater.**
D. Under parts dusky ....5. **Sooty Shearwater.**
D. Under parts white.......................3. **Greater Shearwater.**
  E. Tail forked for over a half inch[3] ....8. **Leach's Petrel.**
  E. Tail square. (**F.**)
F. Upper tail coverts white; nails flat and obtuse[4]..9. **Wilson's Petrel.**
F. Upper tail coverts tipped with black; nails hooked, acute[5] ....
.... 7. **Stormy Petrel.**

1. **Fulmar** (86. *Fúlmarus glaciális*). *Light phase.* — A large white bird with slaty-gray mantle and nearly black wing quills; the tail the color of the back. *Dark phase.* — A nearly uniform dark, slaty-gray bird. This bird is a constant attendant upon fishermen on their trips to the fishing banks, living upon the offal which is

Fulmar

thrown overboard and which they secure while swimming.
The statements made in the general description about the
position of the wings while scudding will enable one to distinguish the fulmars from the gulls. (Noddy.)

Length, 19; wing, 13 (12–14); tail, 4½; tarsus, 2; culmen, 1½. North
Atlantic, south in winter to Massachusetts, casually to New Jersey. The
Lesser Fulmar (86ᵃ. *F. g. minor*) is a similar bird, but much smaller.
Wing, 12; culmen, 1½. The same distribution.

2. **Cory's Shearwater** (88. *Puffinus borealis*). — A rare shearwater, with the wings and tail nearly black, the back somewhat ashy, and the under parts white, with a slight grayish tint
on the breast. The under tail coverts are white, mottled with
grayish, and the sides of head and neck are somewhat lighter
than the back; bill yellowish.

Length, 21; wing, 14 (13½–14½); tail, 6½; tarsus, 2½; culmen, 2½.
Known only by specimens from off the coasts of Massachusetts south to
Long Island.

3. **Greater Shearwater** (89. *Puffinus gravis*). — A sooty-black
or almost black-backed shearwater, with the under parts almost

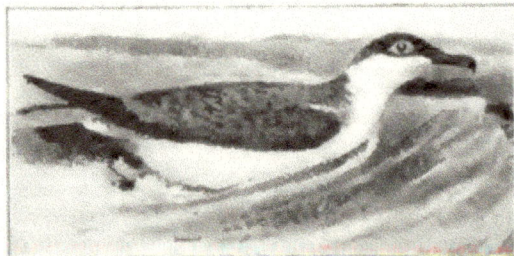

white; shading from
white on the breast
to ashy-gray on the
under tail coverts;
bill blackish. (Hag-
don.)

Length, 20; wing, 12½
(11½–13); tail, 5¾; tar-
sus, 2¾; culmen, 1½. At-
lantic Ocean from Cape

Greater Shearwater

Horn to Cape of Good Hope, north to the Arctic Circle.

4. **Audubon's Shearwater** (92. *Puffinus auduboni*). — A small
shearwater, with all the upper parts from forehead to tail a
sooty-black, and the under parts white. There is a patch of
sooty on the flanks and under tail feathers, and some grayish
on the sides of the breast. This bird is abundant and breeds
in the West Indies.

Length, $11\frac{1}{2}$; wing, 8 ($7\frac{1}{3}$–$8\frac{1}{2}$); tail, $3\frac{1}{4}$; tarsus, $1\frac{1}{2}$; culmen, $1\frac{1}{4}$. Warmer parts of the Atlantic, north casually to Long Island. The **Manx Shearwater** (90. *Púffinus púffinus*) is much like the last, but larger. Length, 14; wing, $8\frac{1}{2}$–$9\frac{1}{4}$; tail, 4; tarsus, $1\frac{3}{4}$; culmen, $1\frac{3}{8}$. A European species, accidental on the North American coast.

Sooty Shearwater

5. **Sooty Shearwater** (94. *Púffinus stricklandi*). — A sooty-black shearwater with the under parts somewhat grayer and the bill blackish. (Black Hagdon.)

Length, 17; wing, $11\frac{3}{4}$ ($11\frac{1}{4}$–12); tail, 4; tarsus, $2\frac{1}{8}$; culmen, $1\frac{3}{4}$. Atlantic Ocean; breeding south of the equator, and migrating north in summer to South Carolina and northward.

6. **Black-capped Petrel** (98. *Æstrélata hasitàta*). — A rare, southern, blackish-brown-backed petrel, with all lower parts and base of tail white. The otherwise white head is distinctly capped with black and marked with a bar of black back of the eye. The tip of tail and the primaries are darker than the back. The *young* has the black of the head more or less connected and continuous down the back neck.

Black-capped Petrel

Length, 15; wing, $11\frac{1}{2}$; tail, 5; tarsus, $1\frac{3}{8}$; culmen, $1\frac{3}{8}$. Warmer portions of the Atlantic Ocean, straying to different sections from Florida to Ontario.

7. **Stormy Petrel** (104. *Procellària pelágica*). — A very small, square-tailed, sooty-black petrel, with white upper tail coverts,

having the longer feathers black tipped. The under tail
coverts are mixed with whitish, and the bill and feet are black.
The common stormy petrel of the Atlantic near
Europe.

Length, 5½; wing, 4¾ (4½-5); tail, 2½; tarsus, ⅞;
culmen, ¼. Atlantic Ocean, south over the
American side to the Newfoundland Banks.

8. **Leach's Petrel** (106. *Oceanódroma
leucórhoa*). — A fork-tailed, sooty-
brown petrel, with white upper tail coverts and black
bill and feet. The
forking of the tail
is over ½ inch.

Length, 8; wing,
6¼ (6-6½); tail, 3½;
tarsus, ⅞; culmen, ⅝.
Northern oceans,
south in America to

Stormy Petrel

California and Virginia : breeding from Maine northward.

9. **Wilson's Petrel** (109. *Oceanites oceanicus*). — A square-
tailed, sooty-brown
petrel, with white
upper tail coverts
and a white bar on
the wings at the
edge of the wing
coverts. The webs
of the feet are most-
ly yellow, and the
under tail coverts
somewhat grayish.
This is the common
small petrel of the
Atlantic Ocean, in
our summer, its

Leach's Petrel

breeding time being the southern summer and its breeding home the southern seas.

Length, 7½; wing, 6 (5¾-6¼); tail, 3; tarsus, 1¼; culmen, ½. North Atlantic ocean, and oceans of the southern hemisphere. The **White - bellied Petrel** (110. *Cymódroma grallaria*) is a small, long-legged, blackish-gray petrel with the lower breast and belly abrupt-

Wilson's Petrel

ly white. The upper tail coverts and the bases of all tail feathers, except the middle pair, are also white. Length, 8; wing, 6¼; tail, 3; tarsus, 1½; culmen, ½. Tropical oceans; accidental on the coast of Florida.

FAMILY LVIII. ALBATROSSES (DIOMEDÈIDÆ)

The albatrosses are large ocean birds of the southern hemisphere, with very great expanse of wings and power of flight. These birds have rarely, if ever, been found on our eastern coasts; four species visit our Pacific coast. They are rarely found near shore, being able, seemingly, to remain on the wing without ever tiring. Two records are given of two of the species.

Wandering Albatross

1. The **Wandering Albatross** (80. 1. *Diomèdea éxulans*) is a large species of dusky to white color, according to age.

Length, 50; wing, 28. Reported from the western coast of Florida.

2. The **Yellow-nosed Albatross** (83. *Thalassogeron culminátus*) is a brownish-backed, white-bellied species.

Length, 36; wing, 18; tail, 8½; tarsus, 3¼; culmen, 4⅜. Reported from the Gulf of St. Lawrence.

ORDER XVI. LONG-WINGED SWIMMERS (LONGIPÉNNES)

An order of swimming birds, with very long, pointed wings, open nostrils, and a small hind toe or none. These birds show great power of sustained flight as well as of swimming.

FAMILY LIX. SKIMMERS (RYNCHÓPIDÆ)

A small family of but three similar, sea-skimming birds, one of which is found frequently on our southern coasts.

1. **Black Skimmer** (80. *Rynchops nigra*). — A short-tailed, long-winged, short-legged, black-backed, white-bellied sea-bird, with a peculiar, long, knife-like bill. These birds skim over the surface of the water with the lower mandible so buried beneath the waves as to "plow the main" for their food, which consists of small sea animals. They feed chiefly during the dusk of the evening and at night; during the daytime they are usually found resting on the sand bars. Their notes are very hoarse, somewhat resembling the croaking of some herons.

Black Skimmer

Length, 18; wing, 15 (14–16); tail, 5; tarsus, 1½; culmen, 2¼–2½. Warmer parts of America; breeding as far north as New Jersey, and wandering to the Bay of Fundy.

FAMILY LX. TERNS AND GULLS (LÁRIDÆ)

A large family (100 species) of birds, divided about equally between the two subfamilies.

The **Terns** are noisy, shrill-voiced, nearly white, swallow-like birds, generally much smaller than the gulls. They have, usually, notched or forked tails, while those of the gulls are even. The terns are almost entirely confined to the coasts; they are most abundant on islands and are numerous on the shores of fresh-water lakes. The gulls are less common except near salt water, and are generally found out at sea far from shore. Terns are readily distinguished from other birds when in the air, but it is almost impossible to determine the species without having them in hand. Terns can easily be separated from gulls by the position of the head while flying. Gulls hold their heads in line with the body, while terns hold theirs pointing downwards.

Gulls are hoarse-voiced, large, long-winged, sea and shore birds, usually with square tails. They are good swimmers, spending much of their time on the water. In this they differ from the terns, which are much of the time on the wing. Gulls procure their food by gathering it from the surface of the water with their strongly hooked bills. Terns plunge downward into the water from the air, often disappearing beneath the surface. Gulls have a varied diet, — mammals, birds, eggs, and fish. Terns live mainly on fish, though some eat insects. The nests of both gulls and terns are almost always on the ground.

Key to the Subfamilies

* Bill more or less hooked, the culmen much curved near tip ;[1] tail about square (No. 25 has a forked tail with the outer feathers rounded at tip, white under parts, and wing over 10 long); colors generally white with a darker, usually grayish mantle on the back (*young* birds have much mottled browns and white) ....Gulls, p. 328. 1

* Bill not hooked ; culmen slightly but evenly curved from end to end [1]
(No. 1 is merely curved near tip [2]) ; tail decidedly forked (No. 1 has
a doubly-rounded tail ; the outer feathers are about 2 inches and
the middle ones about ½ inch shorter than the longest ones)........
....**Terns**, below.

TERNS (SUBFAMILY STERNINÆ)

Characteristics given on p. 321.

Key to the Species

* Tail doubly rounded,[3] the outer feathers about 2 inches, and the middle
    ones ½ inch shorter than the longest ones...............12. **Noddy.**
* Tail decidedly forked. (**A.**)
    A. Tail with the outer feathers broad and rounded ;[4] front toes but
       little more than half webbed ; plumage dark......11. **Black Tern.**
    A. Tail with the outer feathers acutely pointed, and in most cases
       narrow ; front toes well webbed ;[5] plumage light. (**B.**)
    B. Bill dark and stout, its depth at base over ⅓ the length of the cul-
       men[2]....................................... 1. **Gull-billed Tern.**
    B. Bill less stout, usually slender. (**C.**)
       C. Wing, 15 or more long ; tail forked for less than ¼ its length......
          ....2. **Caspian Tern.**
       C. Wing, 14–15 ; tail forked for about ½ its length....3. **Royal Tern.**
       C. Wing less than 13 long. (**D.**)
    D. Head decidedly crested ; wing, 11–13 long..........4. **Cabot's Tern.**
    D. Head but little if at all crested. (**E.**)
       E. Wing under 7 long ; back pearl-gray...............9. **Least Tern.**
       E. Wing, 10¼–12¼ long ; back sooty-black ; inner webs of quills dusky.
          ....10. **Sooty Tern.**
       E. Wing, 8–12 long ; back in adult pearl-gray. (**F.**)
    F. Outer tail feathers with the inner web dusky, outer web white.......
       ....5. **Forster's Tern.**
    F. Outer tail feathers with both webs white...........8. **Roseate Tern.**
    F. Outer tail feathers with inner web white, outer web dusky. (**G.**)
       G. Bill red with a blackened tip ; tail but little more than ½ the length
       of the wing.....................................6. **Common Tern.**
       G. Bill red throughout ; tail over ⅔ the length of the wing...........
       ....7. **Arctic Tern.**

     1   2   3   4   5

1. **Gull-billed Tern** (63. *Gelochelidon nilótica*). — A southern, black-capped, black-billed, black-footed tern, with the upper parts, including the wings, a light pearl-gray, and the lower parts white; tail forked 1½ inches, nearly white. In winter, this, like most terns, loses its black cap: the crown is white, space in front of eyes black-ish, and back of them grayish. A common tern on the southern coast, feeding extensively on insects. The voice has a harshness similar to that of the gulls. (Marsh Tern.)

Gull-billed Tern

Length, 14; wing 12 (11¾-12¼); tail, 5½; tarsus, 1¼; culmen, 1⅜. Nearly throughout the world, in North America chiefly along the Atlantic and Gulf coasts; breeding north to New Jersey, and wandering casually to Massachusetts.

2. **Caspian Tern** (64. *Stérna tschegráva*). — A very large, red-billed tern, with the back of the neck, tail, and under parts white, back and wings pearl-gray, and the primaries slaty-black, with silvery outer webs. In spring, it has a black cap, but after the breeding season and in winter, the top of the head is merely streaked with black. The *young* has the pearl-gray back, and tail spotted or barred with brownish-black, and the head streaked black and white. This is a tern of world-wide distribution, but is not common in North America. It is in appearance, when seen on the wing, almost identical with the next.

Length, 21; wing, 16 (15-17½); tail, 6, forked, 1½; tarsus, 1⅜; culmen, 2½-3½. In North America, breeding locally from Virginia, the Great Lakes, and Texas northward; migrating through the interior as well as along the coast, and probably wintering beyond our borders.

3. **Royal Tern** (65. *Stérna máxima*). — A very large, some-what crested tern, with the back and wings pearl-gray, the

outer web and tips of primaries blackish, and the rest of the plumage white. In the breeding season, there is a black cap, but during the rest of the year the head is streaked black and white. This is much like the last species, but in all ages and seasons, the royal tern can be distinguished by the inner web of the primaries which is white, at least on the inner half. This is a common, strong, and powerful tern of the southern coasts, and is nearly as large as any gull; so the student may distinguish the gulls from the terns by noting the difference in the position of the heads of the species when in flight. The gull's head is in line with the body, the tern's points toward the earth.

Length, 19; wing, 14½ (14-15); tail, 7, forked, 3½; tarsus, 1¾; culmen, 2½; America, chiefly tropical; breeding north to Virginia; wandering to Massachusetts and the Great Lakes, and wintering from the Gulf coast southward.

4. **Cabot's Tern** (67. *Stérna sandvicénsis acuflávida*). — A southern, crested, pearl-gray-backed, white-bellied tern, with a

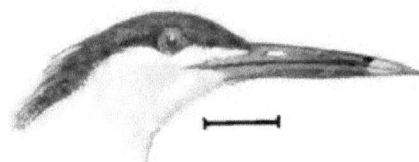

Cabot's Tern

large, yellow-tipped, black bill, and black feet. In the breeding season, the whole top of the head and crest is black, but during the rest of the year the crown is white, somewhat spotted with black, and the crest black streaked with white. The *young* has the pearl-gray back spotted with blackish, the slaty-gray tail short, and the bill nearly all black. (Sandwich Tern.)

Length, 15; wing, 12½; tail, 6, forked over 2; tarsus, 1; culmen, 2½. America, chiefly tropical; breeding along the Gulf coast, and along the Atlantic north to South Carolina; wandering north to New England, and wintering from Key West to Central America.

5. **Forster's Tern** (69. *Stérna fórsteri*). — A medium-sized tern, with wings and back pearl-gray, rump and all under parts white, and bill blackish at tip and dull orange at base. The tail is light colored, and the inner webs of the tail feathers are always darker than the outer ones. In summer, the whole top

of the head is black, but in winter the crown is white spotted
with black, and the side of the head is marked with a large
black spot surrounding the eye. The *young* has a mottled back
and short tail.

Length, 15; wing, 10 (9½–10½); tail, 5–8, forked, 2–5; tarsus, ⅞; culmen, 1⅛. North America; breeding north to Virginia, Illinois, Manitoba,
and California; wandering to Massachusetts, and wintering south to
Brazil. The **Trudeau's Tern** (68. *Sterna trudeaui*) of southern South
America has been seen a few times in the eastern United States. It is a
pale, pearl-gray tern, with the head and under surface of wings white, and
tail and rump lighter than the body. A narrow bar of slate color begins in
front of the eye, passes through it, and curves downward toward the back
of the head. Length, 16; wing, 10¼; tail, 4½–6½; tarsus, 1; culmen, 1¼.

6. Common Tern (70. *Sterna hirundo*). — A pearl-gray-backed,
white-throated tern, with a pale, pearl-gray breast and belly, and
a deeply forked tail. In summer,
the whole top of the head is
black, and the bill is red except
the end third, which is black;
but in winter, the front part
of the head is white, the bill
mainly black, and even the under parts change from pearl-gray
to white. The outer webs of the
outer tail feathers are gray, and
the inner webs white. The *young*
is somewhat mottled, and has a
short tail. On the islands of our
coast this tern was a very common bird, until fashion de-
manded it as an ornament for
ladies' hats; at present it is out

Common Tern

of fashion, but the bird has become almost extinct. (Sea
Swallow; Wilson's Tern.)

Length, 14½; wing, 10¼ (9¾–11¾); tail, 6, forked, 3½; tarsus, ¾; culmen, 1⅛. Northern hemisphere; in North America. mainly east of the
Plains; breeding from Florida and Texas to the Arctic coast, and wintering from Virginia southward.

7. **Arctic Tern** (71. *Stérna paradisæa*). — This is almost exactly like the last, but the tail is somewhat larger, and the bill decidedly redder. Mr. Brewster says the usual cry of the Arctic tern is shriller and more pig-like.

Length, 15½ ; wing, 10–11 ; tail, 6–8½, forked, 4½ ; tarsus, ⅝ ; culmen, 1¼. Northern hemisphere ; breeding from Massachusetts northward, and wintering south to Virginia and California.

8. **Roseate Tern** (72. *Stérna doùgalli*). — A rare, black-billed, white-tailed tern, with back and wings pearl-gray, and the white under parts often delicately pink-tinted. The bill is slightly reddish at base, especially in *young* birds. In summer, the whole top of the head is black, but in winter, the front of the head is white with black streaking; the under parts in winter are pure white. Mr. Chapman says this species "is a less excitable, wilder bird than *hirúndo* [No. 6.], and its single harsh note, *cack*, may be distinctly heard above the uproar of common terns, as it hovers somewhat in the background."

Length, 15½ ; wing, 9½ (9¼–9¾) ; tail, 7¼, forked 4 ; tarsus, ⅞ ; culmen, 1½. Tropical regions generally ; breeding north on the Atlantic coast, rarely to Maine, and wintering south of the United States.

9. **Least Tern** (74. *Stérna antillàrum*). — A very small, rare tern, with the back, wings, and tail pearl-gray, the under parts white, the forehead white, and the bill mainly yellow. There is a black cap extending forward past the eyes, and the outer webs of the outer primaries are black. The *young* is somewhat mottled and has a blackish bill.

Least Tern

Length, 9 ; wing, 6⅝ ; tail, 3½, forked, nearly 2 ; tarsus, ⅝ ; culmen, 1⅛. Northern South America, and north to New England, Minnesota, and California ; breeding mainly throughout. Casual to Labrador.

10. **Sooty Tern** (75. *Stérna fuliginòsa*). — A large tern, with nearly all the upper surface black, and the lower surface white. The tail is deeply forked, and the bill and feet are

black. The outer tail feathers are white. with brownish on the terminal half of the inner web.

Length, 16; wing, 12; tail, 7¼, forked, 3½; tarsus, 1; culmen, 1¾. Tropical regions generally; breeding in North America rarely north to North

Sooty Tern

Carolina; wandering to New England, and wintering south of the United States. The **Bridled Tern** (76. *Sterna anæthetus*) has the two outer tail feathers wholly white. It is a tropical tern; casual in Florida. Wing, 10½; tail, 6½; tarsus. ⅞; culmen, 1½.

Bridled Tern

11. **Black Tern** (77. *Hydrochelidon nigra surinaménsis*). — A small, short-tailed, black tern. with the back, wings, and tail somewhat lighter and more slate colored, and the under tail coverts white. The *young* (also the adult in winter) has the front head and under parts mainly white, and the back and wings pearl-gray. This is an insect-eating bird. and is often found far from large bodies of water, and occasionally on the driest of open plains.

Black Tern

Length, 10; wing, 8¼; tail, 3¾, forked, ⅞; culmen, 1. America, from

Alaska to Brazil; breeding in the interior from Illinois to Alaska, and migrating through all parts of the eastern United States.

12. **Noddy** (79. *Ánous stólidus*). — A southern, dark brown, almost black tern, with a whitish crown and a rounded tail. The *young* lacks the whitish crown, but has more or less of a white line over the eye. This is a common summer visitor in the South Atlantic and Gulf States, and breeds in Florida.

Noddy

Length, 15; wing, 10¼ (10-10⅜); tail, 6; tarsus, 1; culmen, 1¾. Tropical regions generally; in America from Brazil to the southern United States.

GULLS (SUBFAMILY LARINÆ)

Characteristics given on p. 321

Key to the Species

* Hind toe minute or wanting (much less than ¼ inch long); tail slightly notched or even................................14. **Kittiwake.**
* Hind toe small. (**A.**)
    A. Tail forked about 1 inch; tail feathers rounded at tip...........
    ....24. **Sabine's Gull.**
    A. Tail even. (**B.**)
B. Adults pure white; tarsus rough behind and less than the middle toe and nail in length; wing, 13-14 long.............13. **Ivory Gull.**
B. Adults with a darker mantle; tarsus not very rough, and equal to or greater than the middle toe and nail in length. (**C.**)
    C. Wing, 8-9¼ long. The **Little Gull** (60-1 *Larus minutus*) of Europe has been found once on Long Island.
    C. Wing over 9¼ long. (**D.**)
D. Wing, 10-10¾ long; bill black and slender....23. **Bonaparte's Gull.**
D. Wing, 10¾-12 long; bill red, with usually a dark band near tip......
    ....22. **Franklin's Gull.**
D. Wing, 12-13¼ long; outer primary black........21. **Laughing Gull.**
D. Wing over 13½ long. (**E.**)
    E. Primaries pearl-gray, fading to white at tip, no black. (**H.**)

E. Primaries pearl-gray, tipped with white but having distinct gray
spaces on the outer webs...................17. **Kumlien's Gull.**

E. Primaries with white tips and dusky or black spaces near tips (in
young sometimes all dark). (**F.**)

F. Shafts of the primaries white through the dark spaces in adult ; wing.
17-20 long ; back dark slaty-black.....18. **Great Black-backed Gull.**

F. Shafts dark like the spaces. (**G.**)

  G. Wing, 15¼ or more long.......................19. **Herring Gull.**

  G. Wing less than 15¼ long....................20. **Ring-billed Gull.**

    H. Wing over 16½ long ; culmen over 2.............. 15. **Glaucous Gull.**

    H. Wing under 16½ long ; culmen under 2............16. **Iceland Gull.**

13. Ivory Gull (39. *Gàvia àlba*). — A large, rare, northern,
pure white gull with
black feet and yel-
low bill. The *young*
has some gray patches
on different parts of
the body, but espe-
cially at the tips of
the tail feathers and
primaries; sometimes
the wing coverts have
black spots at their
tips.

Ivory Gull

  Length, 15-20 ; wing,
13¼ ; tail, 5½ ; tarsus, 1½ ; culmen, 1¾. Arctic regions ; south in the Atlantic to about the border of the United States.

Kittiwake

14. Kittiwake (40. *Rissa tridáctyla*). — A three-
toed, white gull, with pearl-gray mantle, black
tips to the outer primaries, yellowish bill,
and black feet. The hind toe is represented by a little knob. The
third to the fifth primary
have white tips be-
yond the black. In
winter the top of the
head and the back of
the neck are tinged

with pearl-gray, but there is a darker spot around the eye. The *young* has the back of the neck and lesser wing coverts black. The name is derived from the bird's cry, *kitti-aa, kitti-aa.*

Length, 17; wing, 12¼; tail, 4½; tarsus, 1¼; culmen, 1½. Arctic regions, south in eastern America, in winter to the Great Lakes and the Middle States.

15. **Glaucous Gull** (42. *Làrus glaùcus*). — A very large, northern, nearly white gull, with yellow bill, a light pearl-gray mantle, and white tips; no black anywhere in any plumage. *Young* much mottled ashy and buffy. (Burgomaster.)

Glaucous Gull

Length, 30; wing, 18 (16½–18¾); tail, 8; tarsus, 2⅞; culmen, 2½. Arctic regions; breeding in America from Labrador northward, and south in winter to the Great Lakes and Long Island.

16. **Iceland Gull** (43. *Làrus leucópterus*). — A large, northern, almost white gull, much like the last in coloring, but in its movements and feeding more like the herring gull (No. 19). The mantle is pale pearl-gray, and there are no dark tips to the primaries.

Length, 25; wing, 15½ (14¾–16½); tail, 6¼; tarsus, 2¼; culmen, 1⅞. Arctic regions; south in winter to the Great Lakes and Long Island, sometimes still farther.

17. **Kumlien's Gull** (45. *Làrus kumlieni*). — Similar to the last two, but with the primaries distinctly marked with ashy-gray. The first primary has a white tip with ashy-gray outer web; the second, with only a part of the outer web ashy-gray; the third and fourth have little gray on the outer webs, but some on both webs near the tips.

Length, 24; wing, 16 (15–17); tail, 6½; tarsus, 2¼; culmen, 1¾. Atlantic coast of North America, south in winter to Massachusetts.

18. **Great Black-backed Gull** (47. *Lárus márinus*). — A very large, very shy, black-mantled, white gull, with white tips to all the wing quills. The head and neck are streaked with grayish in winter. The *young* is much mottled with black, browns, buffs, and white. (Saddle-back.)

Length, 30; wing, 18½ (17½–19½); tail, 8; tarsus, 3; culmen, 2½. North Atlantic; breeding in America from the Bay of Fundy northward, and south in winter to Long Island, and sometimes farther.

19. **American Herring Gull** (51ª. *Lárus argentátus smithsónianus*). — A very common, large gull, with dark pearl-gray mantle, and the head, tail, and lower parts white. The ends of the outer primaries are mainly black, but with round white spots near their tips. The *adult* in winter has grayish streaks on head and neck. The *young* is much mottled, ashy, black and buff. This gull is less exclusively marine than most others, as it is found on rivers and in harbors. It shows but little fear of man.

American Herring Gull

Length, 24; wing, 17¼ (15¼-17½); tail, 7½; tarsus, 2½; culmen, 2¼. North America; breeding from northern New York, Minnesota and northward, and wintering from Nova Scotia to Cuba. The **European Herring Gull** (51. *Lärus argentätus*) is occasionally seen in eastern North America. It is somewhat smaller, and the black spot on the first primary is either broken or entirely absent.

20. **Ring-billed Gull** (54. *Lärus delawarénsis*). — A large, white-headed gull, with pearl-gray mantle, white belly, white tail; the tips of the six outer primaries white, and back of the tip black for a less and less distance. The bill is greenish-yellow with a dark ring-like band in front of the nostril. The *young* is very much mottled, with blackish and grayish colors nearly everywhere.

Length, 19; wing, 14¾ (13¾-15¼); tail, 6; tarsus, 2½; culmen, 1⅝. North America at large, more common in the interior; breeding from Minnesota and Newfoundland northward, and wintering from Long Island to Mexico.

21. **Laughing Gull** (58. *Lärus atricilla*). — A rather large, black-headed gull, with dark pearl-gray mantle, the lower neck, breast, belly, and tail white, and the primaries, except the small tips of the inner ones, black. In winter, the head and throat are white, with more or less of grayish tints. Its notes sound "like the odd and excited laughter of an Indian squaw." (Black-headed Gull.)

Laughing Gull

Length, 16½; wing, 13; tail, 5; tarsus, 2; culmen, 1⅝. Atlantic and Gulf coasts of the United States; breeding from Texas to Maine, and wintering from South Carolina to northern South America.

22. Franklin's Gull (59. *Lárus franklinii*). — A western, small, black-headed gull, pearl-gray mantle, and the lower parts and the tail white. The whole head and throat are sooty-black, and the lower parts are often rosy tinted. The first primary is mainly white, but the outer web is black except at the tip; the second has a black mark on the inner web, and a black strip on the outer web near the tip; the third to the sixth are tipped with white. In winter, the head and neck are white. The *young* is much marked with grays and browns. This gull is not found on the Atlantic coast.

Length, 14; wing, 11¼; tail, 4½; tarsus, 1⅔; culmen, 1¼. Interior North America, chiefly from the Rocky Mountains to the Mississippi River; breeding from Iowa northward, and wintering from the Southern States to Peru.

23. Bonaparte's Gull (60. *Lárus philadélphia*). — A small, black-billed, almost black-headed, white-tailed, white-bellied gull, with the wings and back pearl-gray and the first three primaries tipped with black, the next three with small, white tips and three large black spaces. In winter the head and throat are white. The *young* has the back varying from brownish to pearl-gray, the tail banded with black and white, and the head tinted with grayish.

Length, 13; wing, 10¼; tail, 4; tarsus, 1⅜; culmen, 1⅛. North America generally; breeding mainly north of the United States, and wintering from the Middle States southward to the Gulf.

Sabine's Gull

24. Sabine's Gull (62. *Xēma sabínii*). — A very rare, northern, winter-visiting, small, tern-like gull, with a pure white, slightly forked tail. The head and neck in winter (the only season the bird is seen in the United States) mainly white, with a varying number of blackish marks on the back and sides of head; back and wings dark pearl-gray; under parts, except throat, white; first primary black, with the inner half of the inner web white except at the tip; the next three tipped with white; the secondaries tipped with white. In summer the whole head and throat are slate-colored.

Length, 13½; wing, 10¾ (10⅛–11⅞); tail, 4¼, forked, ¾; tarsus, 1¼; culmen, 1. Arctic regions; south in winter to New York, Great Lakes, and Great Salt Lakes; casual in Kansas and the Bahama Islands.

FAMILY LXI. SKUAS AND JAEGERS (STERCORARIIDÆ)

A small family (6 species) of mainly dark-colored, rather long-tailed, long-winged, swift-flying, swimming birds, with the central tail feathers abruptly projecting beyond the others. These birds are hawk-like in the form of their bills[1] and in their actions; they chase the terns and smaller gulls and snatch from them the fish and other prey which they have caught. Although good swimmers, they seem unable to dive. The bill has a large, cere-like covering to the nostrils.

1

Key to the Species

* Wing over 15 long; culmen over 1⅞; tarsus, 2¼–2¾ .... 1. **Skua.**
* Wing, 13½–15 long; culmen under 1⅞; tarsus, 1⅞–2¼ .... 2. **Pomarine Jaeger.**
* Wing not over 13½ long; tarsus not over 1⅞; central tail feathers acute. (A.)
  - **A.** Scaly cere over the nostril more than half the length of the culmen; central tail feathers projecting less than 5 inches beyond the others.................................. 3. **Parasitic Jaeger.**
  - **A.** Scaly cere less than half the length of the culmen; central tail feathers in the adult projecting over 6 beyond the others........ .... 4. **Long-tailed Jaeger.**

1. **Skua** (35. *Megalestris skua*). — A northern, large, stout-bodied, dark-brown sea-bird, with a nearly even tail having all feathers broad at tip; the under parts are somewhat lighter than the upper ones, and the neck is streaked with whitish.

The shafts of the tail feathers, and the shafts and the basal portions of the inner vanes of the wing quills, are white. The *young* is somewhat streaked
with yellowish,
especially about
the head and
neck.

Length, 22;
wing, 16(15¾-16⅛);
tail, 6; tarsus, 2⅝;
culmen, 2⅛. The
coasts and islands
of the North At-
lantic, south in
America to North
Carolina, but very
rare.

Skua

2. **Pomarine Jaeger** (36. *Stercoràrius pomàrinus*). — In usual or *light phase*, a large jaeger with cap, wings, back, and tail blackish-brown, back of neck yellow, and the lower parts white with many streaks and bars of brown, especially on the breast and sides. *Dark phase.* — A dark brown to black bird with the lower parts somewhat lighter, the bill dark greenish, and the feet black. The central projecting tail feathers have rounded tips.

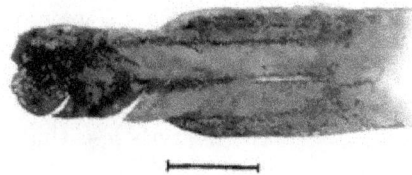

Length, 22; wing, 13¾ (13½-14); tail, 5½-9; tarsus, 2; culmen, 1⅜. Arctic regions; south in winter to Africa, Australia, and probably South America. Found on inland waters as well as seas.

3. **Parasitic Jaeger** (37. *Stercoràrius parasíticus*). — A smaller bird, but similar in coloring to the last, with the brown of the back not so black-ish. It occurs in a light and a dark phase. The middle tail

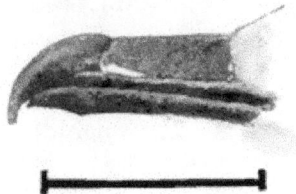

feathers of this and the next species are pointed. The best method of distinguishing this species from the last is by the difference in size and the acute instead of rounded ends to the central tail feathers. To separate it from the next compare the length of the horny covering to the nostrils, with that of the top of bill or culmen; in this species it is always more than half; in the next, less than half. In the adult, the length of tail enables one to separate them.

Length, 18; wing, 12½ (11¾–13½); tail, 5 (young), 8½ (adult); tarsus, 1¾; culmen, 1¾. Northern regions; breeding in high latitudes, and wintering in America from New York and California to South America. Migrates through the Lake region as well as along the coasts.

4. **Long-tailed Jaeger** (38. *Stercorarius longicaúdus*). — This is another bird like the last two, having the same coloring and occurring in the light and dark phases. In the mature birds of this species, the central tail feathers are much longer. This bird, as stated above, has the horny cere which covers the nostrils less than half as long as the culmen. In *young* birds before the full length of the tail is attained, the species can be distinguished only by noting the length of the cere. See the illustration.

Long-tailed Jaeger

Length, 22; wing, 12¼ (11½–13); tail, 6 (young), 11–15 (adult); tarsus, 1¾; culmen, 1¼. Northern regions; breeding in high latitudes, and migrating mainly along the coasts to the Gulf of Mexico and the West Indies.

ORDER XVII. DIVING BIRDS (PYGÓPODES)

This is preëminently the order of water birds; all species are at home only in the water, and all species swim and dive with perfect ease. The legs are situated at the tail end of the body; so in attempting to stand, the birds hold the body in an erect position, and the tarsus and tail are often used as partial supports. These birds are very awkward in their movements on land, their method of progression being by a shuffling motion.

FAMILY LXII. AUKS, PUFFINS, ETC. (ÁLCIDÆ)

A family (30 species) of short-necked, marine divers with peculiar, short bills and three full-webbed toes. The appendages to the bill, which are numerous and remarkable, are shed after the breeding season, and so are practically never observed in the United States, as the nesting grounds are in the far north. These birds differ from the other divers in the use of their wings as an additional aid in swimming under water. They breed, often in immense colonies, in cold regions, and migrate southward in winter. Most species are strong flyers, and all are wonderful swimmers. All the species belong to the northern hemisphere, and more than half are found along the Pacific Ocean. They feed exclusively upon animal matter, and are mainly silent birds.

Key to the Species

* Bill light-colored, and more than an inch high at base. (**F.**)
* Bill dark-colored, and less than an inch high at base. (**A.**)
  A. Culmen about $\frac{1}{2}$ inch long; wing under $5\frac{1}{2}$ long.......8. **Dovekie.**
  A. Culmen, 1 or more long. (**B.**)
 B. Wing, $5\frac{1}{2}-7\frac{1}{4}$ long; nostril overhung by a horny scale. (**E.**)
 B. Wing, $7\frac{1}{4}-9$ long; nostril more or less completely hidden by dense, velvety feathers. (**C.**)
  C. Tail of pointed feathers; bill nearly an inch high at base and much flattened sideways...................... 7. **Razor-billed Auk.**
  C. Tail of rounded feathers; bill less than $\frac{3}{4}$ inch high at base. (**D.**)

D. Culmen over 1½ long.................................... 5. **Murre.**

D. Culmen less than 1½ long..................6. **Brünnich's Murre.**

   E. Greater wing coverts wholly white.........4. **Mandt's Guillemot.**

   E. Greater wing coverts black at base...........3. **Black Guillemot.**

   F. Upper parts, including a band around throat, brownish-black ; belly white............ ....2. **Puffin.**

   F. Upper parts a glossy blue-black ; belly grayish-brown ; head of the adult with crests of yellow feathers...............1. **Tufted Puffin.**

   1. **Tufted Puffin** (12. *Lunda cirrhata*). — A bird similar in form to the next, with the upper parts a glossy blue-black and

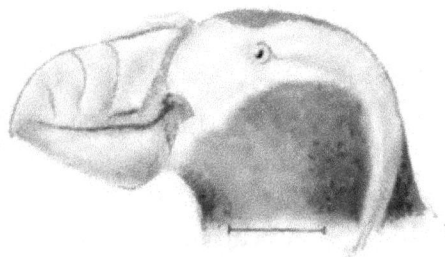

the lower ones grayish-brown. The head is furnished with two crests of yellow, silky feathers above the eyes, and the face portion of the head is white. *Young* lacks crests, white face, and the grooves of the bill.

Tufted Puffin

Length, 15 ; wing, 7¾ ; tail, 2¾ ; tarsus, 1¼ ; culmen, 2⅜. North Pacific ; accidental on coast of Maine.

   2. **Puffin** (13. *Fratercula arctica*). — A very stout-billed diver, with the upper parts, including a band around the neck, brownish-black, breast and belly white, and the sides of the head grayish-white. The bill in life, especially during the breeding season, is peculiarly ridged and of bright red, blue, and white colors. Breeding birds have a horny spine over the eye. (Sea Parrot.)

   Length, 13 ; wing, 6¼ (6-6¾); tail, 2¼ ; tarsus, 1 ; culmen, 1⅞. North Atlan-

Puffin

tic, on coasts and islands; breeding from the Bay of Fundy northward. also south to Long Island, and rarely farther south.

3. **Black Guillemot** (27. *Cépphus grylle*). — In winter, a mottled, grayish-black-backed, white-bellied "sea pigeon," with sooty-black wings marked with a white blotch, formed by the terminal half of the greater wing coverts. The back has the feathers more or less tipped with white. In summer, it is a sooty-black bird, with the same white patch on the wings. These birds fly rapidly in a straight line just above the surface of the waves, but are usually found, in small flocks, swimming or diving in the water.

Length, 13; wing, 6¼ (6–7); tail, 2; tarsus, 1¼; culmen, 1¼. Northern Atlantic Ocean on both shores; in America breeding from Maine to Newfoundland, and wintering south to Philadelphia.

4. **Mandt's Guillemot** (28. *Cépphus mándtii*).—Similar to the last in habits, size, and markings, but the white blotch on the wing is larger, including the bases as well as the tips of the greater wing coverts.

Mandt's Guillemot

Length, 13; wing, 6¼ (5½–7¼); tail, 2; tarsus, 1¼; culmen, 1¼. Arctic regions; in America breeding from Labrador northward, and wintering south to Massachusetts.

5. **Murre** (30. *Uria tróile*). — An auk-like bird, with the upper parts from bill to tail a sooty-black and the lower parts white, excepting a brownish band across the lower neck in summer, which in winter is lacking. The head is more brownish and the back, wings, and tail are more blackish. There are white tips to the secondary quills, making a band across the wing. In winter the throat is somewhat tinted with brown and the belly marked with black. (Common Guillemot.)

Length. 17; wing 8 (7¼-8½); tail. 2¼; tarsus. 1½; culmen, 1¼. Coasts and islands of the North Atlantic; breeding from the Gulf of St. Lawrence northward, and wintering south to southern New England.

Murre

6. **Brünnich's Murre** (31. *Uria lómvia*). — A bird similar to the last, but with a smaller and shorter bill and a slightly longer wing. In breeding plumage there is some difference of color, but in winter, when found in the waters off our eastern shores, the difference in length of bill is the distinguishing mark. (Thick-billed Murre.)

Length. 17; wing, 8½ (7½-8¾); tail. 2¼; tarsus. 1½; culmen, 1¼. Coasts and islands of the North Atlantic; breeding from the Gulf of St. Lawrence northward, and wintering south to New Jersey.

7. **Razor-billed Auk** (32. *Alca tórda*). — A short. high, thin-billed auk, with the upper parts generally sooty-black, and the lower parts white. The black bill is crossed by a white band. there is a white line from the bill to the eye, and a line is formed on the wings by the white tips of the secondaries. The bill is flattened sidewise, whence the bird derives the name of razor-bill. It has the habit when on the water of turning its tail almost directly upward. (Tinker.)

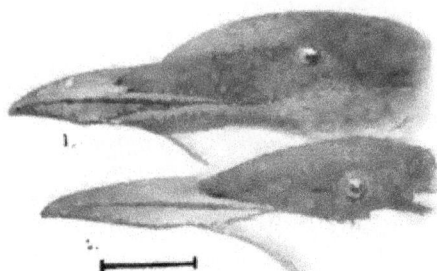

1. Brünnich's Murre   2. Murre

Length, 17; wing, 8¼ (7¼-8½); tail, 3½; tarsus, 1¼; culmen, 1¼. Coasts and islands of the North Atlantic; in America breeding from Maine northward, and south in winter, casually to North Carolina.

8. **Dovekie** (34. *Alle alle*). — A small, short-billed, sooty-backed, white-bellied bird, with white tips to the secondaries, and some white streaks on the shoulders. The small wings of

Razor-billed Auk   Dovekie

this bird are moved with almost bewildering rapidity, enabling it to fly with great swiftness. It swims with grace and ease, and dives, like all of the order, with great expertness. (Sea Dove; Little Auk.)

Length, 8½; wing, 4¾ (4½-5); tail, 1½; tarsus, ¾; culmen, ½. Coasts and islands of the North Atlantic; in America breeding far north; south in winter to New Jersey, accidental in Michigan.

FAMILY LXIII. LOONS (URINATÒRIDÆ)

A small family (5 species) of large, heavy, long-necked, short-tailed, diving birds, with the legs situated at the tail end

of the body. There are four toes, the three in front being full
webbed. In summer, all species when adult have the dark
back regularly spotted with
nearly square white blotches.
They are all migratory, breeding, with one exception, in the
Arctic regions, but found in
the United States in winter.
These birds, like all the divers, are exceedingly clumsy
on land, which they seldom
visit except for breeding purposes; but in the water their
powers of swimming and diving are only equaled by the
grebes. They are also strong
and rapid flyers. In their
migrations, they keep at a
considerable height and are
usually seen in small flocks.

Loon

In pursuit of fish, which forms their only food, they move
through the water by the aid of their feet alone. In this
they are like the grebes, but unlike the auks.

Key to the Species

* Wing, 13–16 long; tarsus, 3–3½; culmen, 2⅜–3½.............1. **Loon**.
* Wing, 10–13¼; tarsus, 2¼–3; culmen, 2–2⅝. (**A**.)
  * **A.** Adult in summer, throat black; adult in winter and young, no
    white spots on the back, but grayish margins to the feathers......
    ....2. **Black-throated Loon**.
  * **A.** Adult in summer, throat gray with a triangular, chestnut patch;
    adult in winter and young, back distinctly spotted with white.....
    ....3. **Red-throated Loon**.

1. **Loon** (7. *Urinator imber*). — *Adult* in summer, a very large,
greenish-black-headed, black-throated loon, with the breast and
belly white. The back and wings are greenish-black, with
many nearly square, white spots. There are spaces on the

sides of the neck and breast, streaked with white, and on the sides of the body and under the tail spotted with white. *Adult* in winter and *young*, a loon with all upper parts blackish, the feathers edged with grayish, but with no white spots; all under parts white, with some grayish on the throat. Birds in the United States can be found with all grades of white spotting on the back. This is the only species of loon breeding in the states and thus the only one to be found at all seasons.

Length, 28–36; wing, 13–15¼; tarsus, 3–3½; culmen, 2¾–3½. Northern hemisphere; breeding from the northern range of states northward, and wintering south to the Gulf of Mexico and Lower California.

2. **Black-throated Loon** (9. *Urinator arcticus*). —A bird similar to the last, but ranging much farther north. *Adult* in winter and *young*, having upper parts, including wings and tail with the feathers, blackish at their centers and grayish along their borders; no white spots on the back. This is practically the winter appearance of the last, so the difference in size must be noted to determine the species. The absence of white spots separates it from the next. In summer the *adult* can be separated from the last by the ashy head, and from the next by the black throat and absence of chestnut color.

Black-throated Loon

Length, 26–29; wing, 12–13¼; tarsus, 2⅝; culmen, 2⅛–2⅜. Northern hemisphere; breeding north of the United States, and south in winter, casually to the northern states east of the Rocky Mountains.

3. **Red-throated Loon** (11. *Urinator lümme*). — This is the smallest of our loons. It is found in the Northern States from October to May, and irregularly south, in winter, to South Carolina.

Red-throated Loon

In winter, this bird in all stages has the square white spots on the back. This separates it from the last species, and its much smaller size distinguishes it from the first. It derives its name from a triangular, chestnut spot on the neck.

All our loons are wild, wary birds. The Pacific loons of the western coast are "tamer than any other water fowl I have seen. . . . They constantly swam around the vessels . . . and all their motions, both on and under the clear water, could be studied to as much advantage as if the birds had been placed in artificial tanks for the purpose. Now two or three would ride lightly over the surface with neck gracefully curved, propelled with idle strokes of their broad paddles . . . while their flashing eyes, first directed upward, then peering into the depths below, sought for some attractive morsel. In an instant, with a peculiar motion impossible to describe, they would disappear beneath the surface, and shoot with marvelous swiftness through the limpid element, transfix on their arrow-like bill an unlucky fish, and lightly rise to the surface again." (Dr. Coues.)

Length, **24–27**; wing, 10–11½; tarsus, 2½; culmen, 2–2¼. **Northern** hemisphere; breeding from Manitoba and New Brunswick northward, and wintering **south to South** Carolina.

FAMILY LXIV. GREBES (PODICIPIDÆ)

A family (30 species) of **fresh-** and salt-water **diving birds** of general distribution throughout the world; **five species are found in** the **region** covered **by this book,** and **only six in** **North** America. The grebes **are** long-necked **divers with** straight, slender bills, and with the feathers **of the** under parts 'of a peculiar satiny texture. The **three front toes have** lobed membranes along **their sides. The heads in many species are** furnished, in the breeding season, with **brightly colored crests,** ruffs, etc. **These are lost after** the nesting **is over,** thus producing **seasonal differences** so great **as** to make the birds appear as separate species. These head appendages, **and** their **erect position, give the birds a most** grotesque appearance.

They have the power, when alarmed, of sinking quietly backwards into the water and then swimming almost any distance
with only the tip of the bill above the surface. Like all of
the divers, their food consists mainly of fish, which they are
able to catch under water by their rapid swimming, using their
feet alone for propulsion. In this they differ from the auks,
which use both legs and wings.

Key to the Species

* Bill stout and somewhat hooked, its length not quite twice its greatest
  depth at base [1] .... 6. **Pied-billed Grebe.**
* Bill straight and more slender, its length more
  than twice its depth at base.[2] (**A.**)

  A. Culmen more than 2½ long ....
  .... 1. **Western Grebe.**

  A. Culmen, 1½–2½ long; wing over 6 long .... 2. **Holbœll's Grebe.**

  A. Culmen, ½–1½ long; wing under 6 long. (**B.**)
  B. Wing under 4¼ long .... 5. **St. Domingo Grebe.**
  B. Wing, 4¼–6 long. (**C.**)

  C. Bill flattened sidewise and thus higher than wide at base ....
  .... 3. **Horned Grebe.**

  C. Bill wider than high at base
  ... 4. **American Eared Grebe.**

1. **Western Grebe** (1. *Æch-
móphorus occidentàlis*).—A very
large, long, slender-billed, mottled, brownish-backed grebe,
with all the under parts satiny
white. The primaries are chocolate-brown with white bases,
and the secondaries are mostly
white. It has a short crest and
puffy cheeks. This is a common grebe of the extreme west.
The grebes rarely fly to escape
their enemies, but depend upon
their diving and swimming
powers.

Western Grebe

Length, 24-29 ; wing, 8 (7½-8½); tarsus, 3 ; culmen, 2½-3¼. Western North America east to Manitoba, and south to central Mexico.

2. **Holbœll's Grebe** (2. *Colýmbus holbœllii*). — *Adult* in winter, a common. blackish-brown-backed. whitish-bellied grebe. with some pale brownish-red on the sides of the neck. *Young,* a blackish-backed, silvery-bellied grebe, with the neck and sides grayish. *Adult* in summer, a blackish-backed, chocolate-brown-sided, white-bellied grebe. with the crown, small crest, and back of the neck black. There is

Holbœll's Grebe

a silvery-ash patch on the throat, changing to deep, brownish-red on the front and sides of neck to the breast. On the water this is a very graceful bird, swimming and diving with the greatest ease. When flying. and it flies rapidly, the neck and feet are stretched to their full length. (Red-necked Grebe.)

Length, 19 ; wing, 7½ (7¼-8½) ; tarsus, 2¾ ; culmen, 1⅝-2¼. North America ; breeding from about the northern border of the states northward, and wintering south to about the Gulf, at least casually.

3. **Horned Grebe** (3. *Colýmbus aurítus*). — *Adult* in summer, — a very much crested and ruffed grebe. with the top of head, hind neck. and throat black ; stripe and plumes behind the eye chestnut, blackening on the sides : front of neck to breast chestnut ; back and wings blackish ; belly white, and sides

washed with chestnut. *Adult* in winter, a common, slightly crested, grayish-black-backed, silvery-white-bellied grebe, with some grayish tints on throat and breast. This and the pied-billed grebe (No. 6.) are in their winter dress much alike in appearance, and are often mistaken for each other. The horned grebe's bill is straighter and more slender than that of the pied-bill. "When ordinarily swimming, the feet struck out alternately, and the progression was steady; but sometimes both feet struck together, and then the movement was by great bounds, and was evidently calculated to force the bird over an expanse of very weedy water, or through any tangle of weeds or rushes in which it might have found itself." (E. E. Thompson.)

Length, 12½-15½; wing, 5½ (5¼-5¾); tarsus, 1¾; culmen, ⅞. North America; breeding from the northern range of states northward, and wintering south to about the Gulf of Mexico.

4. **American Eared Grebe** (4. *Colymbus nigricollis califòrnicus*). — *Adult* in summer, — a western, black-headed, black-necked, blackish-brown-backed, white-bellied grebe, with conspicuous golden-brown ear tufts and a white blotch on the chocolate-brown wings, formed by the tips of the secondaries. The winter coloring is much the same as that of the last, but the difference of bill (wider than high at base), and the smaller size distinguish the species.

Length, 13; wing, 5¼ (5-5½); tarsus, 1⅝; culmen, ⅞. Northern and western North America (west of the Mississippi in the United States), south to Central America.

5. **St. Domingo Grebe** (5. *Colymbus dominicus*). — An extreme southern, very small, brownish-black-backed grebe, with dusky-mottled, silky-white belly. The crown is deep, glossy, steel-blue, and the sides of head and the neck all around are ashy-gray. There are no decided crests or ruffs.

Length, 9½ ; wing. 3¾ (3½–4¼); tarsus, 1½ ; culmen, ⅜. South America, from Paraguay north to Texas and Lower California, including the West Indies.

Pied-billed Grebe

6. **Pied-billed Grebe** (6. *Podilymbus podiceps*). — A common, small, brownish-black grebe (in summer), with the lower breast and belly nearly white. The front and sides of the neck are lighter than the back, and more nearly brown; there is a black band across the bill at the middle. In winter, the coloring is much the same, but the band across the bill is lacking, and the throat is white. This is our commonest grebe. It can dive head first beneath the water, as well as sink gradually like the other species. (Dab-chick; Dipper; Diedapper; Hell-diver; Waterwitch ; etc.)

Length, 13½ ; wing, 4⅞ (4½–5¼) ; tarsus, 1½ ; culmen, ⅞. America, from the Dominion of Canada to the Argentine Republic, including the West Indies ; breeding nearly throughout.

PART III

THE STUDY OF BIRDS IN THE FIELD

This part is designed to enable any person with moderate patience and energy to become familiar with all conspicuous common birds. The only preparation necessary for its use is the ability to recognize the English sparrow, the robin, and the crow when seen, and to tell the difference between an owl and a hawk. The Keys furnish a guide to two hundred of our most common land birds, helping the pupil to recognize them at sight or by their notes, without shooting a single specimen.

With two mornings each week of the spring and early summer devoted to the pleasant task of seeing and hearing the birds, the learner should, in a few seasons, be sufficiently familiar with them to recognize these common birds at sight. A few birds will always remain unidentified until they are dead and in the hands of an experienced ornithologist. It is easy enough to recognize the family to which they belong — to see that they are finches or wood warblers or vireos or flycatchers, but it is far more difficult to determine the species. These difficulties present themselves mainly with females; but since they perplex even the skilled ornithologist, they must not discourage the beginner.

Progress will seem to be slowest during the first season. It will be harder to learn the first ten birds than any succeeding twenty. At the start it appears difficult to observe any birds with care, but one gradually learns to move and work in such a manner as not to frighten the birds. After some practice the observer notes more peculiarities at a single glance than a minute or two of careful study reveals to a beginner. Practice in this as in everything else renders the work easy, certain, and rapid. At the start few bird voices will be heard; after a little experience, the woods and fields will seem to resound with them. To the beginner the bird notes mean little; to the bird lover they are replete with meaning.

In studying birds in the field, the observer must remember that they are naturally timid and have remarkably sharp eyes and ears; almost invariably they see before they are seen. They desire to investigate, not to be investigated; so, the more careless the learner appears to be, the less he shows that he is studying the birds, and the more strange chirps and whistles he can utter, the nearer he can approach and the better he can observe.

Methods of study. — There are three methods of studying birds. (1) To stay in good bird localities and await the approach of the birds. (2) To walk quietly in field and wood, on the alert, while advancing, for the objects of study. (3) To be driven slowly in a carriage or other conveyance through good bird localities. The first method is by far the best for beginners; the others are very useful after a score or more of common birds are well known. The success of the last method will be a surprise. Birds do not expect observation from the occupants of moving vehicles, and so will act naturally and may be closely approached. The slowest of walking horses should be used.

Locality. — In order to choose a good locality for carrying out the first plan, attention must be given to the fact that some birds are always to be found in forests, some in shrubbery, some in open fields, some near the water, some on the ground, and some in the tree tops. Hence a place which combines as many forms of landscape as possible within the scope of the eye and ear will be the one to select. If a stream of water flows through a wood and then into a field, a covered position near the brook at the point where it issues from the forest will be well adapted for a view of many kinds of bird haunts. Then, if the proper time of day and the proper time of the year are chosen, there will be no lack of birds to study; the danger is rather that there will be so many that they will bewilder the beginner.

Season. — For many reasons the best time of the year to begin work is the spring and early summer. The birds are then most brightly plumaged; they sing most loudly, most sweetly, and most characteristically; it is nesting time, and near their nests the same birds can be seen day after day, and thus can be thoroughly studied; the young birds with their plain tints are not abundant enough to confuse the student, and the females are most of the time hidden from view.

Time of day. — In spring and summer the best time of day is the early morning from sunrise to 10 A.M ; next best is the evening just before sunset. The poorest time of all is the middle of the day. During the cold months the best hours are from noon to about 3 P.M.

Which birds to study. — A beginner should try to determine the names of only those birds that have conspicuous colors or markings. They will, as a rule, be males, and are the birds that have characteristic notes, and those that are especially described in the Keys of this part of the book. When a bird is determined upon for study, it should be closely examined through an opera glass, and as many points as possible should be mentally noted before the book is opened, and even before the opera glass is taken from the eyes. All bird workers first become acquainted with the males, and later learn to recognize the females and young by seeing them associated with the males, and reading such descriptions as are found in Part II. of this book.

Special features to examine. — The points to be first determined are the size as compared to that of the English sparrow and the robin; the length of the bill as compared to the length of the head; the form of the bill, whether stout or slender; the actual and comparative length of the wings and the tail; the colors, markings, etc., of the breast, the back, and the wings; the presence or absence of wing bars, and their color, if present;

the tip of the tail, whether notched, square, or rounded; and the presence or absence of white on the tail feathers (to be seen when the bird is on the wing). Of course any peculiarity of habit of perching or flying, any sounds produced, any position habitually taken, the method of gathering food or of progressing on the ground (walking or hopping), should be observed.

Aids to successful work. — A power of mimicry is a valuable attainment for bird study. By imitating the notes heard, not only will you better remember the sounds, but the birds will try to investigate the source of the notes, and will thus come nearer to you than under any other circumstances. If you are not able to imitate bird sounds, then " squeak " by rapidly kissing the finger; this gives a sound similar to that of a bird in distress, and will usually bring into view many of the birds of the vicinity, especially during the mating season. Bird whistles that can be held in the mouth are useful in lieu of mimicry.

Winter study of birds. — Though for a beginner spring and summer are the best times for study, there are some advantages in winter work which are worthy of mention. There are comparatively few birds to be seen in the winter, and no young to confuse by their nondescript plumage; and there is but little foliage to hide the birds from view. The middle of **the** day is the best time for study during the winter.

Local bird lists. — Obtain all the lists you can of the birds of your locality. The more local the list, the better it will be. Such a list will enable you to know what birds are to be expected at any season.

General hints. — In order to emphasize the important points, a brief résumé is here given.

(1) All your movements must be quiet and not *sudden.* Acquire the habit of investigating without appearing to do so. If you need to get near a bird, do it by imperceptible advances.

(2) Your clothing should be free from bright or sharply contrasted **tints**; and it is better to have **the sun** back of you.

(3) You need an opera glass or a field glass. If this is bright or glossy, cover it with gray cloth, and let this cloth extend about an inch beyond the front lenses. It is well also to have a folding artist's stool, as your patience may be tried by an uncomfortable position. Always carry a notebook and pencil with you and *use them*.

(4) Find a good bird locality and visit it day after day, until you have learned a goodly number of its feathered songsters. Good localities are such as have within easy reach trees, bushes, water, swamp, upland, and lowland.

(5) Begin your investigation in spring just before the leaves expand, and attempt to find the name of one new bird at a time. Let that one be a male with some decided peculiarity of color, marking, note, or habit, or, if possible, all of these.

(6) Accustom yourself to observe and remember many things without removing the opera glass from your eyes. Think at the start of each of the following parts: bill, back, breast, belly, crown, wings, and tail, and observe something peculiar about each. The ability to do this will grow rapidly, and you will soon be surprised at the ease with which you observe.

(7) Try to make sounds similar to those of birds, either chirpings or more elaborate sounds. If you can do no better, hold the finger against the lips, and, by drawing in the breath, make kissing sounds somewhat like those of a bird in distress. This will cause a commotion among the smaller birds, and will frequently bring a number into view. Use a mechanical bird whistle if you can do no better.

(8) The true colors of birds cannot be determined with accuracy when seen against a bright sky. So for color of plumage try to observe the bird when brush or grass or trees are in the background.

Method of using the Keys. — The construction of the Keys for the birds in the bush is on the same plan as the others in the book, but as the Field Keys are especially designed for beginners, who need more cautions and hints than others, the directions are here repeated more minutely, with an illustrative example showing the plan of procedure.

Note first the great divisions of birds into groups as given on page 356. You have to decide, mainly by the size of your bird, which Key contains it. Turn to this Key, read all the statements beginning with stars (*), and choose the one which best describes the bird you are investigating; at the end of the one chosen there is a letter in parenthesis (or possibly the name of a bird and the page where it is described). The letter directs you to the statements under the same letter somewhere below, and from among these statements you must choose the one that best describes the bird you are observing. In order to decide, you must carefully read all the statements. At the end of the chosen one you will find another letter in parenthesis. Turn to the place where this is used and continue as before. *Never refer to any letters or read any statements except those to which you are directed by the letter in parenthesis.* At some stage in your progress you will find, instead of a letter in parenthesis, the name of a bird and the page where it is described. Turn to this page and carefully read the description; if there is an illustration, examine it and compare it with the bird you are studying.

The descriptions of all birds in this book **were** especially written for use in the field, and just such markings as can readily be seen at a short distance are emphasized. Great pains have been taken to form descriptions in sentences **so** connected that they can be readily remembered, **and repeated mentally.** If the bird is seen against a bright sky, some allowance must be made for colors.

Suppose you are observing a bird with the following characteristics: when at rest the head, back, and most of the wings appear black. The spots on the wings and the base of the tail are orange or flame color, and the belly white. Under the wings there is much flame color. It is somewhat smaller than the English sparrow; hence you will find it by the aid of the Key on page 356. **(As it is** sometimes nearly as large as a small English sparrow, it is given also in the Key on page 359.) Read the three statements following the stars. Though your bird is a peculiarly lively one, and is often seen flying from twig to twig, floating downward and darting upward, you conclude that it can hardly be considered as generally on the wing, and as it does not show creeping habits on the trunks and larger limbs, you search for it under the third star, where, in parenthesis, you are directed to read the statements following the A's, of which there are four. Reading these carefully, you find that the second is most satisfactory, and you turn to the K's. Here there are five statements, and the first is seemingly right. The name Redstart is given, with the direction to turn to page 96, where a description of your bird will be found.

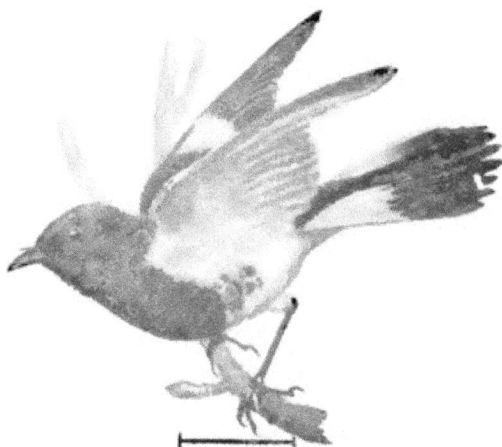

GENERAL KEY TO BIRDS IN THE FIELD

LAND BIRDS

This Key includes a few Water Birds which occasionally do not show their water habits. For the general Key to Water Birds, see page 371.

In the Keys of this section (Part III.) no general attempt has been made to include any but male birds in good plumage, as the introduction of the female and young forms would have increased the number of divisions of the Keys to such an extent as to render them too difficult for the use of beginners. The rarer species are omitted for the same reason.

* Owls. Mainly night-flying birds, of loose plumage and large heads; they have their eyes in a facial disk, and so look forward instead of sideways..Key, p. 369.
* Hawks, Vultures, Kites. and Eagles. Generally large birds, with very hooked bills. These birds in most cases take an erect position in perching ....Key, p. 370.
* Land birds, other than the Birds of Prey. (A.)
  * A. Birds plainly smaller than the English sparrow....... Key below.
  * A. Birds about the size of the English sparrow................p. 359.
  * A. Between the English sparrow and the robin in size........p. 363.
  * A. Birds about the size of the robin........................p. 366.
  * A. Birds larger than the robin............................p. 368.

Key to Birds Smaller than English Sparrows

The numbers refer to the pages where the birds are described.
* Birds seen mainly on the wing. The feet of these birds are small and weak; when at rest they perch on slender things, such as telegraph wires. (S.)
* Creeping birds on tree trunks and larger limbs. (Q.)
* Birds without special creeping habits and not seen constantly on the wing. (A.)
  * A. Birds with conspicuous bright yellow on parts other than the bend of wing or center of the crown and without bright blue or purple. (L.)
  * A. Birds with bright red or flame color, but no distinct lemon-yellow or blue. (K.)
  * A. Birds with either bright purple or blue or slaty-blue (if only slaty-blue, then with no yellow). (I.)
  * A. Birds with none of the above bright colors, except possibly a spot in the crown or at the bend of the wing. (B.)
* B. Very small, plain olive or grayish birds, with no bright colors except in some specimens, a small crown patch; under parts whitish. (H.)
* B. Brown birds, with some cross bars[1] and with the habit of holding the tail erect. (G.)
* B. Birds with the sides of the head and breast white and a conspicuous black throat patch. These birds plainly say chick-ä-dee. (F.)
* B. Upper parts olive; wings and tail blackish; two whitish wing bars;[2] under parts grayish white................Least Flycatcher, p. 165.
* B. Not as above; stout-billed birds.[3][4] (C.)
  * C. Conspicuously streaked, brownish birds. (E.)

1   2   3   4

C. Plain, clay-colored bird, with no conspicuous streaks, but with a white line over the eye............Clay-colored Sparrow, p. 131.

C. Olive-green-backed birds which are generally seen deliberately searching on leaf and twig for insects. **(D.)**

D. Lower parts yellowish; no wing bars.....Philadelphia Vireo, p. 98.

D. Lower parts white; two distinct wing bars;[2] eyes white...........
....White-eyed Vireo, p. 100.

E. Tail notched,[5] and with some yellowish at the base of the blackish feathers; wing also with some yellow... ... Pine Siskin, p. 119.

E. Tail notched;[5] crown chestnut; forehead blackish; a whitish line over the eye.....................Chipping Sparrow, p. 130.

E. Tail not notched, usually rounded,[6] of narrow, sharp-pointed tail feathers.[7] **Grasshopper Sp.**, p. 124. **Henslow's Sp.**, p. 125.
.................................Sharp-tailed Sp., p. 126.

F. Head with a distinct black cap; back ashy.....................
....Chickadee, p. 61. Carolina Chickadee, p. 62.

F. Crown dull brownish; extreme northern......................
.................................Hudsonian Chickadee, p. 62.

G. Tail very short; under parts brown like the back, but lighter.....
.................................Winter Wren, p. 68.

G. Tail longer; under parts grayish; back dark brown, without streaks............................House Wren, p. 68.

G. Upper parts dark cinnamon-brown; a distinct white line over the eye; tail with the outer feathers black, and the central ones barred[1].............................Bewick's Wren, p. 67.

G. Back streaked lengthwise with white or white and black.........
Short-billed Marsh Wren, p. 69. **Long-billed Marsh Wren**, p. 69.

H. Olive-green birds, seen flitting near the tips of twigs and bushes; under parts yellowish-gray; crown usually with a bright spot. .....
.....Golden-crowned Kinglet, p. 56. Ruby-crowned Kinglet, p. 57.

H. Slender, grayish bird, with a long tail..Blue-gray Gnatcatcher, p. 57.

I. Whole plumage blue....................Indigo Bunting, p. 139.

I. Southern bird, with bright blue, green, and red in the plumage....
.................................Painted Bunting, p. 139.

I. Slaty-blue-backed birds. **(J.)**

J. Throat and sides black; lower breast and belly white, a distinct white wing patch....................Black-throated Blue Warbler, p. 82.

J. Throat and belly white; sides streaked with black; two white wing bars[2].. .... Cerulean Warbler, p. 84.

K. With much flame color at base of tail and middle of wing; upper parts black; belly about white ....American Redstart, p. 96.

5 6 7

K. Streaky, winter bird, with red on the crown, and in the male on the breast also.................................**Redpoll, p. 118.**

K. Throat and breast orange flame color; head black striped with flame color......................**Blackburnian Warbler,** p. 86.

K. Very small, olive-green-backed birds, with red or flame color on the crown..
....**Golden-crowned Kinglet,** p. 56.   **Ruby-crowned Kinglet,** p. 57.

K. Bird with crimson, black, yellow, white, and plain brown, in the plumage............................. **European Goldfinch,** p. 120.

L. A yellow-bodied bird, with black wings and tail (in the winter the body is washed with brownish).........**American Goldfinch,** p. 118.

L. Face bright red; back cinnamon-brown; wings with a yellow band ....**European Goldfinch,** p. 120.

L. Bird with some shade of yellow nearly everywhere ....
....**Yellow Warbler,** p. 82.

L. Slender-billed birds, not as above.   **(M.)**

  M. Throat and breast bright yellow, unspotted and unstreaked.   **(P.)**

  M. Breast and belly white or nearly so, with at most a tint of yellow. **(O.)**

  M. Throat and upper breast black; belly white; much of the head yellow; back olive-green....Black-throated **Green Warbler,** p. 87.

  M. Head, neck, and throat bluish-gray, changing to black on the breast; belly yellow; upper parts, including wings and tail, olive-green; no wing bars ....**Mourning Warbler,** p. 92.

  M. Breast yellow, with dark streaks or blotch.   **(N.)**

N. Upper parts grayish-blue, with a golden spot in the middle of the back; two white wing bars.[1]..Parula Warbler, p. 80.

N. Crown black; cheeks chestnut; a broad white wing bar; yellow under parts heavily streaked with black ....**Cape** May Warbler, p. 81.

N. Rump as well as the under parts rich yellow; breast and sides heavily streaked with black; two white wing bars;[1] upper parts dark olive...........................**Magnolia Warbler,** p. 83.

N. Crown chestnut; yellow under parts streaked with chestnut on breast and sides.......**Palm Warbler,** p. 88.   **Yellow Palm** Warbler, p. 89.

  O. Crown and wing patch yellow; chin, throat, and band through eye black; back bluish-gray.......**Golden-winged Warbler,** p. 78.

  O. Back olive-green (abruptly changing to gray on the head of the male); no white wing bars; no black on head and breast........
....**Tennessee Warbler,** p. 80.

  O. Crown yellow; sides chestnut; back and wings streaked with black and yellow................Chestnut-sided Warbler, p. 84.

P. Sides of neck and body with black streaks; back olive-green (spotted with chestnut in the male); two yellow wing bars......... ....
....Prairie Warbler, p. 89.

P. Back olive; head with a peculiar black mask; lower belly white...
.... Maryland Yellow-throat, p. 93.

P. Cap black ; back olive-yellow ; under parts yellow shading to olive on the sides........................ ....Wilson's Warbler, p. 94.

P. Sides streaked with black ; white line over the eye ; two white wing bars ;[1] belly white...............Yellow-throated Warbler, p. 86.

P. Back olive-green, changing to gray on the head and neck ; wing and tail brownish ; no wing bars..............Nashville Warbler, p. 79.

P. Back olive-green; wings slaty-blue; forehead and all under parts bright yellow ; a dark line through eye...Blue-winged Warbler, p. 78.

Q. Slender brown bird, with long tail of sharp-pointed feathers used in climbing ; belly white...................Brown Creeper, p. 63.

Q. Slender bird, with the whole plumage streaked black and white...
.............................Black and White Warbler, p. 76.

Q. Short-tailed birds creeping with the head downward as often as upward. **(R.)**

R. Lower breast and belly reddish-brown. Red-breasted Nuthatch, p. 59.

R. Top of head dark brown............Brown-headed Nuthatch, p. 60.

S. Very small bird, seen hovering over flowers. Hummingbird, p. 166.

S. Larger bird, with mouse-colored back and white belly............
........Rough-winged Swallow, p. 107. Bank Swallow, p. 107.

S. Steel-blue-backed, long-winged bird, with reddish rump..........
......................................Cliff Swallow, p. 105.

S. The flycatchers are so frequently seen on the wing after insects, that they might be looked for here ; they sit on a twig, with depressed tail and quivering wings, till an insect is seen, when they dart out, and after catching their prey, return to the same perch...
..................................Least Flycatcher, p. 165.

Key to Birds about the Size of the English Sparrow

The numbers refer to the pages where the birds are described.

* Birds seen mainly on the wing. The feet of these birds are small and weak ; when at rest they perch on slender things, such as telegraph wires. **(W.)**

* Birds seen creeping along trunks and larger branches of trees. **(V.)**

* Birds neither constantly on the wing nor creeping on tree trunks. **(A.)**

A. Birds with a conspicuous amount of bright yellow, but no red. **(P.)**

A. Birds with rich orange or flame color, but no lemon-yellow. **(O.)**

A. Stout-billed birds, with more or less of distinct red in the plumage. **(N.)**

A. Birds blue in color. — Including breast....Indigo Bunting, p. 139.
— Breast brown.............Bluebird, p. 55.

A. Birds with none of the above bright colors (yellow, flame, red, or blue) in conspicuous amounts. **(B.)**

B. Crested,[2] loud-voiced, gray bird ..Tufted Titmouse, p. 61.

B. Stout-billed birds, without crest. **(F.)**

B. Slender-billed birds, without crest. **(C.)** 2

C. Birds with somewhat barred [1] brown plumage, and with the habit of holding the tail erect. **Bewick's Wren**, p. 67. **Carolina Wren**, p. 66.

C. Crown, throat, upper breast, and sides chestnut................ .... **Bay-breasted Warbler**, p. 85.

C. Black and white streaked ; crown black. **Black-poll Warbler**, p. 85.

C. Walking ground bird, of open fields and pastures, with white tips to the outer tail feathers and a voice which plainly says *dee-dee*, *dee-dee*............................... **American Pipit**, p. 70.

C. Olive-green to olive backed birds, with at most a yellowish tint on the under parts, usually without any yellow. (**D.**)

D. Birds with a thrush-like, spotted breast. (**E.**)

D. Flycatching birds, with the habit of sitting on a perch, with depressed tail and quivering wings, watching for insects. These they capture on the wing with a click of the bill, and then return to the same perch. **Wood Pewee**, p. 163. **Green-crested Flycatcher**, p. 164. **Least Flycatcher**, p. 165.

D. Crown distinctly marked with four black and three buffy stripes ; under parts whitish.................. **Worm-eating Warbler**, p. 77.

E. Crown bright orange, edged with black stripes... **Oven-Bird**, p. 90.

E. Crown olive, like the back.............. ....**Water-Thrush**, p. 90. **Louisiana Water-Thrush**, p. 91.

F. Slate-colored bird, with the belly abruptly white, and the outer tail feathers white .... **The Juncos**, p. 132.

F. Winter bird, mainly white in color, more or less blotched with brownish............................ **Snowflake**, p. 121.

F. Birds with the sides of the head and breast white, and a conspicuous black throat patch ; their notes seem plainly to say *chick-a-dee*...... .... **The Chickadees**, p. 61.

F. Olive-green to olive backed birds, with at most a yellowish tint on the light-colored under parts ; these birds deliberately hunt for insects upon twigs, leaves, and bark. (**M.**)

F. Streaky, brownish birds. (**G.**)

  G. Outer tail feathers conspicuously white. (This can readily be seen when the birds are flying.) (**L.**)

  G. Outer tail feathers not white. (**H.**)

H. Breast grayish to white, unstreaked ; no distinct throat patch. (**J.**)

H. Breast definitely streaky. (**I.**)

H. Breast with a black patch .. ..**European House Sparrow**, p. 120.

H. Breast with an indistinct dark-brown blotch... Tree **Sparrow**, p. 130.

  I. Marsh sparrows, with narrow, sharp-pointed tail feathers [2]....... ....**Sharp-tailed Sparrow**, p. 126. Savanna **Sparrow**, p. 124.

1   2

I. A common, reddish-brown sparrow, with the marks of the breast more or less massed in a blotch at the center; no buffy or creamy band across the breast .... **Song Sparrow**, p. 133.

I. A grayish-brown sparrow, with a sharply streaked, buffy, or creamy band across the breast................ **Lincoln's Sparrow**, p. 134.

J. A salt-marsh sparrow, with a grayish, buffy breast; a yellow spot in front of the eye, and on the bend of the wing; crown not chestnut, but like the back in color.................. **Seaside Sparrow**, p. 126.

J. A western, grayish, clay-colored bird, with but little streakings, and no chestnut on the crown .... **Clay-colored Sparrow**, p. 131.

J. A common town and village sparrow, with ashy crown .... **Female English Sparrow**, p. 120.

J. Sparrows with more or less of chestnut on the crown. (**K.**)

K. Tail rounded;[3] lower parts white, with an ashy band across breast .... **Swamp Sparrow**, p. 134.

K. Tail forked;[4] lower parts dark ashy; a black line through eye.... .... **Chipping Sparrow**, p. 130.

K. Tail forked;[4] lower parts ashy; no dark line through eye; back, bright, reddish-brown; bill reddish........ **Field Sparrow**, p. 131.

L. Sides of head distinctly marked with bands and spots of white, black, and chestnut; under parts white, unstreaked, but with a small black spot on the breast; western.................. **Lark Sparrow**, p. 127.

L. Two white wing bars,[5] and the bend of the wing chestnut; upper parts brownish-gray; under parts white, with the breast and sides distinctly streaked...................... .... **Vesper Sparrow**, p. 123.

L. Under parts buffy; two white wing bars, with a black band between; western winter bird..................... **Smith's Longspur**, p. 122.

L. Wing without distinct wing bars; head, throat, and breast with much black; under parts white, with some dark streaks on the sides of the breast and belly; northern winter bird .. **Lapland Longspur**, p. 121.

M. Crown gray, bordered by blackish, rendering a white line over the eye very distinct; no wing bars; under parts pure white......... .... *............................ **Red-eyed Vireo**, p. 98.

M. Under parts slightly tinted with yellowish; no white line over the eye; no wing bars.............. .... **Warbling Vireo**, p. 99.

M. Head bluish-gray; a white eye ring, two white wing bars[5] .... **Blue-headed Vireo**, p. 99.

N. Streaky winter bird, with red on the crown, and in the male on the breast also; bill not crossed at tip.................. **Redpoll**, p. 118.

N. Bill not crossed at tip; orchard or wood-living bird, with the plumage more or less generally reddish................ **Purple Finch**, p. 116.

3   4   5

N. Bill crossed at tip;[1] pine-woods-living birds, with dull, blood-red plumage .... **The Crossbills,** p. 116.

N. A red-faced bird, with yellow, brown, black, and white in the plumage .... **European Goldfinch,** p. 120.

  O. Throat and breast orange flame color; head black, striped with flame color ....**Blackburnian Warbler,** p. 80.

  O. Whole head, neck, and under parts rich orange...................
....**Prothonotary Warbler,** p. 76.

  O. With much flame color at base of tail and middle of wing; upper parts black; belly about white..........**American Redstart,** p. 96.

P. A yellow-bodied bird, with black wings and tail (in the winter the body has brownish washings)...........**American Goldfinch,** p. 118.

P. A streaky, bluish-gray, slender-billed bird; a yellow spot on rump, crown, and side of breast....................**Myrtle Warbler,** p. 83.

P. Breast yellow, without streaks; belly white or whitish. **(T.)**

P. Breast and belly yellow, unstreaked (the lower belly may be whitish). **(S.)**

P. Belly yellow, but the breast not pure yellow. **(R.)**

P. Breast yellow, with streaks or spots. **(Q.)**

  Q. All under parts yellow, with a necklace of black spots across the breast; upper parts gray...............**Canadian Warbler,** p. 95.

  Q. Crown black; cheeks chestnut; a broad white wing bar; under parts heavily streaked with black.......**Cape May Warbler,** p. 81.

R. Head, neck, and breast bluish-gray; eye ring white..............
......................................Connecticut Warbler, p. 92.

R. Head, neck, and throat bluish-gray, changing to black on the breast; upper parts olive-green; no wing bars.....Mourning Warbler, p. 92.

R. Upper parts dark olive-green; throat, breast, and sides washed with olive-green; two whitish wing bars[2]................................
...............................Yellow-bellied Flycatcher, p. 164.

  S. Crown black, connected below the head with a black throat patch; forehead, sides of head, and belly yellow; back olive-green.......
....**Hooded Warbler,** p. 94.

  S. Crown and side of throat black, but the throat yellow; a curved yellow line over the eye; no wing bars...Kentucky Warbler, p. 91.

  S. Crown like the back olive-green; two whitish wing bars.........
....**Pine Warbler,** p. 88.

T. Stout-billed[3] ground bird, with a black blotch on the throat; chin white; back streaky; breast bright yellow.......Dickcissel, p. 141.

T. Heavy-billed,[4] bright olive-green-backed bird, with two white wing bars[2] and a white eye ring...........Yellow-throated Vireo, p. 99.

1   2   3   4

T. Slender-billed[5] birds. (**U.**)
> **U.** Crown chestnut; yellow under parts streaked with chestnut on breast and sides. ....**Palm Warblers**, p. 88.
> **U.** Back olive; head with a black mask.......................... ....**Maryland Yellow-Throat**, p. 93.
> **U.** Sides streaked with black; a white line over the eye; two white wing bars[2].....................**Yellow-throated Warbler**, p. 86.

V. Slender brown bird, with long, sharp-pointed tail feathers[6]......... ....**Brown Creeper**, p. 63.

V. Slender bird, everywhere streaked with black and white........... ....**Black and White Warbler**, p. 76.

V. Short-tailed creeper, with much black and white in the plumage; creeping head downward as often as upward..................... ....**White-breasted Nuthatch**, p. 59.

W. Iridescent or glossy-backed swallows. (**Z.**)
W. Dull-colored birds, without iridescence. (**X.**)

X. Sooty-brown bird, with very short tail, much shorter than the wings ....**Chimney Swift**, p. 167.

X. Mouse-colored swallows, with not especially short tails. (**Y.**)
> **Y.** Under parts white, with a brownish band across breast......... ....**Bank Swallow**, p. 107.
> **Y.** Throat and breast brownish; belly white..................... ....**Rough-winged Swallow**, p. 107.

Z. Steel-blue-backed swallow, with the throat and breast chestnut; tail deeply forked[7].........................**Barn Swallow**, p. 106.

Z. Back steel-blue; rump chestnut; tail nearly even; head, throat, and breast chestnut.........................**Cliff Swallow**, p. 105.

Z. Blackish-green-backed swallow, with all lower parts white; tail nearly even...................................**Tree Swallow**, p. 106.

Key to Birds between the English Sparrow and Robin in Size

The numbers refer to the pages where the birds are described.

* Creeping birds upon the trunks of trees. (**N.**)
* Birds practically always seen on the wing. (**M.**)
* Ground birds, with slender bills and plainly spotted breasts. (**K.**)
* Decidedly crested, seal-brown birds with yellow tips to the tail feathers ....**Waxwings**, p. 103.
* Not as above. (**A.**)
> **A.** With a conspicuous amount of bright yellow or orange in the plumage. (**J.**)

5   6   7

A. With decided red in the plumage. (**I.**)

A. With decided blue.   (**H.**)

A. Black bird, with a brown head and neck.........**Cowbird, p. 144.**

A. With large amounts of both black and white, but no bright red. (**G.**)

A. Not as above. (**B.**)

B. With head, back, and tail black, and belly chestnut..............
....**Orchard Oriole, p. 147.**

B. Slate-colored bird, with chestnut patch under the tail...**Catbird, p. 65.**

B. Winter bird, mainly white in color, but more or less washed with
brown.......... ....**Snowflake, p. 121.**

B. Ground bird, with pinkish-brown back, white belly, and black crescent on breast............................**Horned Lark, p. 156.**

B. Brown to olive, unstreaked birds, with gray breasts. These birds have
the habit of sitting on a perch, watching for insects, which, when seen,
are captured on the wing with a characteristic click of the bill, the
bird returning to the old perch. (**F.**)

B. Streaked, brownish, heavy-billed, sparrow-like birds. (**C.**)

   C. With acute-pointed tail feathers,[1] and no white anywhere........
....**Bobolink, p. 144.**

   C. Outer feathers of the tail with much white......................
.... **The Longspurs, pp. 121, 122.**

   C. Tail feathers not acute, and the outer ones not white. (**D.**)

D. Head without stripes ; body and wings with much chestnut ; breast
decidedly spotted.......................**Fox Sparrow, p. 135.**

D. Head without stripes ; no chestnut on body or wings ; head and back
blackish streaked ; under parts conspicuously streaked.............
.................................**Red-winged Blackbird, p. 146.**

D. Head decidedly striped ; throat with a distinct patch of white ; breast
grayish. (**E.**)

   E. A yellow spot in front of eye.....**White-throated Sparrow, p. 129.**

   E. No yellow spot in front of eye....**White-crowned Sparrow, p. 128.**

F. Slightly crested bird, with much chestnut on the wings and tail ;
throat and breast pearl-gray ; belly yellow.**Crested Flycatcher, p. 161.**

F. Olive-brown-backed, nearly black-crowned bird, with the under parts
yellowish-white, and the bill black. .. ...**Phœbe, p. 162.**

F. Blackish-olive-backed flycatcher, with the side olive-colored and only
the central line of the lower parts white. **Olive-sided Flycatcher, p. 163.**

F. Grayish-slate-colored bird, with a white band across the tips of the tail
feathers ; belly and throat white ; breast grayish....**Kingbird, p. 160.**

   G. At a distance the bird above given (last **F.**) might be considered
mainly black and white. It can be known by the white tips to the
blackish tail feathers......................**Kingbird, p. 160.**

   G. Entire under parts black ; back of head buffy ; rump white. A
musical bird of meadow and field in spring......**Bobolink, p. 144.**

   G. Head and back black ; belly and outer tail feathers white ; sides
chestnut. A thicket-living bird..................**Towhee, p. 136.**

G. Gray-backed birds, with black wings and tail, and the under parts mainly white ; bill decidedly hooked [2].. ....**Shrikes**, p. 101.

H. Upper parts with much blue ; breast brown ; belly white...........
....**Bluebird**, p. 55.

H. Very stout-billed,[3] dark-blue bird, with black wings and tail. This is a southern bird, found mainly in shrubbery near water............
....**Blue Grosbeak**, p. 138.

I. Whole plumage red ; no crest on head...**Summer Tanager**, p. 109.

I. Head and body red ; wings and tail black.....................
.. ..**Scarlet Tanager**, p. 108.

I. Head and body black ; belly white ; breast rose-color ; wings and tail with white blotches...........**Rose-breasted Grosbeak**, p. 137.

I. A distinctly crested [3] bird.....................**Cardinal**, p. 137.

J. Upper parts olive-green ; throat and breast bright yellow, changing abruptly to white on the lower belly ; a white eye ring............
..................................**Yellow-breasted Chat**, p. 93.

J. Front parts black ; much of breast, belly, and lower back rich orange...............................**Baltimore Oriole**, p. 148.

J. Upper parts and tail olive-green ; under parts yellow..............
....**Orchard Oriole**, p. 147.

J. Upper parts brownish ; under parts dull orange ; wings blackish, with white wing bars..........................**Baltimore Oriole**, p. 148.

J. Slightly crested bird, with brownish-olive back, sulphur-yellow belly, and chestnut edgings on wing and tail feathers ....
.................................**Crested Flycatcher**, p. 161.

J. A streaky, sparrow-like ground bird, with bright-yellow breast and black blotch on throat.......................**Dickcissel**, p. 141.

K. Outer tail feathers with white. A meadow and field, tail-wagging bird, with a dark-olive-brown back..........**American Pipit**, p. 70.

K. Outer tail feathers white tipped. A "*cooing*," southern, brownish-gray bird, with the lower parts wine-tinted...**Ground Dove**, p. 217.

K. A short-tailed, long-winged, "*teetering*" bird, with a slender bill about an inch long [4].................**Spotted Sandpiper**, p. 248.

K. Birds with rather long tails, having the outer feathers without white. (**L.**)

L. Crown reddish-brown, changing gradually to olive on the tail ; breast and sides heavily marked with round, black spots.................
....**Wood Thrush**, p. 51.

L. Crown and back olive, changing gradually to reddish on the tail ; only the breast spotted ; spots wedge-shaped.....................
....**Hermit Thrush**, p. 53.

1   2   3   4

L. Whole back from crown to tip of tail reddish ; upper breast slightly
spotted ; sides white......................Wilson's Thrush, p. 52.
L. Whole back olive ....
........Olive-backed Thrush, p. 53.   Gray-cheeked Thrush, p. 52.
M. A sooty-black, long-winged, but very short-tailed bird...........
......................................Chimney Swift, p. 167.
M. A swallow with a very deeply forked tail [1] and steel-blue upper
parts ; chestnut on throat and upper breast..Barn Swallow, p. 106.
M. All parts more or less shining blue-black...Purple Martin, p. 105.
N. Back black, with a central stripe of white ; wings with round white
spots ; under parts grayish-white ; some red on the head of the *male*.
..........Hairy Woodpecker, p. 173.   Downy Woodpecker, p. 173.
N. Back mottled black, white, and yellowish ; belly greenish-yellow ;
breast with black ; crown (and in the *male* throat also) red........
................................Yellow-bellied Sapsucker, p. 175.

Key to Birds about the Size of the American Robin

The numbers refer to the pages where the birds are described.

* Creeping birds upon tree trunks.   (H.)
* Peculiarly mottled long-winged brown birds, with large mouth, but short
bill ; seen mainly on the wing.   (D.)
* Long-winged, fork-tailed,[1] slender-billed[2] birds ; seen constantly on
the wing and appearing like large swallows, but with harsh voices...
....Terns, pp. 323-327.
* Birds neither fitted for creeping nor seen constantly on the wing.   (A.)
A. Birds with bright red in conspicuous amounts.   (G.)
A. Grayish-slate-colored bird, with chestnut-brown breast, white throat,
and white lower belly ....American Robin, p. 54.
A. Crested birds, mainly blue in color.   (F.)
A. Black bird, without bright red anywhere, but sometimes with rusty
tips to the feathers....................Rusty Blackbird, p. 149.
A. Slate-colored bird, with chestnut blotch under the tail. Catbird, p. 65.
A. Not as above.   (B.)
B. Upper parts uniform in tint, neither streaked nor spotted ; outer tail
feathers either wholly white or distinctly white at tip ; under parts
nearly white.   (E.)
B. Slightly crested bird, with grayish to brownish-olive back, grayish
throat, and sulphur-yellow belly.   There are chestnut edgings to wing
and tail feathers..................... Crested Flycatcher, p. 161.
B. Upper part gray and brown mottled.   (C.)

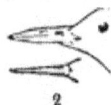

1   2   3

C. Short-billed, short-winged, short-tailed, heavy-bodied ground bird, with striped head and either white or buff throat patch; under parts not yellow...........................**Bob-white**, p. 222.

C. Long-billed, short-tailed meadow bird, with the under parts yellow and a black crescent on the breast...........**Meadowlark**, p. 147.

C. Long-winged birds, with fluffy owl-like plumage. These perch lengthwise on limbs or on the ground, and are much the color of their surroundings; open mouth very large, but culmen short.[3] (**D.**)

D. An evening-flying bird, with a large white spot on the middle of the wing...**Nighthawk**, p. 170.

D. Birds similar to the last, but without the white spot on the wing, and with a white or buffy band across the throat. These birds usually fly near the ground, the nighthawk high in the air ....**Whip-poor-will**, p. 169. **Chuck-will's-widow**, p. 169.

E. All the tail feathers tipped with white...........**Kingbird**, p. 160.

E. Slender birds, with brownish-gray backs, long tails, and long, curved bills[4]...**Cuckoos**, p. 182.

E. Back ashy; tail long, and the outer feathers wholly white. A wonderful song bird, with rather long but nearly straight bill..... ....**Mocking Bird**, p. 64.

E. Upper parts gray; bill decidedly hooked at tip;[5] head with a black stripe on the side extending past the eyes........**Shrikes**, p. 101.

F. Large-headed bird, with long, heavy bill[6] and two bluish bands across the breast..............................**Belted Kingfisher**, p. 179.

F. The bright blue of the tail cross-barred with black; bill only about one inch long; black collar across breast.........**Blue Jay**, p. 153.

G. A conspicuously crested[7] bird, with a black face. The other parts entirely red (*male*) or much red on crest and wings (*female*)..... ....**Cardinal**, p. 137.

G. A winter bird of the Northern States, with no crest, but much rosy-red in the plumage......................**Pine Grosbeak**, p. 115.

G. A black bird, with red on the bend of the wing.................. ....**Red-winged Blackbird**, p. 146.

H. The whole head and neck bright red, back black, belly white, and wings black and white.............**Red-headed Woodpecker**, p. 176.

H. Back distinctly but finely cross-barred with black and white; crown and back neck red in the *male*; belly tinged with red.............. ....**Red-bellied Woodpecker**, p. 177.

H. Back black, marked lengthwise through the center with white; wings black, with many round, white dots......**Hairy Woodpecker**, p. 173.

4   5   6   7

Key to Birds Larger than the Robin

The numbers refer to the pages where the birds are described.

* Birds seen constantly on the wing, and generally near or over the water. (**E**.)

* Mottled-brownish, short-billed[1] ground birds, with feathered legs, walking and scratching like barnyard fowl. (**D**.)

* Crested birds, with more or less of blue in the plumage. (**C**.)

* Black-plumaged birds, not constantly on the wing. (**B**.)

* Not as above. (**A**.)

   A. Long-legged ground birds, with a slender bill,[2] an inch or more long............ ....**Bartramian Sandpiper**, p. 247.

   A. Brown-colored ground birds, with a long tail and a spotted breast.................................**Brown** Thrasher, p. 66.

   A. Loud-voiced, woodpecker-like birds, with much golden color on the under sides of the wings and tail, and a black crescent across the breast ; belly with round, black spots........Flicker, p. 178.

   A. Small-headed, full-breasted, short-billed[3] birds, with reddish breast ; head and neck with metallic tints.............................
   ....**Mourning Dove**, p. 216.   **Passenger** Pigeon, p. 215.

   A. Slender birds, with long, slender, somewhat curved bills,[4] and long tails, having the outer feathers white tipped................
   .... The Cuckoos, p. 182.

   A. Mottled-backed ground birds, with long, straight bill,[5] yellow under parts, and a black crescent on the breast.................
   .... Meadowlark, p. 147.

   A. Very much mottled, short-legged birds, with a white or buffy collar around the throat.............. .....Chuck-will's-widow, p. 169.

B. Black, without iridescence.................The **Crows**, pp. 155, 156.

B. Black, glossy, and iridescent..... ....Purple Grackle, p. 150.

   C. Large-headed bird, with a heavy, long, straight bill,[6] and two bluish bands across the breast................Belted Kingfisher, p. 170.

   C. Tail and wings heavily barred[7] with black ; a black band across breast................................... ... Blue Jay, p. 153.

D. Tail long, and when expanded, fan-shaped ; a ruff of black feathers on the lower part of the side neck... ....Ruffed Grouse, p. 224.

D. Tail extending but little beyond the tips of the wings when closed ; a tuft of feathers higher up on the side neck...........
   ....**Prairie Hen**, p. 225.

1   2   3   4   5   6

E. With square tails;[8] size generally larger than the crow........
....Gulls, pp. 329–333.

E. With forked tails,[9] and usually not larger than the crows........
... .Terns, pp. 323–327.

Key to the Owls

The numbers refer to the pages where the birds are described.

* Owls with conspicuous ear tufts[11] and yellow eyes. (**F.**)
* Owls without ear tufts,[11] and black or yellow eyes. (**A.**)
  A. Large, 12 inches or more long. (**C.**)
  A. Small, less than 12 inches long; back spotted with white. (**B.**)
  B. Ground-burrowing, day owls, of the south and west, with very long legs, nearly naked of feathers ...The Burrowing Owls, pp. 190, 191.
  B. Short-legged owl, less than 9 inches long, with the head streaked, and the back spotted with white... ....Saw-whet Owl, p. 187.
  B. A northern, winter, short-legged owl, nearly a foot long, with both head and back spotted with white.........Richardson's Owl, p. 187.
  C. Eyes black or nearly so. (**E.**)
  C. Eyes distinctly yellow. (**D.**)
  D. A very large, winter owl, with nearly white plumage..............
  .... Snowy Owl, p. 189.
  D. A very large, grayish-mottled owl, with the white lower parts broadly streaked on the breast, and irregularly barred with blackish on the belly and sides............................Great Gray Owl, p. 186.
  D. A medium-sized, day-flying, long-tailed, somewhat hawk-like owl, with the back dark, sooty-brown, and the head and neck much spotted with white.... ....American Hawk Owl, p. 190.
  D. A medium-sized, dull orange to buffy owl, with darker streaks. This owl has short, and usually unnoticed, ear tufts..................
  ....Short-eared Owl, p. 185.
  E. Large owl, with curious, heart-shaped, monkey-like face.[12] This is a spotted and speckled light-colored bird.........Barn Owl, p. 192.
  E. A large, grayish-brown, hooting owl, with the back and breast much barred, and the belly and sides streaked..Barred Owl, p. 186.
  F. A common, small, brownish-gray or reddish owl, less than 12 inches long.............................. ....Screech Owl, p. 188.
  F. Owls over 12 inches long. (**G.**)
  G. A very large, heavy owl, with ear tufts[11] two inches long, and dark, mottled back; the belly is rusty buff, barred with black.........
  ....Great Horned Owl, p. 188.

7 8 9 10 11 12

G. A medium-sized, conspicuously eared owl (ear tufts 1 inch long), with dark brownish back mottled with white and orange. The lower parts buffy, streaked on the breast, and barred on the sides and belly.................... **American Long-eared Owl**, p. 185.

G. A medium-sized, inconspicuously eared owl, with both breast and belly streaked. The general plumage is dull orange to buffy...... ....**Short-eared Owl**, p. 185.

Key to the Hawks, etc.

The numbers refer to the pages where the birds are described.

* Bird of prey, with long, deeply forked tail.[1] **Swallow-tailed Kite**, p. 196.
* Small, less than 14 inches long. (**E.**)
* Large hawks, 14–25 inches long. (**A.**)
* Very large birds of prey, over 25 long. Eagles, p. 205. Vultures, p. 212.

A. Hawk with densely feathered legs ; plumage usually dark-colored.. ....**American** Rough-legged Hawk, p. 204.

A. Tarsus bare for at least one third its length. (**B.**)

B. Plumage with a conspicuous amount of rusty red. (**D.**)

B. Without rusty red. (**C.**)

C. A long-tailed hawk, with the upper tail coverts entirely white. An inhabitant of marshy places................**Marsh Hawk**, p. 198.

C. A fishing hawk, with the head, neck, and lower parts white. This bird is usually seen flying over large bodies of water and frequently dashing down for its fish food...........**American Osprey**, p. 211.

C. All upper parts slate-colored and nearly uniform ; the sides of head with peculiar "mustache" blotches ....**Duck Hawk**, p. 208.

D. Shoulders conspicuously rusty red ; tail black, with about four broad white bands, and white tip..........Red-shouldered **Hawk**, p. 201.

D. Tail rusty red, with a narrow black band near the tip, but the tip white ; upper breast streaked buffy and brown ; lower belly white, without streaks........................ Red-tailed Hawk, p. 200.

D. A blackish-crowned, medium-sized hawk, with a much rounded ashy-gray tail crossed by blackish bands and a white tip................ ... .Cooper's Hawk, p. 199.

D. A medium-sized hawk, with the under parts heavily barred with rusty buff................................Broad-winged **Hawk**, p. 203.

E. A long-tailed, bluish-gray-backed hawk, with the lower parts whitish, barred on the sides and breast with rusty red or brown........... ....**Sharp-shinned** Hawk, p. 198.

E. A very small hawk, with much rusty red on the back and usually on the crown................ **American** Sparrow Hawk, p. 210.

E. A small hawk, with slaty-blue back, a rusty collar on the neck, and about three whitish bars on the tail, and a white tip............. ....Pigeon **Hawk**, p. 208.

General Key to the Groups of Water Birds

This Key is a very general one. The illustrations in Part II. are believed to serve better for the identification of most water birds, seen at a distance, than any field keys that could be prepared. The object of this Key is to state concisely the general characteristics of each group and refer the learner to the pages where descriptions and engravings can be found.

* Shore birds, with round heads, short, pigeon-like bills,[2] short necks, and stout bodies. These are found near both salt and fresh water ponds and streams. None are over 12 inches long. Because of the shortness of the bill, a few of the sandpipers might be looked for here, especially those found on pp. 247 and 248.................................
................................**Plovers**, pp. 229–233. **Turnstone**, p. 228.
* Shore birds, with slender and usually elongated bills,[3] and generally long legs and necks. These are found abundantly on marshes, meadows, and along the shores. The plumage is generally of mottled brown color. The length varies from 6 to 25 inches...............
— Bill long and curved downward...............**Curlews**, p. 249.
— Bill long and curved upward ....
... .**Godwits**, p. 244. **Avocets**, pp. 250, 251.
— Bill straight and of varying length...........................
....**Snipes, Sandpipers**, pp. 237–249. **Phalaropes**, p. 252.
* Reedy marsh birds, with long legs, long toes, and narrow bodies. Plain-colored, generally skulking birds, hiding in the most inaccessible places, and thus difficult to see. They are noisy birds, with penetrating voices of varied character which have been likened to those of pigs, frogs, chickens, etc. The length of the different species varies from 6–15 inches............... **Rails**, p. 255. **Gallinules**, p. 259.
* Swamp birds of large size, with long necks, long, strong bills,[4] and long legs. They are often seen standing on one leg. These are brightly marked and in the breeding season beautifully crested birds.........
..............**Bitterns**, p. 264. **Herons**, pp. 265–270. **Cranes**, p. 261.
* Small swimming birds, which on shore seem much like sandpipers.....
....**Phalaropes**, p. 252.
* Swimming birds, with stout, flattened bodies, large heads, and usually broad, depressed bills.[5] These are generally large birds found swimming in all waters. When flying they move through the air with wonderful velocity .... **Sea Ducks**, pp. 286–296. **River Ducks**, pp. 297–303. **Fish Ducks**, p. 304. **Geese**, p. 280. **Swans**, p. 279.

 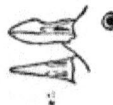 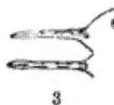  

1   2   3   4   5

* Swimming and diving birds, with almost no tails.   Heads peculiarly
crested in the breeding season.   Body held nearly erect when standing —
    — Neck short..**Auks,** pp. 338–341.
    — Neck long ....**Loons,** p. 342.   **Grebes,** p. 345.
* Very long-winged, flying, and swimming birds, seen usually in the air
over the water along all shores.
    — Smaller birds, with forked tails, and the head so held as to point
    downward when flying ; voices shrill........**Terns,** pp. 323–328.
    — Larger birds, with even tail and the head held in line with the
    body when flying ; voices hoarse........ ...**Gulls,** pp. 329–336.

PART IV

PREPARATION OF BIRD SPECIMENS FOR DISPLAY OR STUDY

Whether it is better to have *skins* or *mounted birds* depends entirely upon the use they are to serve, the number there are to be, and the room at disposal for their preservation.

For beginners in ornithology, mounted birds show far more than skins. A bird properly stuffed, with the mouth slightly open, the wings placed free from the body feathers, and the toes well spread on the perch, can be studied by thousands of beginners and still remain intact. The specimen itself need not be handled, as all the necessary parts of head, bill, wings, and legs can be studied by holding the bird stand in different directions. A bird skin is soon torn to pieces by beginners. They pull the toes apart to see the amount of webbing, move the legs in all directions to examine the tarsus and tibia, raise the wings, and open the mouth. They have no respect for the skin; but the mounted bird they consider a thing of beauty.

The author has hundreds of specimens of mounted birds, which have been studied by thousands of his students in the last twenty years, and they are still in good condition for another twenty years of study, while his bird skins have lasted but a few years. The students much prefer the mounted specimens; indeed, all one need do to insure the birds against careless usage is to warn the students, that, if the mounted birds are harmed by handling, skins will be used instead.

Any moderately ingenious boy or girl can learn to mount birds well by following printed and illustrated instructions. There is an advantage in seeing one specimen prepared by a

373

good taxidermist, but it is better for the student to see this work after he has made a few independent attempts. In any case, the first attempt is certain to be a total failure, and if the first ten are far from successful, it is no cause for discouragement.

It is unfortunate that. although one starts with a thing of beauty. from the moment the mounting operation begins (even if performed by a master). through hours of labor. the specimen looks worse and worse, and less and less like a bird, until just before it is finished. The last five minutes' work once more makes it look alive and beautiful. This is apt to have a discouraging effect upon a student, and the "thing" is often thrown away before the last five minutes of restorative work can accomplish their mission.

The more beautiful the bird taken, the more regret is felt at the loss of the specimen; so the score or more used in first attempts should be birds of no importance, and, if possible, birds whose number needs to be lessened. It is almost universally agreed that the English sparrow belongs to this group, and so the learner should make use of it until success is assured. until at least a half dozen good mounts in different positions have been prepared. This will require a dozen or more specimens, according to the ability of the student. The general directions in this chapter refer to the English sparrow.

Killing the Bird. — Have the specimen killed with "dust" or "No. 12" shot. The dust is smaller and better than No. 12, but cannot always be purchased. Either of these makes such small holes in the skin that there is rarely enough bleeding to injure the plumage. As soon as the bird is shot, the mouth, the nostrils. the vent. and the bleeding shot holes, if there are any, should be plugged with a little cotton and the specimen carefully wrapped in a piece of paper. If a piece of paper is twisted into a cornucopia and the bird slipped into it head first, there will be no danger of ruffling the plumage.

Instruments. — Sharp pocket knife. scissors. pair of pincers with a wire-cutting attachment, **pair of** tweezers, flat file, brad awl, stiff wire in handle, commercial steel pen, stiff brush a fourth of an inch through (No. 4, round, bristle, marking brush), and a two-ounce, large-mouth bottle for arsenic, plainly marked with a POISON label.

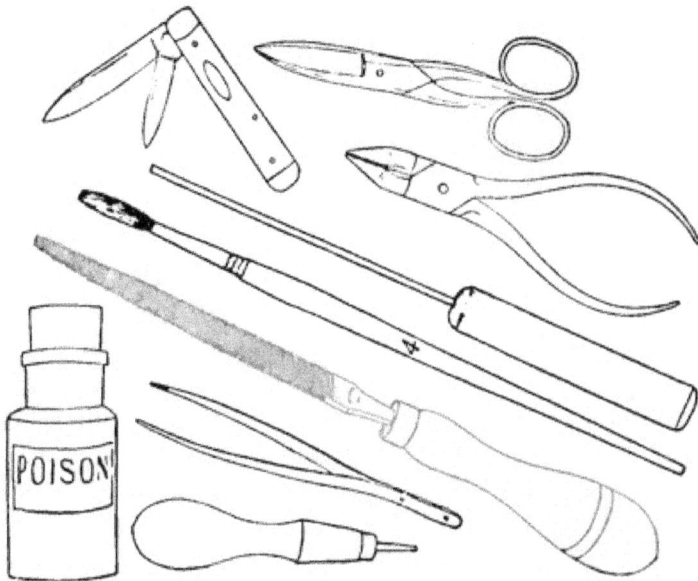

Tools, etc., shown one third size.

Materials. — Two ounces white arsenic and 1 ounce alum mixed together in the bottle with enough water to give the whole the consistency of hasty pudding, 1 pound of good tow (to be obtained from a furniture dealer), a bat of best cotton, black glass eyes a little over $\frac{1}{8}$ inch in diameter (black glass-headed pins of the right size will do), $\frac{1}{2}$ pound of annealed iron wire about No. 22 (Standard Wire Gauge). 2 pounds corn meal, 2 pounds plaster of Paris, 1 pound of good clay, a spool of linen thread No. 40, and bird stands.

Skinning. — Remove all the cotton plugs which were placed in your specimen at the time of shooting and substitute fresh ones. Spread on your table a large newspaper, and you are ready for work.

1. Place the bird with its back on the paper and its head toward your left. With your fingers separate the feathers of the belly from the breastbone to the tail, and thus expose the bare skin which will be found in this region. With your knife cut through the skin from about the lower end of the breastbone back along the middle line of the body to the vent. Especial care must be taken to cut only through the skin and not through the membrane which covers the abdominal cavity.

2. With the left hand lift the edge of skin toward you, and with the side of the knife blade press the flesh from the skin till you reach the knee. The first illustration [1] shows the bird at this stage. If at this or any other time during the skinning process any fluid escapes, the meal is to be used to absorb it.

3. Press the leg up under the skin and thus make the knee project; cut off the leg at this point either with the knife or, better, with a pair of scissors. Reverse the position of the bird and sever the other leg.

4. The next step is a difficult one: the body is to be cut off at the base of the tail, without cutting the skin, loosening the tail feathers, or opening the body cavity so that the entrails can escape. First separate the skin from the body as far back as you well can with the side of the knife and your fingers; place the thumb and first finger of the left hand between the skin and the body near the tail; and, holding the second or third finger above the tail (that is, on the lower side of the bird as you hold it), to feel for the action of the scissors so as not to cut through the skin, cut carefully between the bones and entirely sever the flesh.

5. Held the bird so that it rests with its breastbone on the

table and its belly toward your right, and press the skin away from the back, turning it inside out as you proceed. In the work at this stage you will find that as soon as the skin is partly past the rump it will be well to hold the rump with the right hand and with the fingers of the left gently press the skin from the flesh. You will soon reach the wings, and your specimen will look as in the second illustration.[2]

2.

6. With scissors cut off the wing bones close to the body. At this stage there is danger from profuse bleeding, and the meal must be used very freely. The blood must not be allowed to touch the feathers. Continue the skinning up the neck and over the head. The skinning of the neck is easy, but care must be taken as you press the skin loose from the skull. You will soon have the bird and skin as shown in the third illustration,[3] and the skin of the right ear,

3.

as shown in the figure, is to be carefully pulled from its socket by the aid of the point of the knife; afterwards remove the skin from the other ear.

7. The eyes now come into view, and the membrane which

joins the eyeball and skin is to be carefully cut with the scissors. In this process the eyelids must not be injured nor the eyeballs ruptured. The skin is next pressed from the skull about to the bill. The eyes are now fully exposed and can be readily removed without rupture by the aid of the rounded end of the commercial steel pen.

8. With the scissors cut off the back part of the skull obliquely, as shown in the next figure,[4] and pull away the body, neck, and tongue from the skin. This oblique cutting with the scissors is performed by four cuts, — one across the roof of the mouth, two obliquely upward along the sides of the skull, and the last across the top just above the neck. Next remove the brain with the rounded end of the steel pen. This can often be done without rupturing the surrounding membrane. If the tongue was not pulled out with the neck it must now be removed, together with all the fleshy parts about the base of the skull.

9. Pull the leg and wing bones out the proper distance from the skin and cut away all the flesh possible. The illustration[4] will show how far to pull and what to remove. The base of the tail needs also to be cleaned of superfluous flesh.

10. The skin is now ready to be treated. With the small brush, paint the arsenic mixture over every part of the skin and bones, being especially careful to leave a full supply wherever there is flesh. (See caution in regard to this poison on page 387.)

11. Nearly fill the eye sockets with small, twisted-up wads of cotton, and plaster them even full of clay in about the plastic

condition used by pottery workers. In the center of the clay, on each side, place the head of a mourning pin, or a glass eye. It is well to place a little clay in the top of the skull also, as the neck wire will be much more firmly held in place by this addition.

12. The skin is now to be turned right side out, and the first step — that of getting the skin of the head properly and smoothly over the skull — is difficult for a beginner. Place the thumb of the left hand just where the skull is cut off. and with the tip of the fingers gradually and slowly work the skin upward and backward over the most bulging portions. As soon as you can reach the bill from within the skin take hold of it, and almost immediately the whole skin will be reversed.

13. Take hold of the ends of legs and wings and pull them into place. Shake the whole skin while holding by the bill. Lift up the skin from the skull and thus give it a chance to take its exact old position. The proper adjustment of all feathers depends entirely upon the proper adjustment of the skin. The feathers will come right if you get the skin right. The first arrangement of the eyelids around the glass eyes should now be attended to. The tweezers will be found useful for this purpose.

Stuffing. — 14. The wires needed should first be cut and both ends of each sharpened with the file. A bird should be successfully mounted with the wings closed before any attempt is made to mount one with the wings spread. For the closed wing form, three wires 7 inches long and one 4 inches long are needed. For a spread wing, two additional wing wires about 5 inches long will be necessary.

15. A body is next to be made of tow. It should be in size and shape as nearly as possible like the one taken from the bird. Take a mass of tow in your hand; two or three trials will show how much is needed. Wind this with thread in all possible directions, and at the same time press it into form by the thumb and fingers of your left hand. Compare constantly with the bird's body. If any portion proves too

small, add a little tow to the part and continue the winding. In the end, you should have a very firm, smooth body with thread nearly covering its surface. Pass one of the long wires through this body from the front end; then the protruding end should be passed back and its tip clinched into the body. Next, wind the neck end smoothly with cotton and tie a thread around the part which is to enter the skull. See that the position, length, and size of the neck are like that of the bird. You will now have an object in shape much like the fifth figure.[5] The dotted lines show how the wire goes through the body and is firmly clinched.

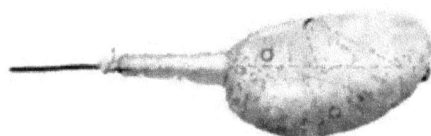

16. The other two wires, 7 inches long, are now to be passed up the legs. Start the wire at the place of the joining of the toes, and slide it along the back of the tarsal bone to the joint; pull the tibia bone through the opening in the skin in such a way that you can get the wire past the joint and along the tibia bone. Both legs are of course to be fixed in the same way.

17. The prepared body is now to be placed in the skin. First introduce the sharp end of the neck wire into the neck, and carefully guide it so that it will enter the skull. Pass it through the skull somewhere near the forehead. Then carefully pull the skin over the body till the tail readily slips past the posterior end of it.

18. The next step is the fastening of the leg wires into the body. Slide the wires up and down the legs till they move freely; then pass them one at a time into the body just where the knee was found on the bird before skinned. The exact place is almost the center lengthwise and one fourth from the lower side, — the spot marked with a small circle on the fifth figure.[5] The wire is to be passed through and back and then clinched. When both leg wires are fastened the bird will appear as shown in the sixth figure.[6]

19. After a little lifting of the skin, pulling out of the wings, and sliding up of the legs, so that the upper end of the tibia bone comes to its proper position against the body, close the skin along the belly and, if necessary, sew it with a stitch or two. The legs extend out straight behind, and in this condition the bird is to be placed on the stand. The two holes for the leg wires should be about one inch apart. These holes are made with the small brad awl. Introduce the leg

wires, and, when the feet rest properly on the crosspiece, bend the wires below so that the bird is held firmly.

20. The bird is now to be given position and form. First bend the leg wires at the heel and knee: then slide down the head so that the neck is not too long, and give the head and neck their proper position and form.

21. The last wire is next to be used to set the tail. Bend it upon itself so as to make a staple-like form about a half inch wide and nearly 2 inches long. Pass this through the skin at the base of and under the tail and then into the body, and bend it, if necessary, so that the tail will rest upon it and hide it. The under tail coverts will hide the wire. Lift the wings, stretch them out, and move them back against the body till you find the feathers taking proper position around them. Pin each in place with about three pins. In fixing the second wing, care must be taken that it matches the one already fastened. Wherever any feathers seem out of place or twisted,

a proper use of the tweezers in lifting the skin and pulling the
twisted feathers will make them all right. Go over the whole
surface of the body — practice will give you the knack —
and get all the plumage in shape. Leave the feathers somewhat open and fluffy, as is natural for this bird. The legs
and neck are to be bent till the position of body suits you.
Your first bird will be apt to have too long a neck and too
much of its legs exposed.
As a rule, the tibiæ of the
sparrows do not show at
all, and even the heel is
well within the feathers.

22. Before putting the
bird away to dry, most
writers on taxidermy ad-
vise the winding of the
whole surface with thread,
so that the shrinking of the
skin over the rough body
will not force the feathers
into poor positions. If this
is done, the thread should
be so lightly drawn as to
barely touch the feathers.
The usual method is to
stick a number of pins into
the body; hook a loop of thread around one of these pins,
and then wind it back and forth from pin to pin in all directions till all the plumage is properly held in place.

If there is a well-formed, smooth body, and each part of the
skin occupies its proper place on that body, the drying of
the skin will not twist or displace the feathers. A well-
mounted bird needs but little, if any, winding. There is apt
to be too much of this work, to the detriment of the specimen.
Let the feathers have a natural, that is, generally, an open
appearance. Watch a caged canary, and see the different

positions it takes, and the frequency with which its feathers are ruffled.

23. The bird, whether wound with thread or not, should be left for several days to dry in some place free from dust. The thread should then be removed, the extra pins pulled out, and the head wire and the pins holding the wings in place cut off as close to the skin as possible so that the feathers will hide them from view. The bird is now finished, and should appear as in the seventh figure.

After preparing two or three good specimens in this position, you will be ready to undertake the mounting of a spread wing. In this case, the two wires, five inches long, are to be passed along the bones of the different joints of the wings. This is to a beginner a difficult process; there seem to be too little flesh and too many bones and joints. The only cautions that can be given are that the wires must be kept straight, the wing so pulled out as to straighten the joints, and the finger and thumb of the left hand must keep the point of the wire within the skin. This wiring of the wings is to be done at the same time as the wiring of the legs (stage 16). Insert the wing

wires into the tow body just before you insert those of the legs (stage 18). The position for the entrance of these wires is found by examining the place where the wings were cut off from the body. The proper locality for both wing and leg wires is shown in the illustration of the tow body (page 380) by small circles. The wing wires need some clinching after being passed through the tow body, though this is not so important as the firm fixing of the leg wires.

When a successful flying bird has been prepared,[8] a slight modification of the plan will enable the student to give a male the strutting position.[9] After this he is ready to undertake the mounting of birds in all kinds of natural attitudes Probably as difficult a one as any is that of gathering food from the ground.[10]

10.

GENERAL HINTS ABOUT THE MOUNTING OF BIRDS

Cleaning blood from feathers. — No matter how small the shot used, there will occasionally be blood spots to be removed. This can be easily done when the blood is fresh. In the worst of cases, with the blood dry and the feathers white, the stains can all be removed if sufficient time and care be given to the work. Wash the spots thoroughly with warm water (and soap also if necessary), and dry with abundance of plaster of Paris. After the moisture has been all absorbed, the plaster is to be completely dusted from the feathers.

Birds difficult to skin. — Birds with large heads and small necks, as the ducks and woodpeckers, will not allow the neck skin to pass over the head. In these cases, the skin of the neck has to be so split open as to allow the head to be skinned and the brain to be removed. Birds with firm, close feathers, as the doves, need to be skinned while perfectly fresh, and with great care, or the plumage will come off from the skin in patches. All large birds are difficult to skin, and many of them need the constant use of the knife to separate the skin from the flesh. All such birds should be suspended from some support by passing a hook (a bent wire nail forms a good one) through the rump. The bird should be suspended as soon as the tail is severed, and then the skinning should be started along the back (stages 4 and 5).

Stuffing for large birds. — Excelsior is the best material for all large bodies, although a little good tow spread over its surface and thoroughly wound down makes it still better.

Legs and wings. — All birds with the tibia exposed should have the leg wire and tibia bone wound together with the proper amount of cotton (stage 16). The wings of large birds are also better if cotton is wound around the bones to take the place of the flesh removed.

Necks. — Most birds need to have the neck stuffed out. This is done with the long wire set in a handle. Either chopped tow or cotton is forced into the mouth and down the throat (stage 21).

Sizes of wire. — The size of wire needed depends more upon the length of the legs and of the neck than upon the size of the body; thus a crane should have much heavier wire than a loon, and the yellow-legs needs as large wire as the blue-winged teal. The sizes given in the following table are those of the "standard wire gauge."

BIRD	No. OF WIRE	BIRD	No. OF WIRE
Hummingbirds	28	Ducks and Long-legged Snipe	16
Kinglets and Warblers	24	Swans and Geese	14
Sparrows	22	Smaller Herons	12
Thrushes	20	Larger Herons	10
Average Snipes	18		

Bird eyes. — Most bird eyes are practically black, and for these, the black glass-headed pins are both good and cheap. A few birds have peculiarly colored eyes, yellow. blue, red,

white, etc., and for these especially made glass bird eyes
should be purchased. Care should be taken that eyes of the
proper color are procured for each bird.

Shot for large birds. — Hunters who merely shoot to kill
generally use shot of large size. The ornithologist, who desires
good specimens or none, will soon find that very small shot
will do. Even ducks, whose plumage is so abundant and close,
can be killed with No. 8 or No. 10 shot. The author once
secured a surf scoter with dust shot. Of course this was possible only under unusual circumstances. The bird flew from
under a bank in a line directly away from the author, and had
reached a distance of only a few feet when aimed at.

Stretching the skin. — The skin should not be stretched. In
skinning large birds, suspended from a hook, the weight of
material, if allowed to hang down upon the neck, will cause
the skin to lengthen, and thus distort the shape. This should
be carefully guarded against. Do not pull the skin from the
flesh, but rather press with the fingers and work with the
knife so that there is no strain upon the skin itself.

Spread wings and crests. — While the bird is drying, some
support must be given to spreading parts, or they will droop in
such a way as to render the bird unsightly. Cotton, held in
place by wire or pins, is the best material for this purpose.

Spreading the tail feathers. — If the tail feathers are to be
widely spread, or, in fact, spread at all, a piece of wire bent
upon itself in the form of a staple, and placed across the tail,
with one prong above and the other below, can be so manipu-
lated with the fingers as to hold the feathers in place till the
bird is dry. Another method is to pass a fine-pointed wire
through all the feathers near their bases (through the hollow
portion), and move each feather along this wire till its position
is satisfactory. A third plan is to reverse the position of
each feather before setting it away to dry. Bring the central
feather below and force each outer pair above the preceding
pair. After the bird is dry the feathers are snapped back into
their proper places.

Birds for study and analysis. — If the birds are to be used in class work, the toes should be carefully spread apart and held in place by pins till dry, so that the amount of growing together of joints and the amount of webbing, if any, can be easily determined. The mouth should also be left slightly open. A good plan is to place a piece of a wooden toothpick between the mandibles, and, if necessary, pass a needle and thread through the nostrils and tie the thread under the bill. To insure the slight elevation of the wings necessary to determine the character of the first primary, place a little cotton under their lower edges before introducing the pins to hold them in place.

Bird stands. — Ground birds and water birds should be placed on flat boards, while most perching birds are better if placed on bird stands, like the one shown in the illustration of the mounted English sparrow (p. 382). If the birds are to be handled by students, the bottoms in all cases should be large enough to preclude the possibility of their being upset.

Finishing the specimen. — After the bird is thoroughly dry, all the projecting wires, pins, etc., are to be carefully cut off. The wire-cutting pincers must work well, so as to completely sever the wires below the surface feathers. Any twisting or bending of the wires in an attempt to break off the parts which the cutting pliers have left will be sure to disturb and injure some part of the specimen.

Bare skin. — The bill, legs, and other parts not covered with feathers will change color and lose brightness; so, sometimes these parts are painted and varnished. In some cases a little of this work is necessary, but generally it is not advisable.

Poison. — Great care must be taken not to get the arsenic into any cut there may be on the hands. If there are any openings in the skin they should be covered with strips of court-plaster. As soon as possible after finishing a bird, the hands should be thoroughly washed and all material carefully removed from under the nails.

Sewing the skin together. — All large birds need to have the skin along the abdomen carefully sewed together.

Determination of sex. — The sex of the specimens should be determined by dissection. The sexual organs are situated between the intestines and the backbone at about the middle of the body, but somewhat on the left side. These organs are large and easily recognized in the spring. The male organs (the testes) are two in number, light, yellowish-white in color, somewhat elongated in form, and, in the English sparrow, in spring, about the size of large peas.[1] The female organs (the ovaries) are clusters of different sized, globular, grape-like parts, united together in a membrane.[2] The English sparrow is a good species to learn from, because the plumage of the head and neck enables one to distinguish the sex before dissection. The best place to cut the body open is along the left side; then by lifting the intestines the sexual organs can be seen. By beginning in the spring and opening a sparrow of each sex each month throughout the year a better knowledge of these organs will be obtained, and an easier recognition of them in all conditions than by any amount of description or illustration.

Cleaning eggs. *Instruments, etc., needed.* — Two or three egg drills of different sizes; two or three blow-pipes of different sizes; a very slender pair of scissors made for the purpose; light spring forceps; a hook formed of a bent needle mounted in a handle; a basin of water; some cotton. The proper instruments can be obtained from any of the dealers in naturalists' supplies.

If the egg is fresh a small hole, a little larger than the point of the smallest blow-pipe, should be carefully drilled in the *side*. Introduce the blow-pipe about a sixteenth of an inch, and blow gently and steadily till the contents are removed. Fill the mouth with water and blow this into and out from the egg so as to rinse it thoroughly. Place the egg with the hole downward on some cotton to dry.

If the egg contains an embryo it will be necessary to drill a larger hole in the side so that the hook, scissors, and forceps can be used as they are needed. Afterwards thoroughly rinse and drain the shell as above described. The work in this case should be performed over the basin of water so as to catch the egg if it slips from the fingers.

As stated in Part I no eggs should ever be gathered till the species of bird has been identified with certainty. On the egg, near the hole, mark with a lead pencil the number of the bird according to the A. O. U. check list. This is the number within the parenthesis next to the scientific name in Part II.

Collecting and preserving nests. — Such well-woven nests as those of the orioles and vireos should be removed from the plant by cutting the branches to which they are attached. These will keep in good condition without any preparation. Others need to be kept from falling to pieces by some artificial support. A good and easy method is to sew loosely with brown thread back and forth over all parts of the nest. Another plan is to make a supporting basket out of annealed wire.

GLOSSARY [1]

Aberrant. Deviating from the usual character.

Acute. Ending in a well-defined angle, usually a sharp one.

Angulated. Forming an angle; applied to the mouth of birds when the direction of the gape suddenly changes at the rictus.

Ashy. A bluish-gray color; about the color of wood ashes.

Axillary plumes. A distinct tuft of feathers, under the wing where it joins the body.

Back. As generally restricted, the upper part of the body of the bird, half way along the mass of flesh (see p. 40); as used in Part II., most of the upper parts, but usually not including wings, tail, or crown.

Barred. With cross bands of distinct colors.

Base. The part of bill or quill attached to the flesh.

Belly. The under parts back of the breast, but not including the tail coverts (see p. 40).

Blotched. Furnished with rounded spots of a different color.

Blunt. Rounded; the opposite of acute.

Booted. Applied to the tarsus of birds when the usual scales along the front are so grown together as to seem continuous.

Bristles. Small hair-like feathers.

Bronzy. Having a metallic appearance like tarnished brass.

Buffy. A light, dull, brownish-yellow.

---

[1] The terms defined in Part I. include all that are generally used in bird descriptions. The words in this glossary are supplementary to those in Part I. and are intended chiefly for use in connection with field study of birds.

Cap. The top of the head when of a distinct color.

Cere. A peculiar covering of the bill of birds of prey extending beyond the nostrils.

Cheek. The side of the head back of and below the eye.

Chin. The part of the head just below the bill (see p. 40).

Claw. The nail of the toe.

Collar. A colored band extending more or less around the lower neck.

Compressed. Flattened sideways.

Conical. Cone-shaped, as the bills of many birds.

Coniferous. Trees which bear cones, as the pines.

Convex. Bulging outward, as the top and bottom ridges of some bills.

Coverts. The small feathers covering the bases of the larger quills of wing and tail.

Creamy. A light pinkish-yellow color like rich cream.

Crest. A tuft of feathers on the top of the head; these can be raised or lowered at will.

Crissum. The under tail coverts (see p. 40).

Crown. The top of the head (see p. 40).

Crustaceans. Animals with jointed covering, as the crab and lobster.

Culmen. The ridge of the upper mandible; as a measure of the length of the bill, it is the straight distance from the feathers on the forehead to the tip of the bill (see p. 14).

Decurved. Bent downward in a regular manner.

Depressed. Flattened at the top and bottom; a depressed bill is wider than high.

Dusky. A dark color of no especial shade.

Erectile. Capable of being raised, as the crest of a bird.

Exserted. Extending beyond the rest, as the central tail feathers of some sandpipers.

Fauna. The animal life of a region.

Flanks. The posterior portion of the sides of a bird (see p. 40).

Forehead. The portion of the head just above the bill (see p. 40).

Forked. Deeply notched as the tails of many birds.

Fulvous. A yellowish-brown.

Fuscous. A dark or blackish-brown of rather indefinite shade.

Gape. The opening of the mouth.

Genus. A closely related group; this close relationship is represented by giving to all members the same scientific name as far as the first word is concerned; this portion being called the generic name.

Graduated tail. One in which the middle pair of feathers is longest, and each successive pair gradually shorter.

Grooves. Furrows.

Habitat. The region or locality inhabited by a species.

Hooked bills. Bills having the point more or less abruptly bent downwards.

Horizontal. Level ; on a line with the horizon.

Horny. Of a material like the finger nail.

Hybrids. An intermediate form between two species caused by inter-breeding.

Impaling. The killing of an animal by striking it on a sharp point.

Inner secondaries. The feathers fastened to the joint of the wing at the elbow. In the illustration these are lengthened.

Inner toe. The inner one of the front toes (see p. 40).

Inserted. Fastened or grown to.

Iridescent. Exhibiting a play of colors like those of the rainbow.

Lobate toes. Those furnished with projecting flaps.

Lobes. Membranous flaps.

Lores. The spaces between the eye and bill, often free from feathers in water birds.

Mandible. One of the two parts of the bill, called upper and lower man-dibles (see p. 14).

Margined toes. Those furnished with a ridge-like border not wide enough to be called lobate.

Marine. Pertaining to salt water.

Membranes. Skin-like parts.

Metallic. Having the appearance of metal, or with the luster of polished metal.

Migratory. Accustomed to move to different countries at different seasons.

Molt. The periodical shedding of feathers.

Mollusks. Soft-bodied animals usually inclosed in shells, as snails and mussels.

Mottled. Marked with different colors in a blotched manner.

Nails. The horny appendages to the toes (see p. 40).

Nape. The part of the head just back of the crown (see p. 40).

Nasal. Pertaining to the nostril.

Nocturnal birds. Those which fly and feed by night.

Nostrils. The external openings in the upper mandible.

Oblique. Slanting or crossing diagonally as the grooves on the bill of the puffin.

Ochraceous. A brownish-orange color; of the color of yellow ocher or a little darker.

Olive. A greenish-brown color like that of pickled olives.

Outer toe. The outside one of the three front toes (see p. 40).

Pectinated nail. A nail furnished with saw-like teeth.

Perching. Lighting or resting on the twigs of plants.

Plumage. The general feathering of the body.

Primary. Any of the quills attached to the outer joint of the wing. The outer one of all is the *first primary* (see p. 23).

Quills. The larger feathers of wings and tail (often restricted to include only the primaries of the wing).

Recurved. Bent backward.

Reticulate. Forming or resembling a network.

Rictal. Pertaining to the rear portion of the mouth.

Rictus. The back or rear of the mouth (see p. 14).

Rufous. Rusty or reddish-brown; the color of the usual red brick.

Rump. The rear portion of the back (see p. 40).

Rusty. A brownish-red; the color of the rust formed on iron.

Scapulars. The tuft of shoulder feathers; the enlarged feathers at the inner part of the wing next the back.

Scutellæ. The nearly square scales along the front of most tarsi. *Scutellate*, having scutella.

Secondaries. The quills of the second joint of the wing (see p. 23).

Sepia-brown. A blackish-brown.

Serration. Saw-tooth-like notches.

Shaft. The midrib of a feather.

Slate color. A dark gray with less bluish than lead color.

Speculum. A bright-colored area on the secondaries of many ducks (S in the cut).

Spinous feathers. Those with thorn-like projecting tips.

Talons. The larger claws or nails of the toes of birds of prey.

Tarsus. The first joint of the leg above the toes.

Tawny. A dark yellow; the color of tanned leather.

Terrestrial. Pertaining to the ground.

Tertials. Usually applied to the inner secondaries if enlarged or peculiarly colored (see p. 23).

Transverse. Turned across; running in a cross direction.

Truncate. With a square tip.

Tubercle. A knob-like projection.

Vane. The whole of a feather excepting the midrib or shaft.

Washings. Tintings.

Webs of toes. The skin-like membranes extending from toe to toe.

Webs of feathers. The spreading portion at either side of the midrib or shaft.

Wing bar. Peculiar strips of color across the base of wings, formed by the tips of the wing coverts.

Zone. A cross-bar on a feather when very wide.

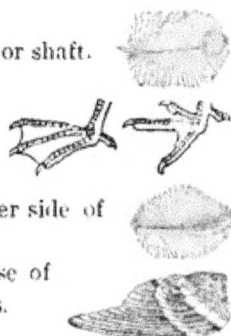

INDEX

TYPOGRAPHY BY J. S. CUSHING & CO., NORWOOD, MASS.

www.ingramcontent.com/pod-product-compliance
Lightning Source LLC
Chambersburg PA
CBHW021349210326
41599CB00011B/805